Molecular
Genetics of
Cancer

The **HUMAN MOLECULAR GENETICS** series

Series Advisors

D.N. Cooper, *Charter Molecular Genetics Laboratory, Thrombosis Research Institute, University of London, UK*

S.E. Humphries, *Division of Cardiovascular Genetics, University College London Medical School, London, UK*

T. Strachan, *Department of Human Genetics, University of Newcastle upon Tyne, Newcastle upon Tyne, UK*

Human Gene Mutation
From Genotype to Phenotype
Functional Analysis of the Human Genome
Molecular Genetics of Cancer

Forthcoming titles

Environmental Mutagenesis
Human Genome Evolution
HLA and MHC: Genes, Molecules and Function

Molecular Genetics of Cancer

John K. Cowell
Haematology and Oncology Unit, Institute of
Child Health, London, UK

SCIENTIFIC
PUBLISHERS

© BIOS Scientific Publishers Limited, 1995

First published in 1995

A CIP catalogue record for this book is available from the British Library.

ISBN 1 872748 09 0

BIOS Scientific Publishers Ltd
9 Newtec Place, Magdalen Road, Oxford OX4 1RE, UK.
Tel. +44 (0) 1865 726286. Fax +44 (0) 1865 246823

DISTRIBUTORS

Australia and New Zealand
  DA Information Services
  648 Whitehorse Road, Mitcham
  Victoria 3132

India
  Viva Books Private Limited
  4346/4C Ansari Road
  New Delhi 110002

Singapore and South East Asia
  Toppan Company (S) PTE Ltd
  38 Liu Fang Road, Jurong
  Singapore 2262

USA and Canada
  Books International Inc
  PO Box 605, Herndon,
  VA 22070

Typeset by Saxon Graphics Ltd, Derby, UK.
Printed by Biddles Ltd, Guildford, UK.

Contents

Contributors

Bishop, D.T. ICRF Genetic Epidemiology Laboratory, Ashley Wing, St James's University Hospital, Beckett Street, Leeds LS9 7TF, UK

Colman, S.D. Department of Pediatric Genetics, University of Florida, College of Medicine, 1600 S.W. Archer Road, Box 100296, Gainesville, FL 32610-0296, USA

Costantini, F. Department of Genetics and Development, Columbia University, The College of Physicians and Surgeons, 701 West 168th Street, New York, NY 10032, USA

Cowell, J.K. Haematology and Oncology Unit, Institute of Child Health, 30 Guilford Street, London WC1N 1EH, UK

D'Agati, V. Department of Genetics and Development, Columbia University, The College of Physicians and Surgeons, 701 West 168th Street, New York, NY 10032, USA

Dunlop, M.G. Edinburgh University Department of Clinical Surgery and MRC Human Genetics Unit, Western General Hospital, Crewe Road, Edinburgh EH4 2XU, UK

Dunn, J.M. The Eye Research Institute of Canada and Visible Genetics Inc., 399 Bathurst St, MP-6, Toronto, Ontario, Canada M5T 2S8

Durbec, P. The National Institute for Medical Research, The Ridgeway, Mill Hill, London NW7 1AA, UK

Gallie, B.L. Division of Immunology and Cancer Research, The Hospital for Sick Children Research Institute, 555 University Avenue, University of Toronto, Toronto, Ontario, Canada M5G 1X8

Haber, D. Massachusetts General Hospital Cancer Center and Harvard Medical School, Boston, MA 02129, USA

Hei, Y.-J. Division of Immunology and Cancer Research, The Hospital for Sick Children Research Institute, 555 University Avenue, University of Toronto, Toronto, Ontario, Canada M5G 1X8

Kaye, F.J. NCI-Navy Oncology Branch, National Cancer Institute, and the Uniformed Services University of the Health Sciences, Bethesda, MD 20889, USA

Kim, Y.W. NCI-Navy Oncology Branch, National Cancer Institute, and the Uniformed Services University of the Health Sciences, Bethesda, MD 20889, USA

Larsson, C. Department of Molecular Medicine, Karolinska Hospital, L6-building, S-171 76 Stockholm, Sweden

Larsson-Blomberg, L. The National Institute for Medical Research, The Ridgeway, Mill Hill, London NW7 1AA, UK

Maheswaran, S. Massachusetts General Hospital Cancer Center and Harvard Medical School, Boston, MA 02129, USA

Mankoo, B. The National Institute for Medical Research, The Ridgeway, Mill Hill, London NW7 1AA, UK

Mostachfi, H. Division of Immunology and Cancer Research, The Hospital for Sick Children Research Institute, 555 University Avenue, University of Toronto, Toronto, Ontario, Canada M5G 1X8

Nordenskjöld, M. Department of Molecular Medicine, Karolinska Hospital, L6-building, S-171 76 Stockholm, Sweden

Otterson, G.A. NCI-Navy Oncology Branch, National Cancer Institute, and the Uniformed Services University of the Health Sciences, Bethesda, MD 20889, USA

Pachnis, V. The National Institute for Medical Research, The Ridgeway, Mill Hill, London NW7 1AA, UK

Schuchardt, A. Department of Genetics and Development, Columbia University, The College of Physicians and Surgeons, 701 West 168th Street, New York, NY 10032, USA

Shapiro, D.N. Department of Experimental Oncology, St Jude Children's Research Hospital, 332 North Lauderdale, Memphis, TN 38101, USA

Soussi, T. Institut de Genetique Moleculaire, Unité 301 INSERM, University P. & M. Curie, 27 rue Juliette Dodu, 75010 Paris, France

Wallace, M.R. Departments of Pediatric Genetics, Biochemistry, and Pathology, Center for Mammalian Genetics, University of Florida, College of Medicine, 1600 S.W. Archer Road, Box 100296, Gainesville, FL 32610-0296, USA

Zedenius, J. Departments of Molecular Medicine and Surgery, Karolinska Hospital, L6-building, S-171 76 Stockholm, Sweden

Abbreviations

ACTH	adrenocorticotrophic hormone
AgT	simian virus 40 large T antigen
APC	adenomatous polyposis coli
APUD	amine precursor uptake and decarboxylation
BCLC	Breast Cancer Linkage Consortium
bHLH	basic-helix–loop–helix
bp	base pair
BWS	Beckwith–Wiedemann syndrome
CASH	Cancer and Steroid Hormone Study
cdk	cyclin-dependent kinase
CHRPE	congenital hypertrophy of the retinal pigment epithelium
CMV	cytomegalovirus
CNS	central nervous system
DCC	deleted in colorectal cancer
DGGE	denaturant gradient gel electrophoresis
DMD	Duchenne muscular dystrophy
DRG	dorsal root ganglia
EBNA	Epstein–Barr nuclear antigen
EBV	Epstein–Barr virus
ELISA	enzyme-linked immunosorbent assay
ENS	enteric nervous system
ESD	esterase D
FAP	familial adenomatous polyposis
FIHP	familial isolated hyperparathyroidism
FMTC	familial medullary thyroid carcinoma
GADD45	growth arrest DNA damage inducible 45
GAP	GTP-ase-activating protein
GH	growth hormone
GI	gastrointestinal
HBV	hepatitis B virus
HCC	hepatocarcinoma
HIV	human immunodeficiency virus
HLH	helix–loop–helix
HNPCC	hereditary non-polyposis colorectal cancer
HPT	hyperparathyroidism
HPV	human papillomavirus
HSCR	Hirschsprung's disease
hsp	heat-shock protein
HSV	herpes simplex virus
IGF2	insulin-like growth factor 2
IGFB4	insulin-like growth factor-binding protein 4
IL	interleukin
kb	kilobase

LCR	ligase chain reaction
LOH	loss of heterozygosity
LSH	loop–sheet–helix
MAP	mitogen-activated protein
MCC	mutated in colon cancer
MCK	murine creatine phosphokinase
MEN	multiple endocrine neoplasia
MTC	medullary thyroid carcinoma
MTS	multiple tumour suppressor
NF1	neurofibromatosis type 1
NF2	neurofibromatosis type 2
NGF	nerve growth factor
NLS	nuclear localization signal
NSCLC	non-small cell lung cancer
NSE	neuron-specific enolase
PALA	N-(phosphonacetyl)-L-aspartate
PCR	polymerase chain reaction
PFGE	pulsed-field gel electrophoresis
pHPT	primary hyperparathyroidism
PNS	peripheral nervous system
PTH	parathyroid hormone
RARA	retinoic acid receptor α
RB	retinoblastoma
RER	replication error
RFLP	restriction fragment length polymorphism
RGC	ribosomal gene cluster
RSV	Rous sarcoma virus
RT	reverse transcriptase
RTK	receptor tyrosine kinase
SCLC	small cell lung cancer
SDS–PAGE	sodium dodecyl sulphate–polyacrylamide gel electrophoresis
SSCP	single-stranded conformation polymorphism
SV40	simian virus 40
TBP	TATA binding protein
TGF	transforming growth factor
WT1	Wilms' tumour
XP	xeroderma pigmentosum

Preface

In the late 1970s and early 1980s, the analysis of the dominantly transforming oncogenes offered new hopes for the understanding of the genetic basis of cancer. Although much has been learned about signal transduction in tumour and normal cells from these studies, the underlying mechanisms of tumour initiation were not forthcoming. With the development of sophisticated gene cloning strategies, the late 80s and early 90s belong to the analysis of the recessive oncogenes, or 'tumour suppressor genes'. This class of genes play an important role in the normal development and differentiation of cells and have been shown to have diverse roles ranging from extracellular signal recognition to DNA repair. The aim of producing this volume was to provide an up-to-date discussion of the cloning and analysis of genes which are critical in the development and progression of human cancers. It has not been possible to review all the different tumour types which show a hereditary predisposition; in particular, the leukaemias, which could fill a volume on their own, have been deliberately excluded in favour of solid tumours.

Due to the clinical importance of the hereditary cancer genes, their study represents a very rapidly moving area in biomedical research. For example, during the relatively short period in which this volume was compiled, one of the genes responsible for hereditary breast cancer was cloned and the location of a second determined. However, by selecting authors directly involved in the cloning and analysis of tumour suppressor genes, the most up-to-date information possible is presented together with insights into how this information is being used in the clinical management of patients with these tumours. The analysis of mutations in tumour suppressor genes is essential for genetic screening and also provides for a fundamental understanding of the function of many of these genes. Thus, wherever possible, the nature of mutations in these genes is reviewed to provide a valuable reference source for these abnormalities.

Finally, the majority of the chapters are written by individuals who are part of the teams involved in the genetic screening of patients with hereditary forms of cancer, and so discuss the relevance of basic research to the counselling of families. This volume is therefore of interest to clinical as well as research scientists. The book has had an accelerated production time with the aim of providing information both current and topical.

John K. Cowell (*London*)

Foreword

Cancer is essentially a somatic evolutionary process. Each individual cancer is an independent realization of this process, and so the existence of genetic mutations that are common to a given type of cancer, such as in the *APC* gene, or that are common to many types of cancers, such as for *p53* and the *RAS* gene family, represent examples of convergent evolution where independent mutation and selection events pick out the same sorts of changes. The clue to the fundamental understanding of cancer is the identification of the genetic mutations that occur in these evolutionary processes and the basis for their selective advantages, namely their functional significance. From this will flow not only new understanding of the biology of tumours but also new approaches to their prevention, early diagnosis and cure, the latter in particular through the identification of novel targets for drug development. The localization and subsequent positional cloning of genes causing inherited cancer susceptibilities, using genomic analysis techniques, has played a major role in the uncovering of key tumour suppressor genes in this evolutionary sequence. Their importance follows from Knudson's seminal hypothesis that genes which are mutated somatically to give rise to key events in the initiation and progression of a tumour will also arise in the germline and then often be a basis for an inherited cancer susceptibility. This is particularly the case for tumour suppressor genes, when what is inherited is a single malfunctioning gene, whereas what is required in the evolution of the tumour is a complete knocking out of function. This requires a second event which leads to the loss of function of the remaining normal version of the tumour suppressor gene. It is this latter process which is uncovered by loss of heterozygosity studies, pioneered by Web Cavanee and his colleagues. This up-to-date and informative book on the molecular genetics of cancer presents a series of chapters illustrating these genetic processes through detailed discussion of particular examples of cancers and oncogenes.

Each chapter has its own author(s) and accompanies a description of the tumour, its pattern of occurrence or inheritance, and then the mutations involved and a discussion of their functions. In some cases, as in breast cancer, the story is only just now unfolding and, while the *BRCA1* gene has been identified, it does not yet fit into the pattern of a typical tumour suppressor gene, and may more often be mutated in sporadic ovarian than breast cancers, suggesting that the story is as yet fairly incomplete. Retinoblastoma is the first and classical example of a Knudsonian tumour suppressor gene whose basic function is being well established. Yet we still do not understand why carriers of a mutant form of the *RB* gene are mainly susceptible to retinoblastomas and not even to breast cancers, which often harbour mutated forms of the retinoblastoma gene. Wilms' tumours are perhaps most interesting because the gene with which they are associated has an interesting developmental role. The neurofibromatosis genes, on the other hand, especially *NF2*, teach us that mutations in a single gene can give rise to an extraordinarily wide range of phenotypes whose basis, in spite of the understanding of the function of the *NF2* gene, is still a major puzzle. The multiple endocrine neoplasia gene has a similarly striking message.

Colon cancer is one of the best developed stories, though why the *APC* gene, which is expressed in all tissues, should be so strikingly associated just with the colon is still something of a mystery. Its probable key role in the control of intercellular adhesion is consistent with the importance of disrupting cell–cell and cell–basement membrane contacts for the early acquisition by epithelial cells of the property of independent growth. The discovery of the mismatch repair gene mutations which underlie the hereditary non-polyposis cancer (HMPCC) syndromes is a remarkable example of perceptive positional cloning, together with functional analysis. But it is important not to let these observations mislead us into believing that selection for increased mutation rates in the somatic development of a tumour is of any importance. These changes, as in the case of *p53*, are most likely to be positively selected for because of a possible role in preventing apoptosis.

The story of the *p53* gene is expertly and thoroughly revealed. It, above all, illustrates the way that analysing the patterns of mutations at the DNA level can tell us about tumour aetiology. The fact that lung cancers have exactly those types of mutations that would be predicted to occur as a result of the carcinogens in cigarette smoke, is perhaps now the best direct evidence for the role of cigarette smoking in lung cancer, if anyone still has any doubts about this. *p53* changes are an extraordinary probe for environmental effects revealing, for example, just those mutations that UV causes in skin cancer and the extraordinary focus of mutations, presumably associated with Aflatoxin B1, in liver cancer. The rhabdomyosarcomas are an intriguing example of a solid tumour in which translocations play a key role, an event that is so strikingly characteristic of leukaemias and lymphomas, but is notably represented hardly at all in carcinomas. The lung cancers tell, of course, the *p53* mutation story but are as yet at a very early stage of their genetic elucidation, and no doubt future such volumes will fill in the story as it develops for other major tumours such as of the prostate and ovaries.

This multi-author volume on the molecular genetics of cancer, appearing in a most timely fashion, provides not only a valuable overview of key areas of the genetics of cancer, but also serves as a valuable reference source for much of the latest work in this fascinating field.

Walter Bodmer
(Imperial Cancer Research Fund, London, UK)

Retinoblastoma: for the next generation

Brenda L. Gallie, Yong-Jiang Hei, Homa Mostachfi and
James M. Dunn

The prototype tumour suppressor gene, *RB1*, was cloned 9 years ago, in 1986. Despite the clear benefit to families with retinoblastoma of identification of the mutation in *RB1* predisposing to disease, and the apparent economic benefits, relatively few mutations have been identified and applied in clinical medicine. This is because the gene is large, no 'hot spots' have been identified and the technologies applied to the task have, to date, been research tools, poorly suited for scaling up to clinical service. We anticipate that this problem will soon be solved.

Once *RB1* was cloned, it was recognized that the protein product, RB, interacts with other nuclear proteins of the cell cycle, and the mimics expressed by DNA tumour viruses, which achieve transformation of mammalian cells in part by inactivating RB. Many groups have rapidly defined the interactions and changes in RB during the cell cycle. However, these studies have not identified why children with *RB1* mutations get retinoblastoma. Indeed, initiation of cancer in the absence of RB appears to depend, instead, on its developmental function. Tissues destined for terminal differentiation may be critically dependent on RB in order to respond to differentiation signals with permanent cessation of proliferation. The transgenic and 'knockout' transgenic mice support this role for RB, but do not yet explain why children get retinoblastoma.

1.1 History

1.1.1 Inheritance

Retinoblastoma has been the prototype for the genetic basis of cancer since the first reported family illustrated the autosomal dominant transmission of predisposition to retinoblastoma (Newton, 1902). Review of many families subsequently revealed that all bilateral retinoblastoma patients and approximately 4–5% of unilateral retinoblastoma patients carry a mutant allele of the *RB1* gene (Bonaiti-Pellie and Briard-Guillemot, 1981; Draper *et al.*, 1992), which predisposes them to development of retinal tumours in infancy.

1.1.2 Two-hit hypothesis

Knudson (1971) analysed child age at diagnosis of bilateral retinoblastoma tumours compared to unilateral tumours (85% non-hereditary), and formulated his pivotal hypothesis: retinoblastoma is induced by two mutations, the first (M_1) being present in the germline of hereditary cases, but only in the somatic retinal cell destined to become the tumour in the non-hereditary cases. Both hereditary and non-hereditary retinoblastoma tumours develop only when the second mutation (M_2) occurs in a retinal cell that already carries M_1. The concept of tumour suppressor genes was launched with the idea that M_1 and M_2 are the two alleles of the predisposing gene, RB1 (Comings, 1973).

1.1.3 Locus

The locus of RB1 was established at chromosome 13q1.4 by cytogenetic documentation of large deletions in the rare children with retinoblastoma and other developmental abnormalities (Yunis and Ramsay, 1978). Linkage studies with the isoenzymes of esterase D (ESD) confirmed that the RB1 gene at 13q1.4 was responsible for the hereditary predisposition to retinoblastoma (Sparkes et al., 1980; Connolly et al., 1983).

1.1.4 Loss of heterozygosity

The prediction that M_1 and M_2 are the two alleles of RB1 was proven correct, first by the observation that retinoblastoma tumours from individuals heterozygous for ESD often showed only one allele of this closely linked isoenzyme (Godbout et al., 1983) and, subsequently, by the recognition that large chromosomal regions surrounding RB1 showed similar loss of heterozygosity (LOH), detected by polymorphic molecular markers, in retinoblastoma tumours (Cavenee et al., 1983). Both hereditary and non-hereditary retinoblastoma tumours showed LOH, confirming Knudson's hypothesis. The observation of LOH became the classical way to identify chromosomal regions encompassing tumour suppressor genes involved in the induction of other specific types of cancer.

1.1.5 RB1

The RB1 gene was cloned by identifying chromosome 13 markers that were completely missing from a few retinoblastoma tumours (Dryja et al., 1986), and identifying a cDNA clone within the deleted region (Friend et al., 1986). RB1 spans 27 exons and 180 kb of genomic DNA, and contains two very large introns (Toguchida et al., 1993). The messenger RNA is 4.7 kb with 3 kb of coding sequence. The promoter of the RB1 gene has been defined and is contained within a 1.5-kb fragment of genomic DNA (Gill et al., 1994).

1.1.6 Origin of RB1 mutations

Only 10–15% of *RB1* germline mutations are inherited. The majority arise in the first affected child in the family as a new mutation. These new mutations are much more likely to occur on the paternal than the maternal chromosome, perhaps because sperm are at greater risk of mutagenesis than ova (Dryja *et al.*, 1989; Zhu *et al.*, 1989).

1.1.7 Non-RB1 mutations in retinoblastoma tumours

Although both alleles of *RB1* are mutated in all retinoblastoma tumours, this change alone may not be sufficient for malignancy. All retinoblastoma tumours show further karyotypic abnormalities (Squire *et al.*, 1984, 1985). The iso(6p) rearrangement is present in 70% of retinoblastoma tumours but is almost undocumented in other tumours (Squire *et al.*, 1984). The i(6p) usually results in four copies of 6p, or low-level 'amplification'. Some interesting genes related to cell cycle control are encoded on 6p, for example cyclin D3 (Inaba *et al.*, 1992), but their roles in retinoblastoma tumour induction are not known.

The proto-oncogene, *MYCN*, named for its frequent amplification in neuroblastoma, is occasionally also amplified in retinoblastoma tumours. However, the level of expression of *MYCN* in most retinoblastoma tumours is similar to the level of expression in normal fetal retina. The 'DEAD' box gene, *DDX1*, with homology to RNA helicase proteins, is frequently expressed and co-amplified with *MYCN* in retinoblastoma tumours (Godbout and Squire, 1993; Godbout *et al.*, 1994). Also amplified or overexpressed in some retinoblastoma lines is the α-subunit of mitochondrial ATP synthase (Godbout *et al.*, 1993). The roles of these proteins in progression of malignancy are undefined.

Retinal tumours, called retinomas, that do not progress to malignancy are documented in clear association with *RB1* gene mutations (Gallie *et al.*, 1982a,b). Rarely, these have been observed, after years of dormancy, to acquire foci of full malignancy, necessitating treatment (Eagle *et al.*, 1989). Retinomas have not been molecularly studied since the tissue is not generally available. We hypothesize that both *RB1* alleles are mutant in retinomas but that further genetic changes necessary for full malignancy are missing; if one cell of a retinoma gains further mutations, retinoblastoma results.

1.2 Mutations of RB1

1.2.1 Impact of molecular knowledge of RB1 on health care

Without knowledge of the *RB1* mutation in the proband or identification of the mutant chromosome in familial cases, all infant relatives, including offspring, siblings, and first cousins, of a bilateral proband are subjected to intense clinical surveillance, in order to detect and treat tiny retinoblastoma tumours before they damage vision. If disease prediction can be made much more accurate by molecular studies, the normal children can be spared the testing, while scarce resources are effectively applied to those with mutations and at very high risk of developing tumours (Cowell, 1989). For familial retinoblastoma, polymorphic markers can provide accurate evaluation as to whether the mutant chromosome has been inherited (Wiggs *et al.*, 1988; Scheffer *et al.*, 1989; Onadim *et al.*, 1990); however, only 15% of all cases have a family history (Draper *et al.*, 1992) so, for the usual family with one proband, linkage studies using polymorphic markers are of limited use. Accurate tumour prediction in all families awaits an efficient and inexpensive method to detect new mutations in *RB1*.

1.2.2 Mutation identification

Effect of mutations. The vast majority of *RB1* mutations result in truncated RB, which is not detected in the retinoblastoma tumour cells (Horowitz *et al.*, 1990). We have reviewed the published mutations that are defined well enough to predict the effect on RB. Of 135 *RB1* mutations in retinoblastoma tumours or patients with retinoblastoma, 99 (73%) predict a truncated protein. Two mutations affected the promoter (Sakai *et al.*, 1991a) and might alter levels of expression of *RB1*. Five large internal deletions, although 'in frame', would ablate known critical domains of RB (Canning and Dryja, 1989; Dunn *et al.*, 1989; Hashimoto *et al.*, 1991; Kato *et al.*, 1994). Sixteen mutations were predicted to affect splicing, but only two groups confirmed the effect of splicing (Dunn *et al.*, 1989; Shimizu *et al.*, 1994). Only 10% of the mutations were small enough to be able to define important regions of RB: eight internal, in-frame deletions and 11 missense mutations would be expected to result in a mutant but stable RB (see *Figure 1.1*). In only a few instances have the effects of these mutations on RB been published (Horowitz *et al.*, 1989; Yandell *et al.*, 1989; Bignon *et al.*, 1990; Kratzke *et al.*, 1994; Paggi *et al.*, 1994).

Methods for detection. About 5% of all retinoblastoma patients show a genetic abnormality detected by study of the karyotype (Yunis and Ramsay, 1978; Motegi, 1981). In the remainder of patients and retinoblastoma tumours, both chromosomes 13 appear entirely normal by microscopic examination. Confirmation of the identity of the newly cloned *RB1* gene was provided by Southern blot analysis of retinoblastoma tumours showing internal deletions (Canning and Dryja, 1989; Shew *et al.*, 1990a,b; Kloss *et al.*, 1991). However, less than 20% of tumours show any abnormality by Southern or Northern blot analysis (Goddard *et al.*, 1988), which will only detect large rearrangements or promoter malfunction.

Several techniques have been effective, although not efficient, in detecting the more common small abnormalities, including nucleotide substitutions. RNase protection of the cDNA generated by reverse transcriptase from the tumour RNA revealed small mutations in 11 samples (Dunn *et al.*, 1988, 1989). Single-stranded

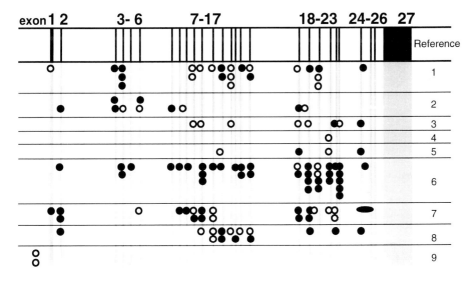

Figure 1.1. Well-characterized *RB1* mutations that have induced retinoblastoma. The mutations are distributed throughout the gene, and generally result in truncated, unstable protein, not detectable in tumours. (●) Mutations detectable by fragment analysis of size or copy number of exons (*n*=75); ○ single nucleotide substitution mutations, requiring other methods for detection (*n*=42). Mutations are grouped by the research teams that identified them: 1 (Hogg *et al.*, 1992, 1993; Kratzke *et al.*, 1994; Onadim *et al.*, 1992); 2 (Blanquet *et al.*, 1993); 3 (Dryja *et al.*, 1993; Yandell *et al.*, 1989); 4 (Weir-Thompson *et al.*, 1991); 5 (Kato *et al.*, 1994); 6 (Lohmann *et al.*, 1992, 1994a,b); 7 (Dunn *et al.*, 1989) and unpublished data; 8 (Shimizu *et al.*, 1994); 9 (Sakai *et al.*, 1991a). The methods used to detect mutations introduce a bias in the types of mutations found.

conformation polymorphism (*Figure 1.2a*) has been used to identify 58 *RB1* mutations (Hogg *et al.*, 1992, 1993; Kato *et al.*, 1994; Sachse *et al.*, 1994; Shimizu *et al.*, 1994; Mostachfi, personal communication). By this means, 40–60% of the mutations in tested retinoblastoma samples have been identified. Less efficient was denaturant gradient gel electrophoresis since mutations were found in 10/120 samples tested, although only 11 exons were studied (Blanquet *et al.*, 1993).

Since a significant fraction of *RB1* mutations involve changes in the sizes of exons, two groups have developed 'fragment analysis', consisting of multiplex methods to examine the length of exon/promoter fragments as a screening procedure. Lohmann *et al.* (1994b) found the germline mutation in 27 out of 104 familial or bilateral retinoblastoma patients by relatively efficient means, grouping exon fragments into 12 sets. We have developed a similar strategy, and included control fragments from other chromosomes in order to assess also the copy number of each fragment. An efficient first screening for mutations that alter exon size or quantity is anticipated to identify 50% of the mutations (*Figure 1.1*). This can be followed by techniques which are less efficient but more sensitive.

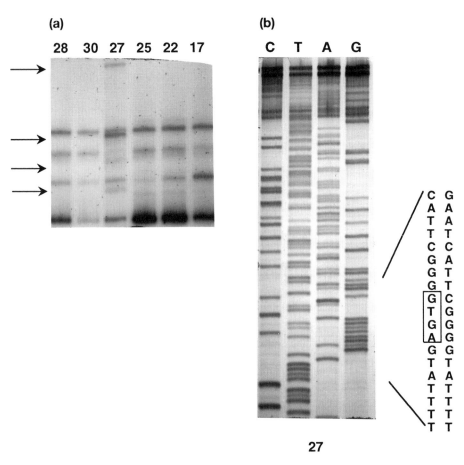

Figure 1.2. Detection of *RB1* mutations. (a) SSCP reveals a series of abnormal bands (arrows) in exon 24 of the patient sample, lane 27. (b) Manual sequencing shows a four nucleotide insertion in exon 24, resulting in many subsequent abnormal bands in the sequence. (a and b courtesy of J. Cowell.) (c) Automated sequencing (Pharmacia A.L.F. sequencer) shows heterozygosity for a G→A substitution ('Y') resulting in Trp195STOP in exon 6 of *RB1* in the blood of a child with bilateral retinoblastoma. The child's tumour showed homozygosity for the mutation.

Once abnormal fragments or patterns have been detected, sequencing is applied by all groups to identify and describe the precise mutation (*Figure 1.2b*). The observation that C → T mutations in CGA-arginine codons are particularly common suggests that exons containing these codons should be studied first (Cowell *et al.*, 1994). Sequencing of each fragment is also proposed when all the screening methods fail to detect the mutant fragment. Automation and optimization for repetitive sequencing of *RB1* may make this last step feasible (*Figure 1.2c*).

(c)

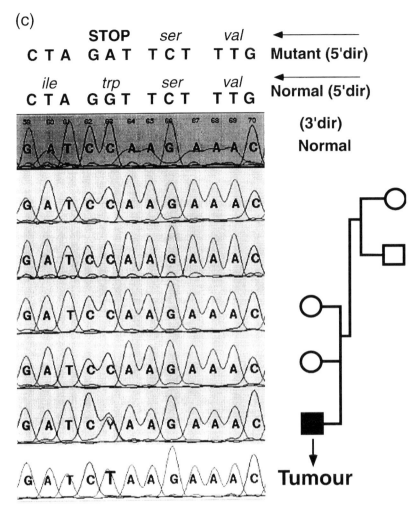

Figure 1.2. Continued.

It would be easier to identify mutations in *RB1* by studying the 3000 bp of coding transcript than by studying the 27 exons of genomic DNA (Dunn *et al.*, 1988). However, RNA from the mutant allele is not generally detectable in the presence of a normal allele, for example, in blood (Dunn *et al.*, 1989). The same mutant allele is readily detectable when retinoblastoma tumour is examined. The inability to detect the mutant allele in the presence of normal RB is presumed to be due to a combination of unstable RNA when translation is incomplete and autoregulation by RB (Gill *et al.*, 1994; Shan *et al.*, 1994). Only when tumour cells are available can RNA, instead of the genomic exons, be studied for detection of the mutation.

RB1 *mutations in other tumour types.* Many tumour types which are not initiated by *RB1* mutations show mutations in this gene (Gallie *et al.*, 1993). These are well documented in small cell lung carcinoma (Harbour *et al.*, 1988; Bignon *et al.*, 1990; Hensel *et al.*, 1990; Mori *et al.*, 1990; Murakami *et al.*, 1991; Kashii *et al.*, 1994), bladder carcinoma (Horowitz *et al.*, 1989, 1990), breast cancer (Lee *et al.*, 1988; T'Ang *et al.*, 1988; Wang *et al.*, 1993b) and osteosarcoma (Wadayama *et al.*, 1994). However, almost every human tumour type can show *RB1* mutations, perhaps associated with tumour progression.

Low penetrance RB1 *mutations.* Although the usual retinoblastoma family has a mean of seven tumours (Lohmann *et al.*, 1994a), some families develop far fewer tumours, demonstrating 'low penetrance' of this dominantly inherited trait. Since it is hard to count the number of tumours accurately, the number of eyes affected has been used to estimate the degree of penetrance. Whereas the usual family has two affected eyes per patient, low penetrance families have many unilateral patients and unaffected gene carriers. *RB1* mutations have been characterized in nine such families (*Figure 1.3*) and all are predicted to result in a stable protein, either due to in-frame deletion (Connolly *et al.*, 1983; Dryja *et al.*, 1993; Lohmann *et al.*, 1994a; Du *et al.*, in preparation; and our unpublished data), missense mutation (Onadim *et al.*, 1992; Kratzke *et al.*, 1994; Lohmann *et al.*, 1994a; and our unpublished data) or promoter mutations resulting in reduced transcription (Sakai *et al.*, 1991b).

Of the 25 identified mutations predicted to result in a stable protein (*Figure 1.3*), four were in tumours other than retinoblastoma and six were present only in the retinoblastoma tumour, not the germline of the family. Fifteen of the 25 were germline mutations associated with retinoblastoma and 9/15 exhibited a low penetrance phenotype. The remaining six germline mutations had deletions or missense mutations within the critical A, B or C domain. Thus, the aetiology of low penetrance retinoblastoma appears to be a 'weak' germline mutation; perhaps LOH for such mutations is insufficient to induce tumours and the M_2 event must be a complete disruption of RB.

1.3 Function of RB

Loss of functional *RB1* leads to malignancy, implicating an inhibitory role for the *RB1* gene product, RB, in cell proliferation (Bernards *et al.*, 1989), as befits a tumour suppressor gene (*Figure 1.4a,b*). Binding of RB to the transforming proteins of several DNA viruses indicated that uncontrolled proliferation required inactivation of RB (*Figure 1.4c*). The transforming abilities of the viral proteins, adenovirus E1a, simian virus 40 (SV40) large T antigen (AgT) and human papillomavirus E7, were shown to be directly correlated with their RB binding abilities, and mutant viral proteins that failed to bind RB also lost their transforming

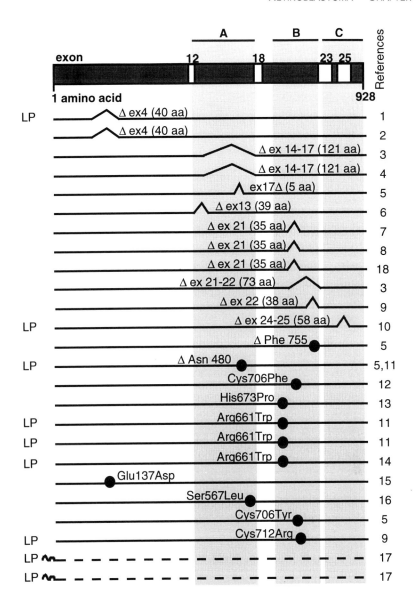

Figure 1.3. The major RB functional domains. A and B (the pocket domain), and C, are indicated at the top. Mutations of *RB1* in tumours or constitutional cells of retinoblastoma patients are indicated as deletions or single amino acid changes. LP, low penetrance family. References: 1 (Dryja *et al.*, 1993); 2 (Hogg *et al.*, 1993); 3 (Hashimoto *et al.*, 1991); 4 (Canning and Dryja, 1989); 5 (Lohmann, *et al.*, 1994a); 6 (Kato *et al.*, 1994); 7 (Horowitz *et al.*, 1990); 8 (Horowitz *et al.*, 1989); 9 and 10 (Gallie, unpublished data); 11 (Lohmann *et al.*, 1992); 12 (Bignon *et al.*, 1990; Hensel *et al.*, 1990); 13 (Sachse *et al.*, 1994); 14 (Onadim *et al.*, 1992; Kratzke *et al.*, 1994); 15 (Blanquet *et al.*, 1993); 16 (Yandell *et al.*, 1989); 17 (Sakai *et al.*, 1991a); 18 (Weir-Thompson *et al.*, 1991).

9

potential (DeCaprio *et al.*, 1988; Whyte *et al.*, 1988; Dyson *et al.*, 1989; Heck *et al.*, 1992). The interaction of RB and the viral oncoproteins appears to involve a region of RB described as a 'pocket', consisting of two non-contiguous regions, termed A and B, between amino acids 393–572 and 646–772, which are frequently affected by mutations or deletions in tumours (Hu *et al.*, 1990) (*Figure 1.3*).

1.3.1 Regulation of RB

RB is a 110-kDa phosphoprotein that localizes to the nucleus. AgT preferentially binds to the underphosphorylated RB (Ludlow *et al.*, 1989), suggesting that this may be the active form of RB. Consistent with this notion, introduction of RB into the osteosarcoma cell line Saos-2, which lacks functional RB, induces growth arrest only in G_1 cells, where RB is in the underphosphorylated form (Goodrich *et al.*, 1991). Phosphorylation of RB by ectopic expression of cyclin A and E renders the protein incapable of suppressing the growth of Saos-2 cells (Hinds *et al.*, 1992). It has also been shown that a mutant RB that can not be phosphorylated has increased activity in repression of transcription (*Figure 1.4d*) (Hamel *et al.*, 1990). Thus, phosphorylation represents a major mechanism for regulation of the function of RB.

RB is underphosphorylated in the G_0/G_1 phases, hyperphosphorylated prior to the G_1/S transition and throughout the S phase, and dephosphorylated at mid-to-late mitosis (Buchkovich *et al.*, 1989; Chen *et al.*, 1989; Ludlow *et al.*, 1989; Mihara *et al.*, 1989). Phosphorylation of RB occurs on multiple serine and threonine amino acids within domains recognized and bound by cyclin-dependent kinases (cdks) (Lees *et al.*, 1991). Early studies demonstrated that purified p34cdc2 was able to associate with, and phosphorylate, RB on sites that are also phosphorylated *in vivo* (Lees *et al.*, 1991; Lin *et al.*, 1991; Hu *et al.*, 1992). However, activation of cdc2 kinase occurs in the G_2/M phase, at which time phosphorylation of RB has already started to decrease. Indeed, the RB kinase may be p33cdk2 (Kato *et al.*, 1993), since the timing of activation of p33cdk2 correlates well with the phosphorylation of RB (Akiyama *et al.*, 1992). Activation of cdk2 requires association with cyclin E or A. More importantly, however, additional G_1 cyclins such as cyclins D1–3 were found to associate with RB, via the LXCXE motif that is present in the viral oncoproteins that bind RB (Dowdy *et al.*, 1993; Ewen *et al.*, 1993). These cyclins can form complexes with, and activate, other cdks (cdk4–6), which may be collectively responsible for the multiple-site phosphorylation of RB in a coordinated manner throughout the cell cycle. Thus, phosphorylation of RB occurs through the action of cyclins D1–3/cdk4–6 complexes in the G1 phase of the cell cycle, a cyclin E/cdk2 complex in the late G1 to early S phase, and a cyclin A/cdk2 complex in the S phase. RB exists in multiple, distinct states of phosphorylation, where subsets of RB with distinct patterns of phosphorylation exhibit different affinities for nuclear tethering (Mittnacht *et al.*, 1994) and interactions with viral oncoproteins and cellular transcription fac-

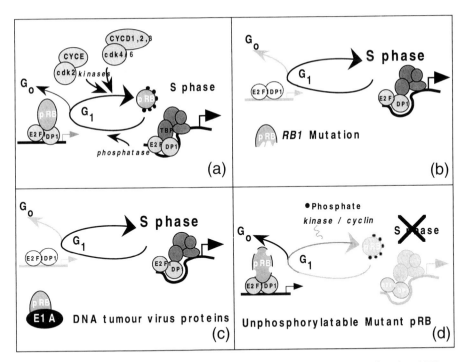

Figure 1.4. Control of the cell cycle by RB (pRB). (a) Normal underphosphorylated RB binds to E2F/DP1 heterodimers to block their activation of transcription. Phosphorylation of RB by cyclin (CYC)E/cdk2 or cyclins D1–3/cdk4-6 complexes releases E2F to interact with TBP and transcription initiation complexes. Genes with binding sites for E2F are activated, promoting entry into the S phase. (b) Mutation of *RB1* usually results in truncation and unstable, undetectable protein. E2F is continuously free to activate S phase, resulting in cell division at a time of development when cells should normally enter G_0. (c) DNA tumour viruses activate the mammalian cell cycle in part by inactivating RB, releasing E2F. (d) The unphosphorylatable mutant (Hamel *et al.*, 1992b) is constitutively active, preventing cells from leaving G_0, despite signals to re-enter the cell cycle.

tors (Templeton, 1992). As a result, fine regulation of cell growth may be achieved by differential phosphorylation of RB.

Since the active form of RB is underphosphorylated, we constructed an RB that was mutant at eight of the potential phosphorylation sites, while remaining functional in all other domains (Hamel *et al.*, 1992a,b) (*Figure 1.4d*). The efficacy of this mutant in controlling the cell cycle has been shown recently by using it to block the benign, but potentially life-threatening, proliferation of the vascular intimal cells, following therapeutic angioplasty (Chang *et al.*, 1995). The unphosphorylatable RB was delivered to the damaged vascular segment by an adenovirus vector and was shown to block serum-stimulated proliferation and entry into the S phase *in vitro*, and DNA synthesis and neointimal proliferation in two different *in vivo* models. Such use of RB offers significant opportunities in many aspects of human disease.

Alteration of RB function is also accomplished by regulation of transcription. Several *cis*-acting elements have been identified in the *RB1* promoter, including binding sites for Sp-1, ATF and E2F-1 (Zacksenhaus *et al.*, 1993; Gill *et al.*, 1994). RB may also negatively autoregulate its own expression (Dunn *et al.*, 1989; Gill *et al.*, 1994; Shan *et al.*, 1994). The protein product of the tumour suppressor gene, *p53*, may also regulate the expression of RB, since a region of the *RB1* promoter, homologous to a single p53 binding site, has been identified (Osifchin *et al.*, 1994).

1.3.2 Interaction of RB with transcription factors

In addition to regulation by phosphorylation, the growth-suppressing function of RB depends on the presence or absence of other factors. For instance, the effect of introduction of RB into retinoblastoma lines and other tumours lacking functional RB varies from complete growth arrest, as in the case of Saos-2 cells (Templeton *et al.*, 1991), to negligible effects in other cell lines, such as C33A cells and retinoblastoma cells (Muncaster *et al.*, 1992), unless highly expressed. Thus, the cell line Saos-2 may express the factors critical for responding strongly to RB, while other malignant cells may have acquired mutations in these factors that prevent RB from blocking the cell cycle.

Several of the cellular transcription factors interacting with RB have been identified: E2F-1, ATF2, PU.1 and Elf-1 (Chellappan *et al.*, 1991; Kim *et al.*, 1992a; Hagemeier *et al.*, 1993; Wang *et al.*, 1993a). The interaction of RB with many transcription factors may be due to the homology between the A and B domains of RB and the general transcription factors TFIID and TFIIB, respectively (Hagemeier *et al.*, 1993).

The best characterized of the above transcription factors, E2F-1, is a member of a family of transcription factors containing the basic-helix–loop–helix (bHLH) motif. E2F was initially identified in adenovirus-infected cells as a DNA binding activity which activated the promoter of adenovirus *E2a* (reviewed in Nevins, 1992). E2F is also known as differentiation-regulated transcription factor 1 (DRTF1), due to an association with a decrease in transcription activity during the differentiation of embryonal carcinoma cells (La-Thangue, 1994). Four different members of the *E2F* gene family (*E2F-1, -2, -3* and *-4*) have been cloned from mammalian cells and share the features of an N-terminal DNA binding domain and a C-terminal transactivation domain which harbours an RB binding site (Kaelin *et al.*, 1992; Shan *et al.*, 1992; Ikeda and Nevins, 1993; Ivey-Hoyle *et al.*, 1993; Beijersbergen *et al.*, 1994). Homologues of *E2F* have recently been cloned from *Drosophila*. Significantly, nine out of 18 amino acids in the putative RB binding site are conserved (Ohtani and Nevins, 1994), suggesting important functional significance of the RB binding site. *Drosophila RB* gene has not been identified.

Stable interaction of E2F-1 with RB requires DP-1, a member of another family of E2F-related proteins (Bandara *et al.*, 1993) (*Figure 1.4a*). E2F-1 and E2F-2 each form heterodimers with DP-1 that bind RB with an increased affinity (Helin *et al.*, 1993). Structurally, DP-1 resembles E2F-1, consisting of a DNA binding domain with bHLH motif and a transactivation domain, but no RB binding site (Bandara *et al.*, 1993). The E2F-1/DP-1 complex also associates with the cyclin A/cdk2 or cyclin E/cdk2 complex (Pagano *et al.*, 1992). Such interactions have been shown to phosphorylate E2F-1 on serines 332 and 337, resulting in the blocking of interactions with RB and DNA (Dynlacht *et al.*, 1994; Krek *et al.*, 1994; Xu *et al.*, 1994).

The DNA binding domain of E2F-1 recognizes a putative consensus E2F binding motif (TTTCGCGC) that has been found in the adenovirus *E2* promoter as well as the promoter of a number of cellular genes including dihydrofolate reductase (Blake and Azizkhan, 1989), *cdc2* (Dalton, 1992; Yamamoto *et al.*, 1994), *MYC* (Robbins *et al.*, 1990; Hamel *et al.*, 1992b; Adnane and Robbins, 1995), *MYCN* (Robbins *et al.*, 1990), thymidine kinase (Dou *et al.*, 1992), thymidylate synthase and DNA polymerase α (Pearson *et al.*, 1991), cyclin A (Pagano *et al.*, 1992), *MYB* (Lam *et al.*, 1994), *RB1* (Gill *et al.*, 1994; Ohtani-Fujita *et al.*, 1994), and *E2F* itself (reviewed in Farnham *et al.*, 1993). The fact that all of the cellular genes containing putative *E2F* elements are cell cycle-related strongly suggests that *E2F* is a critical regulator in cell cycle progression. Such a notion also helps us to understand the mechanism of RB action in viral transformation, cell cycle regulation and malignant progression. The interaction of RB with E2F inhibits E2F-mediated transcription (Hiebert *et al.*, 1992). By binding to, and blocking the action of, E2F on E2F binding sites in promoters, RB can inhibit the expression of genes that are required for cell division and suppress cell growth, while inactivation of RB, by interaction with viral oncoproteins, phosphorylation or mutation, results in the release of E2F to stimulate cell growth and promote transformation or malignancy. In the case of infection by adenoviruses, the LXCXE motif of E1a binds to the A/B domains of RB, displacing E2F-1 (Hiebert *et al.*, 1992; Qian *et al.*, 1992; Ikeda and Nevins, 1993).

However, the E2F sites in the *MYB* promoter may act as negative elements (Lam and Watson, 1993); when the sites are mutated, transcriptional activity increases. Although RB can bind to the E2F element, it is not required for the repressive activity, which is intact in cells with no RB, and is unaffected by reconstitution of these cells with RB, suggesting that factors other than RB may be responsible for such regulation. The RB-related protein, p107, has been shown to play a role in this process (Lam *et al.*, 1994).

Transactivation by E2F-1 involves DNA bending, shown by circular permutation analysis, centred on the E2F element (Huber *et al.*, 1994). The binding of RB to E2F-1 significantly reduced the angle of DNA bending, and shifted the centre of the bend in a 3' direction. In addition, the half-life of dissociation of E2F/DNA was extended from 3 to 30–45 min by RB. The sequence context of

E2F sites, and the heterodimeric E2F partners available, may determine the angle and orientation of bending of E2F sites, allowing an array of positive and negative transactivation effects (Melillo *et al.*, 1994).

RB also positively regulates transcription. RB activates transcription of transforming growth factor β2 (TGF-β2) via an ATF site, although the TGF-β2 promoter also contains Sp-1 and E2F sites (Kim *et al.*, 1992b). The effect is mediated by ATF2 (Kim *et al.*, 1992b), which binds RB immediately C-terminal of the B domain, in a region containing a cluster of phosphorylation sites (Hamel *et al.*, 1992a). Interestingly, the interaction of RB with ATF2 is much reduced by JNK kinase-induced phosphorylation of ATF2 (Gupta *et al.*, 1995). Ectopic expression of RB causes a 3–6-fold increase in expression of an ATF-CAT reporter dependent on the intact ATF site, in six cell lines with or without intact endogenous RB (Park *et al.*, 1994). RB also induced the expression of the TGF-β1 gene in CCL-64 mink lung epithelial cells, but repressed its expression in NIH3T3 cells, suggesting that RB can positively or negatively regulate transcription, dependent on cell type (Kim *et al.*, 1991).

Earlier studies of the mapping of the elements involved in the negative regulation of *c-fos* and *TGF-β1* expression by RB led to the definition of a 'retinoblastoma control element' (RCE) (Robbins *et al.*, 1990; Kim *et al.*, 1991), which was subsequently identified as a site for the binding of the transcription factor Sp-1 (Kim *et al.*, 1992a). Transcription mediated through Sp-1 is stimulated by RB (Udvadia *et al.*, 1993), presumably through the binding of a negative regulator of Sp-1 by RB (Chen *et al.*, 1994).

In addition to the transcription factors described above, a lymphoid-specific Ets transcription factor, Elf-1, was shown to bind RB via the LXCXE motif in Elf-1. This interaction may be important in regulating T-cell function (Wang *et al.*, 1993a).

The proto-oncogenes, *MYC* and *MYCN*, have also been shown to interact with RB *in vitro* (Rustgi *et al.*, 1991). RB and TATA binding protein (TBP) may bind to very similar sites on both MYC and E1A, suggesting that RB may act by interfering with interaction of MYC with TBP; viral transforming proteins may disrupt this effect of RB (Rustgi *et al.*, 1991; Hateboer *et al.*, 1993). Unexpectedly, however, RB was also shown to stimulate MYC-mediated transcription *in vivo* in a cell type-specific manner (Adnane and Robbins, 1995). An interaction of RB with TBP is also suggested by its homology with the basal transcription factors TFIID and TFIIB. In addition to regulation of activated transcription, RB may also regulate basal transcription (Hateboer *et al.*, 1993). When RB was brought into a promoter with Gal4 binding sites, by fusion of RB to the DNA binding domain of Gal4, repression of basal transcription was suggested (Bremner *et al.*, 1995). The suppressive effect was modified by phosphorylation, shown by co-transfection of cells with cyclins E and A.

1.3.3 The RB family of proteins

Other proteins that resemble RB, structurally and functionally, have been identified by virtue of their interaction with adenovirus E1a protein, and are collectively known as 'pocket proteins'. Each contain A and B domains with high homology, and are recognized by the transforming proteins of the DNA tumour viruses (E1a, AgT, and E7). This family includes RB, p107 and p130. It is likely that the pocket proteins are required to function at different times during cell cycle and development, and in different cells (Cobrinik *et al.*, 1993).

p107 (Ewen *et al.*, 1991) binds to E2F in the presence of cyclins, similar to RB but with different cell cycle timing (Cao *et al.*, 1992), and has activity as a transcriptional repressor, similar to RB (Zamanian and La-Thangue, 1993; Zhu *et al.*, 1993). p107 specifically associates with E2F-4, but not the E2F-1, -2 or -3 isoforms (Beijersbergen *et al.*, 1994; Ginsberg *et al.*, 1994). By association with MYC, p107 suppresses MYC-activated transcription; this has not been demonstrated for RB (Gu *et al.*, 1994).

p130 is a 'pocket' protein related more closely to p107 than to RB (Hannon *et al.*, 1993; Li *et al.*, 1993; Mayol *et al.*, 1993). Like RB and p107, p130 binds SV40 AgT, E2F, cyclin A, cyclin E and cyclin D. However, the timing of complex formation with these molecules is apparently different for p130, RB and p107 (Hannon *et al.*, 1993; Li *et al.*, 1993; Mayol *et al.*, 1993), suggesting that each pocket protein may have distinct functions in cell cycle regulation.

p300 is another E1a binding protein which has been shown to be a component of the TBP complex (Abraham *et al.*, 1993), and may regulate cell growth (Moran, 1993). However, structural comparison indicates that p300 may be a member of a family of proteins that bind the cAMP response element-binding protein (CREB) (Arany *et al.*, 1994), and not a pocket protein.

The pocket proteins cannot fully substitute for each other, since mice with no normal *RB1* gene (*RB–/–*) have normal pocket proteins except for RB, but do not survive to birth (see below). However, muscle development in these mice may be normal, suggesting partial substitution for RB, perhaps by p107, in muscle (Schneider *et al.*, 1994).

1.4 RB in cell differentiation and development

Inhibition of cell division is a prerequisite for cell differentiation (Yen *et al.*, 1992). Since the cell cycle proceeds very well with no RB at all, the real function of RB may be to prepare the cell for differentiation. The tissues that are completely dependent on RB for that differentiation may be the tissues that become malignant in the absence of RB, such as retina. Other, less dependent, tissues may achieve differentiation by other means.

1.4.1 Interaction of RB with bHLH proteins

RB was shown to interact directly with a family of bHLH proteins which are products of 'master' genes for muscle differentiation. The best characterized of this family is MyoD, which binds the B domain and C-terminal part of RB through its bHLH domain (Gu *et al.*, 1993). Consistent with a requirement of RB to maintain the terminally differentiated state, muscle cells from *RB–/–* mice can be induced to re-enter the cell cycle by serum stimulation, while normal differentiated muscle cells can not (Schneider *et al.*, 1994). Furthermore, MyoD failed to induce muscle-specific genes in cells lacking functional RB or in cells transfected with AgT (Gu *et al.*, 1993; Maione *et al.*, 1994). Thus, RB may function in conjugation with genes such as *MyoD* to accomplish cell growth arrest and the onset of differentiation. However, maintenance of the undifferentiated state may require inactivation of RB. Indeed, the growth-promoting protein, Id-2, a member of the family of HLH proteins which function by heterodimerization with their partner bHLH proteins, has been shown to bind to hypophosphorylated RB to enhance proliferation specifically and reverse the RB-dependent block of cell cycle progression (Iavarone *et al.*, 1994).

1.4.2 Development and tissue-specific role of RB

RB may play a regulatory role in differentiation. In human promyelocytic leukaemia cells (HL-60) induced to differentiate *in vitro*, RB expression was down-regulated in the early 'pre-commitment' state (Yen *et al.*, 1993a,b), while blocking phosphorylation of RB by a src-kinase inhibitor promoted differentiation (Yen *et al.*, 1994). However, dephosphorylation of RB in cells stimulated to differentiate was significant only after a large portion of the cells had already been arrested and differentiated, suggesting a role of RB in the maintenance rather than induction of differentiation (Yen and Varvayanis, 1994).

The function of RB may be limited normally by the integrated pathways in the different tissues and stages of development within which it acts. By assaying the relative effectiveness of a wild-type AgT and a mutant AgT (defective only in RB binding) in stimulating DNA synthesis, the dependence on RB for suppression of DNA synthesis was found to differ in quiescent versus senescent fibroblasts (Sakamoto *et al.*, 1993). Sodium butyrate caused the quiescent fibroblasts to differentiate and respond like senescent fibroblasts.

1.4.3 RB1 *mutation in mice*

While natural mutation of *RB1* has not been identified in mice, *RB–/–* mice have been produced by homologous recombination (Clarke *et al.*, 1992; Jacks *et al.*, 1992; Lee *et al.*, 1992). The *RB–/–* mice die *in utero* between embryonic day 13.5 and 15.5 as a result of neuronal cell death and haematopoeitic abnormalities, while the heterozygous mice (*RB–/+*) are born normal. Unlike *RB–/+* humans,

$RB-/+$ mice do not develop retinoblastoma, but instead are prone to tumours in other tissues. Ninety per cent of the heterozygous mice developed tumours by 8 months in the intermediary lobe of the pituitary (Chang et al., 1993; Hu et al., 1994); 70% developed medullary thyroid carcinoma (Williams et al., 1994a). The intermediary lobe of the human pituitary is vestigial.

Further analysis of the $RB-/-$ mice revealed that embryonic neuronal cells lacking a functional $RB1$ gene failed to mature properly, had reduced levels of the neuronal markers, TrkA and TrkB, and the low-affinity neurotrophin receptor p75^{LNGFR}, and died by apoptosis shortly after entering the S phase (Lee, 1994). Peculiarly, the development of the lens of the eye was disrupted in the $RB-/-$ embryos with failure of differentiation and $p53$-dependent apoptosis in the lens fibre cells (Morgenbesser et al., 1994). Humans with heterozygous $RB1$ mutations do not have a tendency towards cataract formation.

Since $RB-/-$ mice die before many tissues, such as retina, are fully differentiated, two approaches have been taken to examine the function of $RB1$ in the development of these tissues. First, chimeric mice were created by injecting embryonic stem cells with double knock-out of both $RB1$ alleles into wild-type blastocysts, producing mice containing both wild-type cells and cells with only mutant $RB1$ alleles (Maandag et al., 1994; Williams et al., 1994b). These $RB-/-$ chimeric mice developed normally but exhibited enhancement of two phenotypes of the $RB-/+$ mice: cataract with failure of lens fibre differentiation and a very early and lethal tumour of the intermediary lobe of the pituitary. Human pituitary tumours (anterior, not intermediary lobe) do not show $RB1$ mutations (Woloschak et al., 1994; Zhu et al., 1994).

At the time of death of $RB-/-$ embryos, the retina shows no abnormalities, but normally exhibits very low expression of $RB1$. The chimeric mice demonstrated increased cell death and ectopic mitoses in inner nuclear layers at embryonic day 16.5–18.5, the time when $RB1$ expression is normally turned on in these layers (Szekely et al., 1992). Adult retina appeared normal, with a significantly reduced contribution of $RB-/-$ cells relative to other tissues. Thus, loss of $RB1$ caused cell death in differentiating murine retina, with no evidence of malignancy.

The failure of erythropoiesis noted in the $RB-/-$ mice was not noted in the chimeric mice, suggesting a defect in the environment required for erythropoiesis, rather than in the red cell precursors. Since the $RB-/-$cells contributed to all tissues, RB is dispensible for the achievement and maintenance of the differentiated state of most murine cell types.

The second approach was to knock out $RB1$ function by expressing viral proteins that are known to inactivate RB. SV40 AgT, which binds RB and many other cell cycle regulatory proteins, including p53, caused tumour development when expressed in lens (Mahon et al., 1987) or retina (Windle et al., 1990; al-Ubaidi et al., 1992). The viral protein HPV-16 E7, however, binds only RB. In mice expressing E7 protein in the lens, apoptosis of the lens fibre cells was evident (Fromm et al., 1994; Howes et al., 1994; Pan and Griep, 1994). The apoptosis

17

was apparently p53-dependent, since co-expression of HPV-16 E6 protein, which binds and inactivates p53, blocked the apoptosis. When E7 protein was expressed in the photoreceptor cells, the retina underwent degeneration due to death of photoreceptors by apoptosis. In *p53–/–* mice, inactivation of RB by expression of E7 protein in the retina led to retinal tumours. This suggests that both RB and p53 must be mutant in order for mouse retina to achieve full transformation. However, preliminary studies in our laboratory have revealed few *p53* mutations in human retinoblastoma.

Mice mutant for both *RB1* and *p53* were created by crossing *RB–/+* and *p53–/+* mice (Williams *et al.*, 1994a). *RB–/+; p53+/–* mice, in addition to 100% pituitary tumours and 83% thyroid tumours with LOH for *RB*, also developed a few novel tumours, but 0% lymphoma. *RB–/+; p53–/–* died before 6 months of age and had many novel tumours, such as islet cell, pineal (40%), bronchial hyperplasia (38%) and retinal dysplasia (41%), and tumours characteristic of *p53–/–* mice such as lymphoma (39%). The novel tumours showed LOH for *RB1* but only 1/7 lymphomas showed LOH for *RB1*, suggesting that loss of *p53* was sufficient for lymphoma.

The combined effects of *RB1* and *p53* mutations could be explained by several models: (i) each tumour suppressor gene negatively regulates cell growth and loss of both produces additive effects; (ii) loss of *p53* increases the *RB1* mutation rate; or (iii) absence of *p53* allows survival of *RB–/–* cells, which would otherwise die by apoptosis.

1.5 Therapy for retinoblastoma

Identification of the gene initiating retinoblastoma is close to being widely applied to help families with the disease. Use of *RB1* in therapy for retinoblastoma, however, is in the future and will depend on the detailed understanding of *RB1* function that is beginning to emerge from studies such as those described here. A large amount of basic research is occurring at the present time to characterize in detail the 'housekeeping' role of RB in determination of when a cell should next divide. Loss of this function in cells that are already malignant probably results in faster, more autonomous, growth of the cells and there may well be ways in the future to utilize derivatives of RB in inhibiting tumour growth. The developmental function of *RB1*, which may be to lock certain cell types forever in G_0 (terminal differentiation), is just beginning to be approached, but is highly relevant to our initial question: Why do children get retinoblastoma?

References

Abraham SE, Lobo S, Yaciuk P, Wang HG, Moran E. (1993) p300, and p300-associated proteins, are components of TATA-binding protein (TBP) complexes. *Oncogene* 8: 1639–1647.

Adnane J, Robbins PD. (1995) The retinoblastoma susceptibility gene product regulates Myc-mediated transcription. *Oncogene* **10**: 381–387.

Akiyama T, Ohuchi T, Sumida S, Matsumoto K, Toyoshima K. (1992) Phosphorylation of the retinoblastoma protein by cdk2. *Proc. Natl Acad. Sci. USA* **89**: 7900–7904.

al-Ubaidi MR, Font RL, Quiambao AB, Keener MJ, Liou GI, Overbeek PA, Baehr W. (1992) Bilateral retinal and brain tumors in transgenic mice expressing simian virus 40 large T antigen under control of the human interphotoreceptor retinoid-binding protein promoter. *J. Cell Biol.* **119**: 1681–1687.

Arany I, Rady P, Tyring SK. (1994) Interferon treatment enhances the expression of underphosphorylated (biologically-active) retinoblastoma protein in human papilloma virus-infected cells through the inhibitory TGF beta 1/IFN beta cytokine pathway. *Antiviral. Res.* **23**: 131–141.

Bandara LR, Buck VM, Zamanian M, Johnston LH, La-Thangue NB. (1993) Functional synergy between DP-1 and E2F-1 in the cell cycle-regulating transcription factor DRTF1/E2F. *EMBO J.* **12**: 4317–4324.

Beijersbergen RL, Kerkhoven RM, Zhu L, Carlée L, Voorhoeve PM, Bernards R. (1994) E2F-4, a new member of the E2F gene family, has oncogenic activity and associates with p107 in vivo. *Gene Devel.* **8**: 2680–2690.

Bernards R, Schackleford GM, Gerber MR, Horowitz JM, Friend SH, Schartl M, *et al.* (1989) Structure and expression of the murine retinoblastoma gene and characterization of its encoded protein. *Proc. Natl Acad. Sci. USA* **86**: 6474–6478.

Bignon YJ, Shew JY, Rappolee D, Naylor SL, Lee EY, Schnier J, Lee WH. (1990) A single Cys706 to Phe substitution in the retinoblastoma protein causes the loss of binding to SV40 T antigen. *Cell Growth Differ.* **1**: 647–651.

Blake MC, Azizkhan JC. (1989) Transcription factor E2F is required for efficient expression of the hamster dihydrofolate reductase gene in vitro and in vivo. *Mol. Cell. Biol.* **9**: 4994–5002.

Blanquet V, Turleau C, Gross M-S, Goossens M, Besmond C. (1993) Identification of germline mutations in the RB1 gene by denaturant gradient gel electrophoresis and polymerase chain reaction direct sequencing. *Hum. Mol. Genet.* **2**: 975–979.

Bonaiti-Pellie C, Briard-Guillemot ML. (1981) Segregation analysis in hereditary retinoblastoma. *Hum. Genet.* **57**: 411–419.

Bremner R, Cohen BL, Sopta M, Gallie BL, Phillips RA. (1995) Direct transcriptional repression by pRB or p107, and its reversal by specific cyclins. *Mol. Cell. Biol.* in press.

Buchkovich K, Duffy LA, Harlow E. (1989) The retinoblastoma protein is phosphorylated during specific phases of the cell cycle. *Cell* **58**: 1097–1105.

Canning S, Dryja TP. (1989) Short, direct repeats at the breakpoints of deletions of the retinoblastoma gene. *Proc. Natl Acad. Sci. USA* **86**: 5044–5048.

Cao L, Faha B, Dembski M, Tsai LH, Dyson N. (1992) Independent binding of the retinoblastoma protein and p107 to the transcription factor E2F. *Nature* **355**: 176–179.

Cavenee WK, Dryja TP, Phillips RA, Benedict WF, Godbout R, Gallie BL, *et al.* (1983) Expression of recessive alleles by chromosomal mechanisms in retinoblastoma. *Nature* **305**: 779–784.

Chang CY, Riley DJ, Lee EY, Lee WH. (1993) Quantitative effects of the retinoblastoma gene on mouse development and tissue-specific tumorigenesis. *Cell Growth Differ.* **4**: 1057–1064.

Chang MW, Barr E, Seltzer J, Jiang Y-Q, Nabel GJ, Nabel EG, Parmacek MSL, Lleiden JM. (1995) Cytostatic gene therapy for vascular proliferative disorders with a constitutively active form of the retinoblastoma gene product. *Science* 267: 518–522.

Chellappan SP, Hiebert S, Mudryl M, Horowitz JM, Nevins JR. (1991) The E2F transcription factor is a cellular target for the RB protein. *Cell* 65: 1053–1061.

Chen LI, Nishinaka T, Kwan K, Kitabayashi I, Yokoyama K, Fu YH, *et al.* (1994) The retinoblastoma gene product RB stimulates Sp1-mediated transcription by liberating Sp1 from a negative regulator. *Mol. Cell. Biol.* 14: 4380–4389.

Chen PL, Scully P, Shew JY, Wang JY, Lee WH. (1989) Phosphorylation of the retinoblastoma gene product is modulated during the cell cycle and cellular differentiation. *Cell* 58: 1193–1198.

Clarke AR, Maandag ER, Van Roon M, Van der Lugt NMT, Van der Valk M, Hooper ML, *et al.* (1992) Requirement for a functional Rb-1 gene in murine development. *Nature* 359: 328–330.

Cobrinik D, Whyte P, Peeper DS, Jacks T, Weinberg RA. (1993) Cell cycle-specific association of E2F with the p130 E1A-binding protein. *Genes Devel.* 7: 2392–2404.

Comings DE. (1973) A general theory of carcinogenesis. *Proc. Natl Acad. Sci. USA* 70: 3324–3328.

Connolly MJ, Payne RH, Johnson G, Gallie BL, Alderdice PW, Marshall WH, Lawton RD. (1983) Familial, EsD-linked, retinoblastoma with reduced penetrance and variable expressivity. *Hum. Genet.* 65: 122–124.

Cowell JK. (1989) One hundred years of retinoblastoma research. From the clinic to the gene and back again. *Ophthalmic Paediatr. Genet.* 10: 75–88.

Cowell JK, Smith T, Bia B. (1994) Frequent constitutional C to T mutations in CGA-arginine codons in the RB1 gene produce premature stop codons in patients with bilateral (hereditary) retinoblastoma. *Eur. J. Hum. Genet.* 2: 281–290.

Dalton S. (1992) Cell cycle regulation of the human cdc2 gene. *EMBO J.* 11: 1797–1804.

DeCaprio JA, Ludlow JW, Figge J, Shew JY, Huang CM, Lee WH, *et al.* (1988) SV40 large tumor antigen forms a specific complex with the product of the retinoblastoma susceptibility gene. *Cell* 54: 275–283.

Dou QP, Markell PJ, Pardee AB. (1992) Thymidine kinase transcription is regulated at G1/S phase by a complex that contains retinoblastoma-like protein and a cdc2 kinase. *Proc. Natl Acad. Sci. USA* 89: 3256–3260.

Dowdy SF, Hinds PW, Louie K, Reed SI, Arnold A, Weinberg RA. (1993) Physical interaction of the retinoblastoma protein with human D cyclins. *Cell* 73: 499–511.

Draper GJ, Sanders BM, Brownbill PA, Hawkins MM. (1992) Patterns of risk of hereditary retinoblastoma and applications to genetic counselling. *Br. J. Cancer* 66: 211–219.

Dryja TP, Rapaport JM, Joyce JM, Petersen RA. (1986) Molecular detection of deletions involving band q14 of chromosome 13 in retinoblastomas. *Proc. Natl Acad. Sci. USA* 83: 7391–7394.

Dryja TP, Mukai S, Petersen R, Rapaport JM, Walton D, Yandell DW. (1989) Parental origin of mutations of the retinoblastoma gene. *Nature* 339: 556–558.

Dryja TP, Rappaport J, McGee TL, Nork TM, Schwartz TL. (1993) Molecular etiology of low-penetrance retinoblastoma in two pedigrees. *Am. J. Hum. Genet.* 52: 1122-1128.

Dunn JM, Phillips RA, Becker AJ, Gallie BL. (1988) Identification of germline and somatic mutations affecting the retinoblastoma gene. *Science* 241: 1797–1800.

Dunn JM, Phillips RA, Zhu X, Becker AJ, Gallie BL. (1989) Mutations in the RB1 gene and their effects on transcription. *Mol. Cell. Biol.* **9**: 4594–4602.

Dynlacht BD, Flores O, Lees JA, Harlow E. (1994) Differential regulation of E2F *trans*-activation by cyclin/cdk2 complexes. *Genes Devel.* **8**: 1772–1786.

Dyson N, Howley PM, Münger K, Harlow E. (1989) The human papilloma virus-16 E7 oncoprotein is able to bind to the retinoblastoma gene product. *Science* **242**: 934–937.

Eagle RCJ, Shields JA, Donoso L, Milner RS. (1989) Malignant transformation of spontaneously regressed retinoblastoma, retinoma/retinocytoma variant. *Ophthalmology* **96**: 1389–1395.

Ewen M, Xing Y, Lawrence JB, Livingston D. (1991) Molecular cloning, chromosomal mapping, and expression of the cDNA for p107, a retinoblastoma gene product-related protein. *Cell* **66**: 1155–1164.

Ewen ME, Sluss HK, Sherr CJ, Matsushime H, Kato J, Livingston DM. (1993) Functional interactions of the retinoblastoma protein with mammalian D-type cyclins. *Cell* **73**: 487–497.

Farnham PJ, Slansky JE, Kollmar R. (1993) The role of E2F in the mammalian cell cycle. *Biochim. Biophys. Acta* **1155**: 125–131.

Friend SH, Bernards R, Rogelj S, Weinberg RA, Rapaport JM, Albert DM, Dryja TP. (1986) A human DNA segment with properties of the gene that predisposes to retinoblastoma and osteosarcoma. *Nature* **323**: 643–646.

Fromm L, Shawlot W, Gunning K, Butel JS, Overbeek PA. (1994) The retinoblastoma protein-binding region of simian virus 40 large T antigen alters cell cycle regulation in lenses of transgenic mice. *Mol. Cell. Biol.* **14**: 6743–6754.

Gallie BL, Ellsworth RM, Abramson DH, Phillips RA. (1982a) Retinoma: spontaneous regression of retinoblastoma or benign manifestation of the mutation? *Br. J. Cancer* **45**: 513–521.

Gallie BL, Phillips RA, Ellsworth RM, Abramson DH. (1982b) Significance of retinoma and phthisis bulbi for retinoblastoma. *Ophthalmology* **89**: 1393–1399.

Gallie BL, Muncaster M, Cohen BL, Gill RM, Hamel PA, Phillips RA. (1993) Retinoblastoma mutations: initiation vs. progression of cancer. In: *Osteosarcoma Research Conference 1991* (ed. JF Novak). Hogrefe and Huber, Pittsburgh, PA, pp. 367–373.

Gill RM, Hamel PA, Jiang Z, Zacksenhaus E, Gallie BL, Phillips RA. (1994) Characterization of the human *RB1* promoter and of elements involved in transcriptional regulation. *Cell Growth Differ.* **5**: 467–474.

Ginsberg D, Vairo G, Chittenden T, Xiao Z-X, Xu G, Wydner KL, *et al.* (1994) E2F-4, a new member of the E2F transcription factor family, interacts with p107. *Genes Devel.* **8**: 2665–2679.

Godbout R, Squire J. (1993) Amplification of a DEAD box protein gene in retinoblastoma cell lines. *Proc. Natl Acad. Sci. USA* **90**: 7578–7582.

Godbout R, Dryja TP, Squire J, Gallie BL, Phillips RA. (1983) Somatic inactivation of genes on chromosome 13 is a common event in retinoblastoma. *Nature* **304**: 451–453.

Godbout R, Bisgrove DA, Honoré LH, Day R III. (1993) Amplification of the gene encoding the alpha-subunit of the mitochondrial ATP synthase complex in a human retinoblastoma cell line. *Gene* **123**: 195–201.

Godbout R, Hale M, Bisgrove D. (1994) A human DEAD box protein with partial homology to heterogeneous nuclear ribonucleoprotein U. *Gene* **138**: 243–245.

Goddard AD, Balakier H, Canton M, Dunn J, Squire J, Reyes E, *et al.* (1988) Infrequent genomic rearrangement and normal expression of the putative RB1 gene in retinoblastoma tumors. *Mol. Cell. Biol.* **8**: 2082–2088.

Goodrich DW, Wang NP, Qian Y-W, Lee EY-HP, Lee W-H. (1991) The retinoblastoma gene product regulates progression through the G_1 phase of the cell cycle. *Cell* **67**: 293–302.

Gu W, Schneider JW, Condorelli G, Kaushal S, Mahdavi V, Nadal-Ginard B. (1993) Interaction of myogenic factors and the retinoblastoma protein mediates muscle cell commitment and differentiation. *Cell* **72**: 309–324.

Gu W, Bhatia K, Magrath IT, Dang CV, Dalla-Favera R. (1994) Binding and suppression of the Myc transcriptional activation domain by p107. *Science* **264**: 251–254.

Gupta S, Campbell D, Dérijard B, Davis RJ. (1995) Transcription factor ATF2 regulation by the JNK signal transduction pathway. *Science* **267**: 389–393.

Hagemeier C, Bannister AJ, Cook A, Kouzarides T. (1993) The activation domain of transcription factor PU.1 binds the retinoblastoma (RB) protein and the transcription factor TFIID in vitro: RB shows sequence similarity to TFIID and TFIIB. *Proc. Natl Acad. Sci. USA* **90**: 1580–1584.

Hamel PA, Cohen BL, Sorce LM, Gallie BL, Phillips RA. (1990) Hyperphosphorylation of the retinoblastoma gene product is determined by domains outside the SV40 large-T-binding regions. *Mol. Cell. Biol.* **10**: 6586–6595.

Hamel PA, Gill M, Phillips RA, Gallie BL. (1992a) Regions controlling hyperphosphorylation and conformation of the retinoblastoma gene product are independent of domains required for transcriptional repression. *Oncogene* **7**: 693–701.

Hamel PA, Gill RM, Phillips RA, Gallie BL. (1992b) Transcriptional repression of the E2-containing promoters EIIaE, c-myc, and RB1 by the product of the RB1 gene. *Mol. Cell. Biol.* **12**: 3431–3438.

Hannon GJ, Demetrick D, Beach D. (1993) Isolation of the Rb-related p130 through its interaction with CDK2 and cyclins. *Genes Devel.* **7**: 2378–2391.

Harbour JW, Lai SL, Whang PJ, Gazdar AF, Minna JD, Kaye FJ. (1988) Abnormalities in structure and expression of the human retinoblastoma gene in SCLC. *Science* **241**: 353–357.

Hashimoto T, Takahashi R, Yandell DW, Xu HJ, Hu SX, Gunnell S, Benedict WF. (1991) Characterization of intragenic deletions in two sporadic germinal mutation cases of retinoblastoma resulting in abnormal gene expression. *Oncogene* **6**: 463–469.

Hateboer G, Timmers HT, Rustgi AK, Billaud M, vant-Veer LJ, Bernards R. (1993) TATA-binding protein and the retinoblastoma gene product bind to overlapping epitopes on c-Myc and adenovirus E1A protein. *Proc. Natl Acad. Sci. USA* **90**: 8489–8493.

Heck DV, Yee CL, Howley PM, Munger K. (1992) Efficiency of binding the retinoblastoma protein correlates with the transforming capacity of the E7 oncoproteins of the human papillomaviruses. *Proc. Natl Acad. Sci. USA* **89**: 4442-4446.

Helin K, Wu CL, Fattaey AR, Lees JA, Dynlacht BD, Ngwu C, Harlow E. (1993) Heterodimerization of the transcription factors E2F-1 and DP-1 leads to cooperative trans-activation. *Genes Devel.* **7**: 1850–1861.

Hensel CH, Hsieh CL, Gazdar AF, Johnson BE, Sakaguchi AY, Naylor SL, *et al.* (1990) Altered structure and expression of the human retinoblastoma susceptibility gene in small cell lung cancer. *Cancer Res.* **50**: 3067–3072.

Hiebert SW, Chellapan SP, Horowitz JM, Nevins JR. (1992) The interaction of RB with E2F coincides with an inhibition of the transcriptional activity of E2F. *Genes Devel.* **6:** 177–185.

Hinds PW, Mittnacht S, Dulic V, Arnold A, Reed SI, Weinberg RA. (1992) Regulation of retinoblastoma protein functions by ectopic expression of human cyclins. *Cell* **70:** 993–1006.

Hogg A, Onadim Z, Baird PN, Cowell JK. (1992) Detection of heterozygous mutations in the RB1 gene in retinoblastoma patients using single strand conformation polymorphism (SSCP) analysis and PCR sequencing. *Oncogene* **7:** 1445–1451.

Hogg A, Bia B, Onadim Z, Cowell JK. (1993) Molecular mechanisms of oncogenic mutations in tumors from patients with bilateral and unilateral retinoblastoma. *Proc. Natl Acad. Sci. USA* **90:** 7351–7355.

Horowitz JM, Yandell DW, Park S-H, Canning S, Whyte P, Buchkovich K, et al. (1989) Point mutational inactivation of the retinoblastoma antioncogene. *Science* **243:** 936–940.

Horowitz JM, Park SH, Bogenmann E, Cheng JC, Yandell DW, Kaye FJ, et al. (1990) Frequent inactivation of the retinoblastoma anti-oncogene is restricted to a subset of human tumor cells. *Proc. Natl Acad. Sci. USA* **87:** 2775–2779.

Howes KA, Ransom N, Papermaster DS, Lasudry JGH, Albert DM, Windle JJ. (1994) Apoptosis or retinoblastoma: alternative fates of photoreceptors expressing the HPC-16 E7 gene in the presence or absence of p53. *Genes Devel.* **8:** 1300–1310.

Hu N, Gutsmann A, Herbert DC, Bradley A, Lee WH, Lee EY. (1994) Heterozygous Rb-1 delta 20/+ mice are predisposed to tumors of the pituitary gland with a nearly complete penetrance. *Oncogene* **9:** 1021–1027.

Hu Q, Lees JA, Buchkovich KJ, Harlow E. (1992) The retinoblastoma protein physically associated with the human cdc2 kinase. *Mol. Cell. Biol.* **12:** 971–980.

Hu QJ, Dyson N, Harlow E. (1990) The regions of the retinoblastoma protein needed for binding to adenovirus E1A or SV40 large T antigen are common sites for mutations. *EMBO J.* **9:** 1147–1155.

Huber HE, Goodhart PJ, Huang PS. (1994) Retinoblastoma protein reverses DNA bending by transcription factor E2F. *J. Biol. Chem.* **269:** 6999–7005.

Iavarone A, Garg P, Lasorella A, Hsu J, Israel MA. (1994) The helix-loop-helix protein Id-2 enhances cell proliferation and binds to the retinoblastoma protein. *Genes Devel.* **8:** 1270–1284.

Ikeda MA, Nevins JR. (1993) Identification of distinct roles for separate E1A domains in disruption of E2F complexes. *Mol. Cell. Biol.* **13:** 7029–7035.

Inaba T, Matsushime H, Valentine M, Roussel MF, Sherr CJ, Look AT. (1992) Genomic organization, chromosomal localization, and independent expression of human cyclin D genes. *Genomics* **13:** 565–574.

Ivey-Hoyle M, Conroy R, Huber HE, Goodhart PJ, Oliff A, Heimbrook DC. (1993) Cloning and characterization of E2F-2, a novel protein with the biochemical properties of transcription factor E2F. *Mol. Cell. Biol.* **13:** 7802–7812.

Jacks T, Fazeli A, Schmitt EM, Bronson RT, Goodell MA, Weinberg RA. (1992) Effects of an *Rb* mutation in the mouse. *Nature* **359:** 295–300.

Kaelin WGJ, Krek W, Sellers WR, Ajchenbaum F, Fuchs CS, et al. (1992) Expression cloning of a cDNA encoding a retinoblastoma-binding protein with E2F-like properties. *Cell* **70:** 351–364.

Kashii T, Mizushima Y, Monno S, Nakagawa K, Kobayashi M. (1994) Gene analysis of K-, H-ras, p53, and retinoblastoma susceptibility genes in human lung cancer cell lines by the polymerase chain reaction/single-strand conformation polymorphism method. *J. Cancer Res. Clin. Oncol.* **120**: 143–148.

Kato J, Matsushime H, Hiebert SW, Ewen ME, Sherr CJ. (1993) Direct binding of cyclin D to the retinoblastoma gene product (pRb) and pRb phosphorylation by the cyclin D-dependent kinase CDK4. *Genes Devel.* **7**: 331–342.

Kato MV, Ishizaki K, Toguchida J, Kaneko A, Takayama J, Tanooka H, et al. (1994) Mutations in the retinoblastoma gene and their expression in somatic and tumor cells of patients with hereditary retinoblastoma. *Hum. Mutat.* **3**: 44–51.

Kim SJ, Lee HD, Robbins PD, Busam K, Sporn MB, Roberts AB. (1991) Regulation of transforming growth factor beta 1 gene expression by the product of the retinoblastoma-susceptibility gene. *Proc. Natl Acad. Sci. USA* **88**: 3052–3056.

Kim SJ, Onwuta US, Lee YI, Li R, Botchan MR, Robbins PD. (1992a) The retinoblastoma gene product regulates Sp1-mediated transcription. *Mol. Cell. Biol.* **12**: 2455–2463.

Kim SJ, Wagner S, Liu F, O'Reilly MA, Robbins PD, Green MR. (1992b) Retinoblastoma gene product activates expression of the human TGF-beta 2 gene through transcription factor ATF-2. *Nature* **358**: 331–334.

Kloss K, Wahrisch P, Greger V, Messmer E, Fritze H, Hopping W, et al. (1991) Characterization of deletions at the retinoblastoma locus in patients with bilateral retinoblastoma. *Am. J. Med. Genet.* **39**: 196–200.

Knudson AG. (1971) Mutation and cancer: statistical study of retinoblastoma. *Proc. Natl Acad. Sci. USA* **68**: 820–823.

Kratzke RA, Otterson GA, Hogg A, Coxon AB, Geradts J, Cowell JK, Kaye FJ. (1994) Partial inactivation of the RB product in a family with incomplete penetrance of familial retinoblastoma and benign retinal tumors. *Oncogene* **9**: 1321–1326.

Krek W, Ewen ME, Shirodkar S, Arany Z, Kaelin WG Jr, Livingston DM. (1994) Negative regulation of the growth-promoting transcription factor E2F-1 by a stably bound cyclin A-dependent protein kinase. *Cell* **78**: 161–172.

La-Thangue NB. (1994) DRTF1/E2F: an expanding family of heterodimeric transcription factors implicated in cell-cycle control. *Trends Biochem. Sci.* **19**: 108–114.

Lam EW, Morris JD, Davies R, Crook T, Watson RJ, Vousden KH. (1994) HPV16 E7 oncoprotein deregulates B-myb expression: correlation with targeting of p107/E2F complexes. *EMBO J.* **13**: 871–878.

Lam EW-F, Watson RJ. (1993) E2F regulation of B-myb transcription. *EMBO J.* **12**: 2705–2713.

Lee EY, To H, Shew JY, Bookstein R, Scully P, Lee WH. (1988) Inactivation of the retinoblastoma susceptibility gene in human breast cancers. *Science* **241**: 218–221.

Lee EYHP. (1994) Dual roles of the retinoblastoma protein in cell cycle regulation and neuron differentiation. *Genes Devel.* **8**: 2008.

Lee EYHP, Chang CY, Hu N, Wang YCJ, Lai CC, Herrup K, et al. (1992) Mice deficient for Rb are nonviable and show defects in neurogenesis and haematopoiesis. *Nature* **359**: 288–294.

Lees JA, Buchkovich KJ, Marshak DR, Anderson CW, Harlow E. (1991) The retinoblastoma protein is phosphorylated on multiple sites by human cdc2. *EMBO J.* **10**: 4279–4290.

Li Y, Graham C, Lacy S, Duncan AM, Whyte P. (1993) The adenovirus E1A-associated 130-kD protein is encoded by a member of the retinoblastoma gene family and physically interacts with cyclins A and E. *Genes Devel.* **7**: 2366–2377.

Lin BT, Gruenwald S, Morla AO, Lee WH, Wang JY. (1991) Retinoblastoma cancer suppressor gene product is a substrate of the cell cycle regulator cdc2 kinase. *EMBO J.* **10**: 857–864.

Lohmann D, Horsthemke B, Gillessen KG, Stefani FH, Hofler H. (1992) Detection of small RB1 gene deletions in retinoblastoma by multiplex PCR and high-resolution gel electrophoresis. *Hum. Genet.* **89**: 49–53.

Lohmann DR, Brandt B, Höpping W, Passarge E, Horsthemke B. (1994a) Distinct RB1 gene mutations with low penetrance in hereditary retinoblastoma. *Hum. Genet.* **94**: 349–354.

Lohmann DR, Brandt B, Höpping W, Passarge E, Horsthemke B. (1994b) Spectrum of small length germline mutations in the RB1 gene. *Hum. Mol. Genet.* **3**: 2187–2193.

Ludlow JW, DeCaprio JA, Huang C-M, Lee W-H, Paucha E, Livingston DM. (1989) SV40 large T antigen binds preferentially to an underphosphorylated member of the retinoblastoma susceptibility gene product family. *Cell* **56**: 57–65.

Maandag EC, van der Valk M, Vlaar M, Feltkamp C, O'Brien J, van Roon M, *et al.* (1994) Developmental rescue of an embryonic-lethal mutation in the retinoblastoma gene in chimeric mice. *EMBO J.* **13**: 4260–4268.

Mahon KA, Chepelinsky AB, Khillan JS, Overbeek PA, Piatigorsky J, Westphal H. (1987) Oncogenesis of the lens in transgenic mice. *Science* **235**: 1622–1628.

Maione R, Fimia GM, Holman P, Schaffhausen B, Amati P. (1994) Retinoblastoma antioncogene is involved in the inhibition of myogenesis by polyomavirus large T antigen. *Cell Growth Differ.* **5**: 231–237.

Mayol X, Grana X, Baldi A, Sang N, Hu Q, Giordano A. (1993) Cloning of a new member of the retinoblastoma gene family (pRb2) which binds to the E1A transforming domain. *Oncogene* **8**: 2561–2566

Melillo RM, Helin K, Lowy DR, Schiller JT. (1994) Positive and negative regulation of cell proliferation by E2F-1: influence of protein level and human papillomavirus oncoproteins. *Mol. Cell. Biol.* **14**: 8241–8249.

Mihara K, Cao XR, Yen A, Chandler S, Driscoll B, Murphree AL, *et al.* (1989) Cell cycle-dependent regulation of phosphorylation of the human retinoblastoma gene product. *Science* **246**: 1300–1303.

Mittnacht S, Lees JA, Desai D, Harlow E, Morgan DO, Weinberg RA. (1994) Distinct sub-populations of the retinoblastoma protein show a distinct pattern of phosphorylation. *EMBO J.* **13**: 118–127.

Moran E. (1993) DNA tumor virus transforming proteins and the cell cycle. *Curr. Opin. Genet. Devel.* **3**: 63–70.

Morgenbesser SD, Williams BO, Jacks T, DePinho RA. (1994) p53-dependent apoptosis produced by Rb-deficiency in the developing mouse lens [see Comments]. *Nature* **371**: 72–74.

Mori N, Yokota J, Akiyama T, Sameshima Y, Okamoto A, Mizoguchi H, *et al.* (1990) Variable mutations of the RB gene in small-cell lung carcinoma. *Oncogene* **5**: 1713–1717.

Motegi T. (1981) Lymphocyte chromosome survey in 42 patients with retinoblastoma: effort to detect 13q14 deletion mosaicism. *Hum. Genet.* **58**: 168–173.

Muncaster M, Cohen B, Phillips RA, Gallie BL. (1992) Failure of RB1 to reverse the malignant phenotype of human tumor cell lines. *Cancer Res.* **52**: 654–661.

Murakami Y, Katahira M, Makino R, Hayashi K, Hirohashi S, Sekiya T. (1991) Inactivation of the retinoblastoma gene in a human lung carcinoma cell line detected by single-strand conformation polymorphism analysis of the polymerase chain reaction product of cDNA. *Oncogene* **6:** 37–42.

Nevins JR. (1992) E2F: a link between the Rb tumor suppressor protein and viral oncoproteins. *Science* **258:** 424–429.

Newton RE. (1902) Glioma of retina. A remarkable family history. *Aust. Med. Gazette* **21:** 236–237.

Ohtani K, Nevins JR. (1994) Functional properties of a Drosophila homolog of the E2F1 gene. *Mol. Cell. Biol.* **14:** 1603–1612.

Ohtani-Fujita N, Fujita T, Takahashi R, Robbins PD, Dryja TP, Sakai T. (1994) A silencer element in the retinoblastoma tumor-suppressor gene. *Oncogene* **9:** 1703–1711.

Onadim A, Hogg A, Baird PN, Cowell JK. (1992) Oncogenic point mutation in exon 20 of the RB1 gene in families showing incomplete penetrance and mild expression of the retinoblastoma phenotype. *Proc. Natl Acad. Sci. USA* **89:** 6177–6181.

Onadim ZO, Mitchell CD, Rutland PC, Buckle BG, Jay M, Hungerford JL, *et al.* (1990) Application of intragenic DNA probes in prenatal screening for retinoblastoma gene carriers in the United Kingdom. *Arch. Dis. Child.* **65:** 651–656.

Osifchin NE, Jiang D, Ohtani-Fujita N, Fujita T, Carroza M, Kim SJ, *et al.* (1994) Identification of a p53 binding site in the human retinoblastoma susceptibility gene promoter. *J. Biol. Chem.* **269:** 6383–6389.

Pagano M, Durst M, Joswig S, Draetta G, Jansen DP. (1992) Binding of the human E2F transcription factor to the retinoblastoma protein but not to cyclin A is abolished in HPV-16-immortalized cells. *Oncogene* **7:** 1681–1686.

Paggi MG, Martelli F, Fanciulli M, Felsani A, Sciacchitano S, Varmi M, *et al.* (1994) Defective human retinoblastoma protein identified by lack of interaction with the E1A oncoprotein. *Cancer Res.* **54:** 1098–1104.

Pan H, Griep AE. (1994) Altered cell cycle regulation in the lens of HPV-16 E6 or E7 transgenic mice: implications for tumor suppressor gene function in development. *Genes Devel.* **8:** 1285–1299.

Park K, Choe J, Osifchin NE, Templeton DJ, Robbins PD, Kim SJ. (1994) The human retinoblastoma susceptibility gene promoter is positively autoregulated by its own product. *J. Biol. Chem.* **269:** 6083–6088.

Pearson BE, Nasheuer H-P, Wang TS-F. (1991) Human DNA polymerase α gene: sequences controlling expression in cycling and serum-stimulated cells. *Mol. Cell. Biol.* **11:** 2081–2095.

Qian Y, Luckey C, Horton L, Esser M, Templeton DJ. (1992) Biological function of the retinoblastoma protein requires distinct domains for hyperphosphorylation and transcription factor binding. *Mol. Cell. Biol.* **12:** 5363–5372.

Robbins PD, Horowitz JM, Mulligan RC. (1990) Negative regulation of human c-*fos* expression by the retinoblastoma gene product. *Nature* **346:** 668–671.

Rustgi AK, Dyson N, Bernards R. (1991) Amino-terminal domains of c-myc and N-myc proteins mediate binding to the retinoblastoma gene product. *Nature* **352:** 541–544.

Sachse R, Murakami Y, Shiraishi M, Hayashi K, Sekiya T. (1994) DNA aberrations at the retinoblastoma gene locus in human squamous cell carcinomas of the lung. *Oncogene* **9:** 39–47.

Sakai T, Ohtani N, McGee TL, Robbins PD, Dryja TP. (1991a) Oncogenic germline mutations in Sp1 and ATF sites in the human retinoblastoma gene. *Nature* **353**: 83–86.

Sakai T, Toguchida J, Ohtani N, Yandell DW, Rapaport JM, Dryja TP. (1991b) Allele-specific hypermethylation of the retinoblastoma tumor-suppressor gene. *Am. J. Hum. Genet.* **48**: 880–888.

Sakamoto K, Howard T, Ogryzko V, Xu NZ, Corsico CC, Jones DH, Howard B. (1993) Relative mitogenic activities of wild-type and retinoblastoma binding-defective SV40 T antigens in serum-deprived and senescent human diploid fibroblasts. *Oncogene* **8**: 1887–1893.

Scheffer H, te Meerman GJ, Kruize YC, vanden Berg AH, Penninga DP, Tan KE, der Kinderen DJ, Buys CH. (1989) Linkage analysis of families with hereditary retinoblastoma: nonpenetrance of mutation, revealed by combined use of markers within and flanking the *RB1* gene. *Am. J. Hum. Genet.* **45**: 252–260.

Schneider JW, Gu W, Zhu L, Mahdavi V, Nadal-Ginard B. (1994) Reversal of terminal differentiation mediated by p107 in Rb–/– muscle cells. *Science* **264**: 1467–1471.

Shan B, Zhu X, Chen P-L, Durfee T, Yang Y, Sharp D, Lee W-H. (1992) Molecular cloning of cellular genes encoding retinoblastoma-associated proteins: identification of a gene with properties of the transcription factor E2F. *Mol. Cell. Biol.* **12**: 5620–5631.

Shan B, Chang C-Y, Jones D, Lee W-H. (1994) The transcription factor E2F-1 mediates the autoregulation of RB gene expression. *Mol. Cell. Biol.* **14**: 299–309.

Shew J-Y, Chen P-L, Bookstein R, Lee EY-HP, Lee W-H. (1990a) Deletion of a splice donor site ablates expression of the following exon and produces an unphosphorylated RB protein unable to bind SV40 T antigen. *Cell Growth Differ.* **1**: 17–25.

Shew JY, Lin BT, Chen PL, Tseng BY, Yang FTL, Lee WH. (1990b) C-terminal truncation of the retinoblastoma gene product leads to functional inactivation. *Proc. Natl Acad. Sci. USA* **87**: 6–10.

Shimizu T, Toguchida J, Kato MV, Kaneko A, Ishizaki K, Sasaki MS. (1994) Detection of mutations of the RB1 gene in retinoblastoma patients by using exon-by-exon PCR-SSCP analysis. *Am. J. Hum. Genet.* **54**: 793–800.

Sparkes RS, Sparkes MC, Wilson MG, Towner JW, Benedict W, Murphree AL, Yunis JJ. (1980) Regional assignment of genes for human esterase D and retinoblastoma to chromosome band 13q14. *Science* **208**: 1042–1044.

Squire J, Phillips RA, Boyce S, Godbout R, Rogers B, Gallie BL. (1984) Isochromosome 6p, a unique chromosomal abnormality in retinoblastoma: verification by standard staining techniques, new densitometric methods, and somatic cell hybridization. *Hum. Genet.* **66**: 46–53.

Squire J, Gallie BL, Phillips RA. (1985) A detailed analysis of chromosomal changes in heritable and non-heritable retinoblastoma. *Hum. Genet.* **70**: 291–301.

Szekely L, Jiang W-Q, Bulic-Jakus F, Rosen A, Ringertz N, Klein G, Wiman KG. (1992) Cell type and differentiation dependent heterogeneity in retinoblastoma protein expression in SCID mouse fetuses. *Cell Growth Differ* **3**: 149–156.

T'Ang A, Varley JM, Chakroborty S, Murphree AL, Fung YKT. (1988) Structural rearrangement of the retinoblastoma gene in human breast carcinoma. *Science* **242**: 263–266.

Templeton DJ. (1992) Nuclear binding of purified retinoblastoma gene product is determined by cell cycle-regulated phosphorylation. *Mol. Cell. Biol.* **12**: 435–443.

Templeton DJ, Park SH, Lanier L, Weinberg RA. (1991) Nonfunctional mutants of the retinoblastoma protein are characterized by defects in phosphorylation, viral oncoprotein association, and nuclear tethering. *Proc. Natl Acad. Sci. USA* **88**: 3033–3037.

Toguchida J, McGee TL, Paterson JC, Eagle JR, Tucker S, Yandell DW, Dryja TP. (1993) Complete genomic sequence of the human retinoblastoma susceptibility gene. *Genomics* **17**: 535–543.

Udvadia AJ, Rogers KT, Higgins PD, Murata Y, Martin KH, Humphrey PA, Horowitz JM. (1993) Sp-1 binds promoter elements regulated by the RB protein and Sp-1-mediated transcription is stimulated by RB coexpression. *Proc. Natl Acad. Sci. USA* **90**: 3265–3269.

Wadayama B, Toguchida J, Shimizu T, Ishizaki K, Sasaki MS, Kotoura Y, Yamamuro T. (1994) Mutation spectrum of the retinoblastoma gene in osteosarcomas. *Cancer Res.* **54**: 3042–3048.

Wang CY, Petryniak B, Thompson CB, Kaelin WG, Leiden JM. (1993a) Regulation of the Ets-related transcription factor Elf-1 by binding to the retinoblastoma protein. *Science* **260**: 1330–1335.

Wang NP, To H, Lee WH, Lee EYHP. (1993b) Tumor suppressor activity of RB and p53 genes in human breast carcinoma cells. *Oncogene* **8**: 279–288.

Weir-Thompson E, Condie A, Leonard RC, Prosser J. (1991) A familial RB1 mutation detected by the HOT technique is homozygous in a second primary neoplasm. *Oncogene* **6**: 2353–2356.

Whyte P, Buchkovich KJ, Horowitz JM, Friend SH, Raybuck M, Weinberg RA, Harlow E. (1988) Association between an oncogene and an anti-oncogene: the adenovirus E1A proteins bind to the retinoblastoma gene product. *Nature* **334**: 124–129.

Wiggs J, Nordenskjold M, Yandell D, Rapaport J, Grondin V, Janson M, *et al.* (1988) Prediction of the risk of hereditary retinoblastoma, using DNA polymorphisms within the retinoblastoma gene. *N. Engl. J. Med.* **318**: 151–157.

Williams BO, Remington L, Albert DM, Mukai S, Bronson RT, Jacks T. (1994a) Cooperative tumorigenic effects of germline mutations in *Rb* and *p53*. *Nature Genet.* **7**: 480–484.

Williams BO, Schmitt EM, Remington L, Bronson RT, Albert DM, Weinberg RA, Jacks T. (1994b) Extensive contribution of Rb-deficient cells to adult chimeric mice with limited histopathological consequences. *EMBO J.* **13**: 4251–4259.

Windle JJ, Albert DM, OBrien JM, Marcus DM, Disteche CM, Bernards R, Mellon PL. (1990) Retinoblastoma in transgenic mice. *Nature* **343**: 665–669.

Woloschak M, Roberts JL, Post KD. (1994) Loss of heterozygosity at the retinoblastoma locus in human pituitary tumors. *Cancer* **74**: 693–696.

Xu M, Sheppard K-A, Peng C-Y, Yee AS, Piwnica-Worms H. (1994) Cyclin A/CDK2 binds directly to E2F-1 and inhibits the DNA-binding activity of E2F-1/DP-1 by phosphorylation. *Mol. Cell. Biol.* **14**: 8420–8431.

Yamamoto M, Yoshida M, Ono K, Fujita T, Ohtani-Fujita N, Sakai T, Nikaido T. (1994) Effect of tumor suppressors on cell cycle regulatory genes: RB suppresses p34cdc2 expression and normal p53 suppresses cyclin A expression. *Exp. Cell Res.* **210**: 94–101.

Yandell DW, Campbell TA, Dayton SH, Petersen R, Walton D, Little JB, *et al.*
(1989) Oncogenic point mutations in the human retinoblastoma gene: their
application to genetic counseling. *N. Engl. J. Med.* **321**: 1689–1694.

Yen A, Varvayanis S. (1994) Late dephosphorylation of the RB protein in G2 during
the process of induced cell differentiation. *Exp. Cell Res.* **214**: 250–257.

Yen A, Chandler S, Forbes ME, Fung YK, TAng A, Pearson R. (1992) Coupled
down-regulation of the RB retinoblastoma and c-myc genes antecedes cell
differentiation: possible role of RB as a 'status quo' gene. *Eur. J. Cell. Biol.* **57**:
210–221.

Yen A, Coles M, Varvayanis S. (1993a) 1,25-Dihydroxy vitamin D3 and 12-O-
tetradecanoyl phorbol-13-acetate synergistically induce monocytic cell
differentiation: FOS and RB expression. *J. Cell Physiol.* **156**: 198–203.

Yen A, Forbes ME, Varvayanis S, Tykocinski ML, Groger RK, Platko JD. (1993b)
C-FMS dependent HL-60 cell differentiation and regulation of RB gene expression.
J. Cell Physiol. **157**: 379–391.

Yen A, Soong S, Kwon HJ, Yoshida M, Beppu T, Varvayanis S. (1994) Enhanced cell
differentiation when RB is hypophosphorylated and down-regulated by radicicol, a
SRC-kinase inhibitor. *Exp. Cell Res.* **214**: 163–171.

Yunis JJ, Ramsay N. (1978) Retinoblastoma and subband deletion of chromosome 13.
Am. J. Dis. Child. **132**: 161–163.

Zacksenhaus E, Gill RM, Phillips RA, Gallie BL. (1993) Molecular cloning and
characterization of the mouse RB1 promoter. *Oncogene* **8**: 2343–2351.

Zamanian M, La-Thangue NB. (1993) Transcriptional repression by the Rb-related
protein p107. *Mol. Biol. Cell* **4**: 389–396.

Zhu J, Leon SP, Beggs AH, Busque L, Gilliland DG, Black PM. (1994) Human
pituitary adenomas show no loss of heterozygosity at the retinoblastoma gene locus.
J. Clin. Endocrinol. Metab. **78**: 922–927.

Zhu L, van-den-Heuvel S, Helin K, Fattaey A, Ewen M, Livingston D, *et al.* (1993)
Inhibition of cell proliferation by p107, a relative of the retinoblastoma protein.
Genes Devel. **7**: 1111–1125.

Zhu X, Dunn JM, Goddard AD, Paton KE, Becker AJ, Phillips RA, Gallie BL. (1989)
Preferential germline mutation of the paternal allele in retinoblastoma. *Nature* **340**:
312–313.

The genetics of Wilms' tumour

Shyamala Maheswaran and Daniel Haber

2.1 Introduction

Wilms' tumour is a paediatric kidney cancer, affecting children under 5 years of age, with an incidence of approximately 1/10 000 children. The tumour is comprised of embryonal cells, with a characteristic 'triphasic' histology, including blastemal, epithelial and stromal components (Bennington and Beckwith, 1975). Some tumours show elements of further differentiation, such as skeletal muscle, leading to the concept that Wilms' tumour originates from embryonic precursor cells that have retained some of their multilineage potential.

Wilms' tumour can present as a sporadic tumour or, rarely, within a genetically susceptible family or in association with a congenital syndrome. Approximately 10% of Wilms' tumours are bilateral at the time of presentation, requiring special precautions in clinical management. Most tumours are treated by surgical excision of the affected kidney, followed by chemotherapy and radiation therapy in cases with adverse prognostic features (Grundy *et al.*, 1989). At the time of surgery, the contralateral kidney is examined for evidence of a second tumour, in which case a more conservative kidney resection is undertaken. With current treatment modalities, approximately 80% of children with Wilms' tumour are cured of their disease.

The genetic aetiology of Wilms' tumour was first addressed by Knudson and Strong in their now classic study of paediatric cancers such as retinoblastoma, Wilms' tumour and neuroblastoma (Knudson, 1971; Knudson and Strong, 1972). Knudson observed that children with bilateral Wilms' tumours developed their cancers at an earlier age than those with unilateral tumours. By analogy with retinoblastoma, he proposed that two 'rate-limiting' genetic hits were required for tumorigenesis; and that children with unilateral tumours had the rare occurrence of these two independent hits in a single kidney precursor cell. In contrast, children with bilateral tumours had inherited the first hit, present in all somatic cells; only one more genetic hit in a kidney precursor cell was therefore required for tumorigenesis. The requirement for only a single additional 'rate-limiting' genetic lesion in these children explained the earlier onset as well as the multifocal or bilateral nature of their cancer. The subsequent discovery of tumour suppressor genes readily identified the two genetic hits as the inactivation of two alleles of a critical growth inhibitory gene. However, unlike the relatively simple genetics of retinoblastoma, Wilms' tumour presents a number of unresolved issues. Despite a 10% incidence of bilateral Wilms' tumours, familial transmission of genetic susceptibility appears to be rare, estimated at 1% of Wilms' tumour cases (Cochran and Froggatt, 1967); hence, the great majority of bilateral tumours must represent *de novo* germline mutations. In addition, multiple genetic loci have been implicated in Wilms' tumorigenesis but only one tumour suppressor gene, *WT1*, has been isolated (Call *et al.*, 1990; Gessler *et al.*, 1990). The identities of the other Wilms' tumour genes and their potential interactions with each other are unknown.

2.2 Genetic loci implicated in Wilms' tumour

Two loci on chromosome 11 have been identified as the site of genes involved in both germline and somatic events leading to Wilms' tumorigenesis. A third locus on chromosome 16 has been implicated in tumour progression, but not in genetic susceptibility. Another, yet unmapped, locus appears to be responsible for some cases of familial susceptibility to Wilms' tumour.

2.2.1 Chromosome 11p13

An early observation was the association between aniridia, a malformation of the iris, and Wilms' tumour (Miller *et al.*, 1964). These two conditions were observed to arise together with far greater than expected incidence. Subsequently, WAGR syndrome (an acronym for Wilms' tumour, aniridia, genito-urinary malformations and mental retardation) was found to be associated with a gross, cytogenetically visible deletion of one 11p13 allele (Riccardi *et al.*, 1978). This deletion is now known to encompass the Wilms' tumour suppressor gene *WT1*, along with the aniridia gene, *Pax 6* (Ton *et al.*, 1991b). In Wilms' tumour specimens, allelic losses at 11p13 were demonstrated by loss of heterozygosity (LOH) at polymorphic markers (Mannens *et al.*, 1988). Thus, 11p13 was demonstrated to be the site of a gene whose disruption occurred in both the germline of genetically susceptible children and somatic cells of children with sporadic Wilms' tumour. *WT1* was identified within the 11p13 locus, and mutational and functional analyses confirmed its characterization as a tumour suppressor gene (see Section 2.3).

2.2.2 Chromosome 11p15

A second genetic locus, closer to the telomere of chromosome 11, was implicated in Wilms' tumour by the observation of allelic losses that excluded 11p13. Thus, a distinct genetic locus harboured a gene whose loss conferred a growth advantage, consistent with another tumour suppressor gene. Analysis of children with Beckwith–Wiedemann syndrome (BWS) also demonstrated germline abnormalities at the 11p15 locus (*Figure 2.1*). This congenital syndrome is characterized by asymmetric organomegaly, often involving the tongue and abdominal viscera, umbilical hernia and neonatal hypoglycaemia, along with predisposition to a number of paediatric cancers, including Wilms' tumour, adrenocortical carcinoma and hepatoblastoma (Wiedemann, 1964; Beckwith, 1969). Rare families with genetic transmission of BWS demonstrated linkage to chromosome 11p15

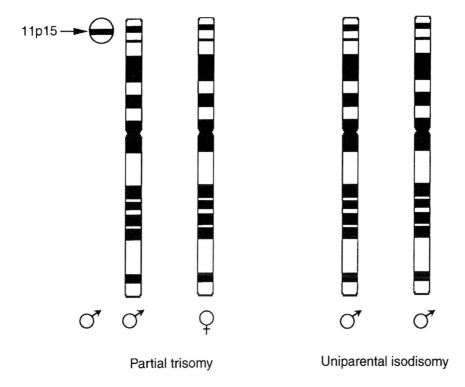

Partial trisomy   Uniparental isodisomy

Figure 2.1. Germline abnormalities in BWS (11p15 locus). Patients with BWS have been found to have an increased dosage of the paternally derived 11p15 locus, implicating an imprinted gene in the aetiology of this syndrome. In some cases, partial trisomy is observed at 11p15, with the duplicated chromosome fragment derived from the father. In other cases, both chromosomes 11 appear grossly normal but molecular markers identify both of them as paternal in origin, with no maternally inherited chromosome 11 (uniparental isodisomy). The identity of the target gene at chromosome 11 is unknown, although the *IGF2* gene, which maps to that locus and is imprinted in humans, is a potential candidate. Although the incidence of Wilms' tumour is increased in BWS, allelic losses at 11p15 in tumour specimens appear to implicate a distinct tumour suppressor gene, referred to as *WT2*.

(Koufos *et al.*, 1989; Ping *et al.*, 1989), and some individuals with *de novo* BWS were found to have partial trisomy at 11p15 (Waziri *et al.*, 1983; Turleau *et al.*, 1984). In these cases, the duplicated chromosome fragment was derived from the paternal allele. More recently, a number of BWS cases that were grossly diploid at 11p15 were found in fact to have inherited both copies of chromosome 11 from the father, and none from the mother (so-called uniparental isodisomy) (Grundy *et al.*, 1991; Henry *et al.*, 1991). These unusual genetic mechanisms are highly suggestive of genomic imprinting, a phenomenon by which only the allele derived from one parent is expressed in the child. In the case of BWS, normal expression of a paternally derived allele would result in increased gene expression in a child with two alleles inherited from the father. Much interest has focused on insulin-like growth factor 2 (*IGF2*), a growth inducing gene, located at 11p15, and shown to be imprinted as predicted for the *BWS* gene (DeChiara *et al.*, 1991; Giannoukakis *et al.*, 1993; Ohlsson *et al.*, 1993). However, direct proof that overexpression of *IGF2* is the cause of BWS remains to be demonstrated. Similarly, the effect of *IGF2* expression on Wilms' tumour growth is unclear. Tumours have been shown to lose or 'relax' the imprint at *IGF2*, allowing expression of both alleles (Ogawa *et al.*, 1993; Rainier *et al.*, 1993; Weksberg *et al.*, 1993). However, the pattern and distribution of allelic losses at 11p15 suggest that another, more conventional tumour suppressor gene may also reside at that Wilms' tumour locus.

2.2.3 Other Wilms' tumour loci

Large families with hereditary transmission of Wilms' tumour susceptibility are rare. In three well-characterized pedigrees, however, linkage analysis of tumour susceptibility excluded both loci on chromosome 11 (Grundy *et al.*, 1988; Huff *et al.*, 1988; Schwartz *et al.*, 1991). Thus, another gene capable of transmitting Wilms' tumour susceptibility – presumably without affecting reproductive ability – remains to be mapped. In addition, LOH at chromosome 16 has been demonstrated in Wilms' tumour specimens. This locus has not been implicated in genetic susceptibility to Wilms' tumour but clinical correlation studies suggest that its disruption may confer a worse clinical prognosis (Maw *et al.*, 1992).

2.3 The *WT1* tumour suppressor gene

2.3.1 Structural properties of WT1

The discovery and mapping of homozygous deletions within the 11p13 locus led to the identification of *WT1*, the tumour suppressor gene targeted by disruption of this genetic locus (Call *et al.*, 1990; Gessler *et al.*, 1990; Haber *et al.*, 1990). *Figure 2.2* gives a schematic representation of the *WT1* tumour suppressor gene. *WT1* encodes a transcription factor of the Cys-His zinc finger class. The four DNA-

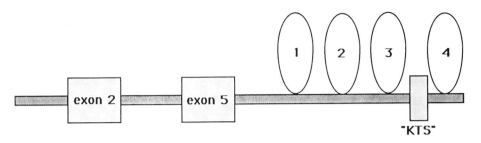

Figure 2.2. Schematic representation of the *WT1* tumour suppressor gene. *WT1* encodes a zinc finger protein with properties of a transcriptional repressor. The amino terminal has transactivational properties, while the four zinc finger domains at the carboxy terminal determine DNA binding specificity. Distinct WT1 isoforms are encoded by alternatively spliced transcripts. Two alternative splices are found in constant proportion in all *WT1*-expressing tissues: alternative splice I is encoded by exon 5, between the transactivation and DNA binding domains, and alternative splice II results from the use of an alternative splice donor site in exon 9 (zinc finger 3). The three amino acids encoded by alternative splice II (KTS) disrupt the spacing between zinc fingers 3 and 4, altering the DNA binding specificity of *WT1*. In ~ 10% of Wilms' tumour specimens, mRNA splicing results in the deletion of exon 2 from the *WT1* transcript, an aberrant mRNA species that is not seen in normal *WT1*-expressing tissues. This in-frame deletion within the *WT1* transcriptional repression domain converts the protien to a potent transcriptional activator.

DNA-binding zinc finger domains have a high degree of homology to those of the early growth response gene 1 (*EGR1*, also known as *NGF1A, Krox 24, Zif 268*). *In vitro* synthesized WT1 zinc finger proteins will also bind to the *EGR1* DNA consensus sequence 5'-GCGGGGGCG-3' (Rauscher III *et al.*, 1990), although additional DNA recognition sites may be recognized by purified WT1 protein (Bickmore *et al.*, 1992; Wang *et al.*, 1993). The amino terminal of *WT1* encodes a transactivation domain rich in prolines and glutamines. This domain appears to mediate transcriptional repression in transient transfection assays, leading to the suggestion that the WT1 protein may be a repressor of genes containing the *EGR1* consensus in their promoters (Madden *et al.*, 1992). Many promoters, including those of many growth factors and growth-inducing genes, appear to be repressed by WT1 in transient transfection assays (Drummond *et al.*, 1992; Gashler *et al.*, 1992; Madden *et al.*, 1992; Wang *et al.*, 1992). This effect is modulated by the presence of the p53 tumour suppressor protein, with which the WT1 protein forms a complex (Maheswaran *et al.*, 1993). In the absence of p53, WT1 appears to act as a potent transcriptional activator, suggesting that its transactivational properties may be complex and depend upon cellular context and different target genes. However, the effect of WT1 on endogenous native genes remains to be elucidated. In addition the *WT1* gene is subject to extensive alternative mRNA splicing

(Haber *et al.*, 1991). Four isoforms of WT1 are encoded in constant proportion in all *WT1*-expressing tissues. Although the exact nature of the functions mediated by these isoforms is unknown, one alternative splicing variant results in the disruption of the *WT1* zinc finger domain, altering its DNA binding specificity and, presumably, the identity of its target genes (Rauscher III *et al.*, 1990; Bickmore *et al.*, 1992; Drummond *et al.*, 1994).

2.3.2 Developmental role of WT1

The normal expression pattern of *WT1* is consistent with a role in the formation and differentiation of specific tissues. In the kidney, both in humans and in the mouse, *WT1* is expressed in cells of the condensing mesenchyme, particularly in S-shaped bodies that form a precursor to the mature glomerulus (Pritchard–Jones *et al.*, 1990). In the mouse, *WT1* expression peaks at the time of birth and rapidly declines as the kidney matures (Buckler *et al.*, 1991). Similarly, in humans, *WT1* expression is high in the fetal kidney and barely detectable in the adult organ (Haber *et al.*, 1990). This striking developmental pattern is not seen in other tissues expressing *WT1* [i.e. the mesothelial lining of abdominal viscera, splenic stroma, pericardial lining, Sertoli cells of the testis, granulosa cells of the ovary and muscle cells of the uterus (Pelletier *et al.*, 1991a; Armstrong *et al.*, 1992; Park *et al.*, 1993a)]. In all these tissues, *WT1* expression persists in the adult form, suggesting a different biological role to that played in the kidney. To date, *WT1* mutations have only been noted in Wilms' tumours and in mesothelially derived tumours (see Section 2.3.3).

The role of *WT1* in development is best demonstrated by its inactivation in the mouse. A mouse mutant, Sey/Dey, with a hemizygous deletion of *WT1* analogous to the human *WAGR* deletion, displays no abnormalities in genito-urinary development (Glaser *et al.*, 1990). However, homozygous inactivation of *WT1* leads to the complete failure of kidney and gonadal development (Kreidberg *et al.*, 1993). In addition, cardiac and diaphragmatic malformations are noted, presumably resulting from abnormalities of their mesothelial lining normally expressing *WT1*, and resulting in embryonic lethality. Mice with one deleted *WT1* allele have no increased incidence of malignancy, pointing to potential differences between human and mouse *WT1*-regulated pathways. In contrast, rat nephroblastomas are histologically similar to human Wilms' tumours, and the role of *WT1* in their genesis is under study (Hasgekar *et al.*, 1989).

2.3.3 Inactivation of WT1 in Wilms' and mesothelial tumours

As predicted by the Knudson model and by the genetics of Wilms' tumour, *WT1* is inactivated by mutations in a subset of Wilms' tumour, and in the germline of genetically susceptible children. Mutations consist of premature terminations as well as missense mutations, and no mutational 'hot spots' have been demonstrated (Haber *et al.*, 1990; Cowell *et al.*, 1991; Huff *et al.*, 1991; Ton *et al.*, 1991a; Little *et al.*, 1992; Pelletier *et al.*, 1992b; Coppes *et al.*, 1993). Mutations within the zinc finger domains result in the abolishment of DNA binding ability (Haber *et al.*, 1990; Pelletier *et al.*, 1991c), while some point mutations in the transactivation domain appear to convert the WT1 protein from a transcriptional repressor to a transcriptional activator of its target promoters (Haber *et al.*, 1993; Park *et al.*, 1993a,b). Approximately 10% of sporadic Wilms' tumours contain a homozygous mutation in *WT1*, and these mutations appear to arise early, being present in nephrogenic rests that may constitute genetic precursors of Wilms' tumour (Park *et al.*, 1993c).

Germline mutations in *WT1* are rare and appear to confer a high degree of genetic susceptibility, as demonstrated for other tumour suppressor genes. In most cases, a germline point mutation in *WT1* is followed by somatic losses of heterozygosity affecting the remaining *WT1* allele (Huff *et al.*, 1991; Pelletier *et al.*, 1991b). However, most children with WAGR syndrome, who harbour an initial germline deletion encompassing *WT1*, appear to have a somatic point mutation affecting the remaining *WT1* allele (Baird *et al.*, 1992; Brown *et al.*, 1992; Park *et al.*, 1993b) (*Figure 2.3*). Loss of one *WT1* allele in the germline, either by a point mutation or by WAGR syndrome, also appears to confer varying degrees of genito-urinary malformations, including hypospadias, undescended testes and renal malformations (Pelletier *et al.*, 1991b). These developmental abnormalities appear to be more prevalent in boys than in girls with germline *WT1* mutations.

Not all *WT1* mutations are equivalent. Germline *WT1* mutations, particularly point mutations within zinc finger 3 of *WT1*, confer a severe phenotype consisting of renal mesangeal sclerosis and pseudohermaphroditism, as well as predisposition to Wilms' tumour – the so-called Denys–Drash syndrome (Pelletier *et al.*, 1991c) (*Figure 2.3*). The severity of this phenotype suggests that some *WT1* mutations may act as 'dominant negatives', causing more severe developmental defects than those induced by simple loss of function. The Wilms' tumours of most Denys–Drash patients show loss of the remaining *WT1* allele, suggesting that this second genetic hit does confer additional growth advantage to the tumour. However, a significant fraction of *WT1* mutations in Wilms' tumours are heterozygous, and most of these involve the zinc finger domain. One such mutant has been shown capable of transforming primary kidney cells in cooperation with the adenovirus oncogene *E1A* (Haber *et al.*, 1992). The mechanism underlying this dominant negative effect has not yet been elucidated.

Additional mechanisms of inactivating *WT1* have recently been described. Approximately 10% of Wilms' tumours express significant levels of an aberrantly spliced *WT1* transcript (Haber *et al.*, 1993). This transcript results from an in-frame deletion of exon 2, within the transactivation domain of *WT1* (see *Figure 2.2*). Like point mutations in this domain, the aberrant *WT1* splice form converts the WT1 protein from a transcriptional repressor to a transactivator of its DNA target. No mutations within *WT1* exon 2 or its surrounding introns have been found in Wilms' tumours expressing this aberrant splice form, suggesting that it

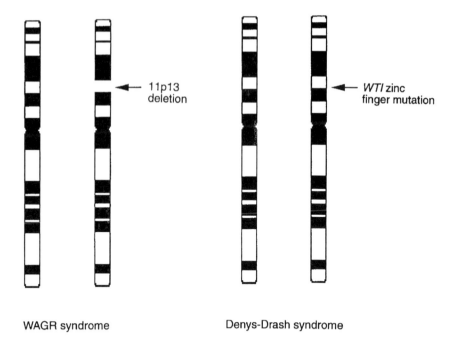

WAGR syndrome   Denys-Drash syndrome

Figure 2.3. Germline abnormalities at the 11p13 locus. In WAGR syndrome, a gross deletion in band p13 of one chromosome 11 results in hemizygosity for the Wilms' tumour suppressor gene *WT1*, the aniridia gene *Pax 6*, and other surrounding genes, including catalase and B-follicle-stimulating hormone. Affected children have abnormalities of the iris, resulting from a reduced dosage of *Pax 6*, varying degrees of genito-urinary anomalies, presumably associated with a reduced dosage of *WT1*, and mental retardation, whose aetiology is unknown. Loss of the remaining *WT1* allele in somatic kidney cells, typically by a point mutation, leads to the development of Wilms' tumour. In Denys–Drash syndrome, a *WT1* point mutation is present in one germline allele, typically a missense mutation within zinc finger 3. This mutation results in a dysfunctional protein, a so-called 'dominant negative', whose developmental effects are more severe than those seen in children with loss of one *WT1* allele, i.e. WAGR syndrome. Children with Denys–Drash syndrome have pseudohermaphroditism and renal mesangial sclerosis, suggesting that the dysfunctional WT1 protein disrupts both sexual and renal developmental pathways. Wilms' tumours arising in children with Denys–Drash syndrome show loss of the remaining *WT1* allele, typically by a gross chromosomal rearrangement, detected by loss of heterozygosity at 11p13.

may result from abnormalities in a distinct regulatory gene required for correct splicing of *WT1* mRNA. In addition to mutational and splicing abnormalities, a specific chromosomal translocation involving *WT1* has recently been reported in desmoplastic small round cell tumour; a rare, paediatric cancer derived from primitive mesothelial cells (Ladanyi and Gerald, 1994). This translocation results in the fusion of the putative transactivatory domain of the Ewing's sarcoma gene

EWS to zinc fingers 2–4 of *WT1*. The resulting chimeric protein may also have deregulated transactivational activity on *WT1* target genes.

2.3.4 Functional properties of the WT1 protein

The transactivational properties of WT1 are undoubtedly essential for its function as a tumour suppressor – a concept supported by mutational studies in Wilms' and mesothelial tumours. Re-introduction of wild-type *WT1* into a Wilms' tumour cell line expressing an endogenous *WT1* transcript with a deletion of exon 2 results in growth suppression (Haber *et al.*, 1993). This effect is consistent with cell fusion and chromosome transfer experiments (Weissman *et al.*, 1987), and suggests that while *WT1* inactivation is an early event in Wilms' tumorigenesis, restoring wild-type WT1 function is sufficient to inhibit malignant growth. An understanding of these properties of *WT1* will be required to define its role in normal organ development and cellular growth control.

References

Armstrong J, Pritchard-Jones K, Bickmore W, Hastie N, Bard J. (1992) The expression of the Wilms' tumour gene, WT1, in the developing mammalian embryo. *Mech. Devel.* **40**: 85–97.

Baird P, Groves N, Haber D, Housman D, Cowell J. (1992) Identification of mutations in the WT1 gene in tumours from patients with the WAGR syndrome. *Oncogene* **7**: 2141–2149.

Beckwith J. (1969) Macroglossia, omphalocele, adrenal cytomegaly, gigantism and hyperplastic visceromegaly. *Birth Defects* **5**: 188–196.

Bennington J, Beckwith J. (1975) Tumours of the kidney, renal pelvis and ureter. In: *Atlas of Tumor Pathology*, Series 2, Fascile 12. Armed Forces Institute of Pathology, Washington, DC.

Bickmore W, Oghene K, Little M, Seawright A, vanHeyningen V, Hastie N. (1992) Modulation of DNA binding specificity by alternative splicing of the Wilms tumour wt1 gene transcript. *Science* **257**: 235–237.

Brown K, Watson J, Poirier V, Mott M, Berry P, Maitland N. (1992) Inactivation of the remaining allele of the WT1 gene in a Wilms' tumour from a WAGR patient. *Oncogene* **7**: 763–768.

Buckler A, Pelletier J, Haber D, Glaser T, Housman D. (1991) Isolation, characterization, and expression of the murine Wilms' tumour gene (WT1) during kidney development. *Mol. Cell Biol.* **11**: 1707–1712.

Call K, Glaser T, Ito C, Buckler A, Pelletier J, Haber D, Rose E, Kral A, Yeger H, Lewis W, Jones C, Housman D. (1990) Isolation and characterization of a zinc finger polypeptide gene at the human chromosome 11 Wilms' tumour locus. *Cell* **60**: 509–520.

Cochran W, Froggatt P. (1967) Bilateral nephroblastoma in two sisters. *J. Urol.* **97**: 216–220.

Coppes M, Liefers G, Paul P, Yeger H, Williams B. (1993) Homozygous somatic WT1 point mutations in sporadic unilateral Wilms tumour. *Proc. Natl Acad. Sci. USA* **90**: 1416–1419.

Cowell J, Wadey R, Haber D, Call K, Housman D, Prichard J. (1991) Structural rearrangements of the WT1 gene in Wilms' tumour cells. *Oncogene* **6**: 595–599.

DeChiara T, Roberson E, Efstradiatis A. (1991) Parental imprinting of the mouse insulin-like growth factor II gene. *Cell* **64**: 849–859.

Drummond I, Badden S, Rohwer-Nutter P, Bell G, Sukhatme V, Rauscher III F. (1992) Repression of the insulin-like growth factor II gene by the Wilms tumor suppressor WT1. *Science* **257**: 674–678.

Drummond I, Ruprecht H, Rohwer-Nutter P, Lopez-Guisa J, Madden S, Rauscher III F, Sukhatme V. (1994) DNA recognition by splicing variants of the Wilms' tumor suppressor, WT1. *Mol. Cell Biol.* **14**: 3800–3809.

Gashler A, Bonthron D, Madden S, Rauscher III F, Collins T, Sukhatme V. (1992) Human platelet-derived growth factor A chain is transcriptionally repressed by the Wilms tumour suppressor WT1. *Proc. Natl Acad. Sci. USA* **89**: 10984–10988.

Gessler M, Poustka A, Cavenee W, Neve R, Orkin S, Bruns G. (1990) Homozygous deletion in Wilms tumours of a zinc-finger gene identified by chromosome jumping. *Nature* **343**: 774–778.

Giannoukakis N, Deal C, Paquette J, Goodyer C, Polychronakos C. (1993) Parental genomic imprinting of the human IGF2 gene. *Nature Genet.* **4**: 98–100.

Glaser T, Lane J, Housman D. (1990) A mouse model of the Aniridia–Wilms' tumour deletion syndrome. *Science* **250**: 823–827.

Grundy P, Koufos A, Morgan K, Li F, Meadows A, Cavenee W. (1988) Familial predisposition to Wilms' tumour does not map to the short arm of chromosome 11. *Nature* **336**: 374–376.

Grundy P, Breslow N, Green D, Sharples K, Evans A, D'Angio G. (1989) Prognostic factors for children with recurrent Wilms' tumor: results from the second and third National Wilms' Tumor Study. *J. Clin. Oncol.* **7**: 638–647.

Grundy P, Telzerow P, Haber D, Li F, Paterson M, Garber J. (1991) Chromosome 11 uniparental isodisomy in a child with hemihypertrophy and embryonal neoplasms. *Lancet* **338**: 1079–1080.

Haber D, Buckler A, Glaser T, Call K, Pelletier J, Sohn R, Douglass E, Housman D. (1990) An internal deletion within an 11p13 zinc finger gene contributes to the development of Wilms' tumor. *Cell* **61**: 1257–1269.

Haber D, Sohn R, Buckler A, Pelletier J, Call K, Housman D. (1991) Alternative splicing and genomic structure of the Wilms tumor gene WT1. *Proc. Natl Acad. Sci. USA* **88**: 9618–9622.

Haber D, Timmers H, Pelletier J, Sharp P, Housman D. (1992) A dominant mutation in the Wilms tumor gene WT1 cooperates with the viral oncogene E1A in transformation of primary kidney cells. *Proc. Natl Acad. Sci. USA* **89**: 6010–6014.

Haber D, Park S, Maheswaran S, Englert C, Re G, Hazen–Martin D, Sens D, Garvin A. (1993) WT1-mediated growth suppression of Wilms tumor cells expressing a WT1 splicing variant. *Science* **262**: 2057–2059.

Hasgekar N, Pendse A, Lalitha V. (1989). Rat renal mesenchymal tumor as an experimental model for human congenital mesoblastic nephroma. *Pediatr. Pathol.* **9**: 131–139.

Henry I, Bonaiti-Pellie C, Chehensse V, Beldjord C, Schwartz C, Utermann G, Junien C. (1991) Uniparental paternal disomy in a genetic cancer-predisposing syndrome. *Nature* **351**: 665–667.

Huff V, Compton D, Chao L, Strong L, Geiser C, Saunders G. (1988) Lack of linkage of familial Wilms' tumour to chromosomal band 11p13. *Nature* **336**: 377–378.

Huff V, Miwa H, Haber D, Call K, Housman D, Strong L, Saunders G. (1991) Evidence for WT1 as a Wilms tumour (WT) gene: intragenic germinal deletion in bilateral WT. *Am. J. Hum. Genet.* **48**: 997–1003.

Knudson A. (1971) Mutation and cancer: statistical study of retinoblastoma. *Proc. Natl Acad. Sci. USA* **68**: 820–823.

Knudson A, Strong L. (1972) Mutation and cancer: a model for Wilms tumor of the kidney. *J. Natl Cancer Inst.* **48**: 313–324.

Koufos A, Grundy P, Morgan K, Aleck K, Hadro T, Lampkin B, Kalbakji A, Cavenee W. (1989) Familial Wiedemann-Beckwith syndrome and a second Wilm's tumor locus both map to 11p15.5. *Am. J. Hum. Genet.* **44**: 711–719.

Kreidberg J, Sariola H, Loring J, Maeda M, Pelletier J, Housman D, Jaenisch R. (1993) WT1 is required for early kidney development. *Cell* **74**: 679–691.

Ladanyi M, Gerald W. (1994) Fusion of the EWS and WT1 genes in the desmoplastic small round cell tumor. *Cancer Res.* **54**: 2837–2840.

Little M, Prosser J, Condie A, Smith P, vanHeyningen V, Hastie N. (1992) Zinc finger point mutations within the WT1 gene in Wilms tumor patients. *Proc. Natl Acad. Sci. USA* **89**: 4791–4795.

Madden S, Cook D, Morris J, Gashler A, Sukhatme V, Rauscher III F. (1992) Transcriptional repression mediated by the WT1 Wilms tumor gene product. *Science* **253**: 1550–1553.

Maheswaran S, Park S, Bernard A, Morris J, Rauscher III F, Hill D, Haber D. (1993) Physical and functional interaction between WT1 and p53 proteins. *Proc. Natl Acad. Sci. USA* **90**: 5100–5104

Mannens M, Slater R, Heyting C, Bliek J, de Kraker J, Coad N, de Pagter–Holthuizen P, Pearson P. (1988) Molecular nature of genetic changes resulting in loss of heterozygosity for chromosome 11 in Wilms' tumor. *Hum. Genet.* **81**: 41–48.

Maw MA, Grundy PE, Millow JW. (1992) A third Wilms' tumor locus on chromosome 16q. (1992) *Cancer Res.* **52**: 3094–3098.

Miller R, Fraumeni J, Manning M. (1964) Association of Wilms' tumor with aniridia, hemihypertrophy and other congenital malformations. *N. Engl. J. Med.* **270**: 922–927.

Ogawa O, Becroft D, Morison I, Ecles M, Skeen J, Mauger D, Reeve A. (1993). Constitutional relaxation of insulin-like growth factor II gene imprinting associated with Wilms' tumour and gigantism. *Nature Genet.* **5**: 408–412

Ohlsson R, Nystrom A, Pfeifer–Ohlsson S, Tohonen V, Hedborg F, Schofield, P, Flam F, Ekstrom T. (1993) IGF2 is parentally imprinted during human embryogenesis and in the Beckwith–Wiedemann syndrome. *Nature Genet.* **4**: 94–97.

Park S, Schalling M, Bernard A, Maheswaran S, Shipley G, Roberts D, Fletcher J, Shipman R, Rheinwald J, Demetri G, Griffin J, Minden M, Housman D, Haber D. (1993a) The Wilms tumour gene WT1 is expressed in murine mesoderm-derived tissues and mutated in a human mesothelioma. *Nature Genet.* **4**: 415–420.

Park S, Tomlinson G, Nisen P, Haber D. (1993b) Altered trans-activational properties of a mutated WT1 gene product in a WAGR-associated Wilms' tumor. *Cancer Res.* **53**: 4757–4760.

Park S, Bernard A, Bove K, Sens D, Hazen–Martin D, Garvin A, Haber D. (1993c) Inactivation of WT1 in nephrogenic rests, genetic precursors to Wilms' tumour. *Nature Genet.* **5**: 363–367.

Pelletier J, Schalling M, Buckler A, Rogers A, Haber D, Housman D. (1991a) Expression of the Wilms' tumor gene WT1 in the murine urogenital system. *Genes Devel.* **5**: 1345–1356.

Pelletier J, Bruening W, Li F, Haber D, Glaser T, Housman D. (1991b) WT1 mutations contribute to abnormal genital system development and hereditary Wilms' tumour. *Nature* **353**: 431–434.

Pelletier J, Bruening W, Kashtan C, Mauer S, Manivel J, Striegel J, Houghton D, Junien C, Habib R, Fouser L, Fine R, Silverman B, Haber D, Housman D. (1991c) Germline mutations in the Wilms' tumor suppressor gene are associated with abnormal urogenital development in Denys–Drash syndrome. *Cell* **67**: 437–447.

Ping J, Reeve A, Law D, Young M, Boehnke M, Feinberg A. (1989) Genetic linkage of Beckwith–Wiedemann syndrome to 11p15. *Am. J. Hum. Genet.* **44**: 720–723.

Pritchard–Jones K, Fleming S, Davidson D, Bickmore W, Porteous D, Gosden C, Bard J, Buckler A, Pelletier J, Housman D, van Heyningen V, Hastie N. (1990) The candidate Wilms' tumour gene is involved in genitourinary development. *Nature* **346**: 194–197.

Rainier S, Johnson L, Dobry C, Ping A, Grundy P, Feinberg A. (1993) Relaxation of imprinted genes in human cancer. *Nature* **362**: 747–749.

Rauscher III F, Morris J, Tournay O, Cook D, Curran T. (1990) Binding of the Wilms' tumor locus zinc finger protein to the EGR-1 consensus sequence. *Science* **250**: 1259–1262.

Riccardi V, Sujansky E, Smith A, Francke U. (1978) Chromosomal imbalance in the aniridia–Wilms' tumor association: 11p interstitial deletion. *Pediatrics* **61**: 604–610.

Schwartz C, Haber D, Stanton V, Strong L, Skolnick M, Housman D. (1991) Familial predisposition to Wilms tumor does not segregate with the WT1 gene. *Genomics* **10**: 927–930.

Ton C, Huff V, Call K, Cohn S, Strong L, Housman D, Saunders G. (1991a) Smallest region of overlap in Wilms' tumor deletions uniquely implicates an 11p13 zinc finger gene as the disease locus. *Genomics* **10**: 293–297.

Ton C, Hirronen M, Miwa H, Weil M, Monaghan P, Jordan T, van Heyningen V, Hastie N, Meigers-Heijboer H, Drechsler M, Royer-Pokora B, Collins F, Swaroop A, Strong LC, Saunders G. (1991b) Positional cloning and characterization of a paired box- and homeobox-containing gene from the aniridia region. *Cell* **67**: 1059–1074.

Turleau C, de Grouchy J, Nihoul-Fekete C, Chavin-Colin F, Junien C. (1984) Del 11p13/nephroblastoma without aniridia. *Hum. Genet.* **67**: 455–456.

Wang Z, Madden S, Deuel T, Rauscher III F. (1992) The Wilms' tumor gene product, WT1, represses transcription of the platelet-derived growth factor A-chain gene. *J. Biol. Chem.* **267**: 21999–22002.

Wang Z-Y, Qiu Q-Q, Deuel T. (1993) The Wilms' tumor gene product WT1 activates or suppresses transcription through separate functional domains. *J. Biol. Chem.* **268**: 9172–9175.

Waziri M, Patil SR, Hanson JW, Bartley JA. (1983) Abnormality of chromosones 11 in patients with features of Beckwith–Wiedemann syndrome. *J. Pediatr.* **102**: 873–876.

Weissman B, Saxon P, Pasquale S, Jones G, Geiser A, Stanbridge E. (1987) Introduction of a normal chromosome 11 into a Wilms' tumour cell line controls its tumorigenic expression. *Science* **236**: 175–180.

Weksberg R, Shen D, Fei YL, Song Q, Squire J. (1993) Disruption of insulin-like growth factor 2 imprinting in Beckwith–Wiedemann syndrome. *Nature Genet.* **5**: 143–149.

Wiedemann H. (1964) Complex familial malformation with umbilical hernia and macroglossia – a novel syndrome? *J. Genet. Hum.* **13**: 223–232.

Molecular genetics of neurofibromatosis types 1 and 2

Steven D. Colman and Margaret R. Wallace

3.1 History of neurofibromatosis types 1 and 2

In 1882, Friedrich von Recklinghausen first described a clinical syndrome characterized by nerve-derived tumours called neurofibromas. This syndrome was subsequently called 'von Recklinghausen neurofibromatosis', and is now referred to as neurofibromatosis type 1 (NF1) (Warkany, 1981; Riccardi, 1992). NF1 is one of the most frequently occurring human autosomal dominant diseases (1/3500 individuals). The hereditary nature of NF1 was not fully recognized until Borberg (1951) and Crowe *et al.* (1956) determined that the mode of inheritance is autosomal dominant. Thirty-five years after NF1 was first described, neurosurgeon Harvey Cushing concluded that another syndrome, characterized by bilateral acoustic neuromas (more accurately, vestibular schwannomas), was a form of neurofibromatosis (Eldridge, 1990). He reached this conclusion after noting pathological similarities between tumours seen in this disorder and those seen in NF1. This classification persisted for many years and, although the suggestion that these two diseases were separate entities was made as early as 1930, the distinction did not become widely accepted until the 1980s (Short *et al.*, 1994). This second disease, which is now known to be genetically and phenotypically distinct from NF1, is referred to as neurofibromatosis type 2 (NF2), an autosomal dominant disorder with a frequency of 1/40 000 births. The hallmark features of NF2 are the formation of bilateral vestibular schwannomas, subsequent hearing loss and premature death. Together, NF1 and NF2 represent the vast majority of the neurofibromatoses; in fact, it is possible that the other, rarer reported types of NF are actually phenotypic variants of NF1 or NF2.

3.2 Clinical features of NF1

Clinical and pathological findings of NF1, briefly summarized below, have been reviewed in detail elsewhere (Mulvihill *et al.*, 1990; Rubenstein and Korf, 1990; Riccardi, 1992; Gutmann and Collins, 1995; Huson and Hughes, 1994). Biochemical and cell biological analyses, predominantly conducted prior to the cloning of the NF1 gene, are also reviewed elsewhere (Rubenstein *et al.*, 1986; Wallace and Collins, 1991; Riccardi, 1992; Gutmann and Collins, 1995). NF1 diagnosis is currently based solely on clinical examination – there is no biochemical test, and direct DNA analysis is not routinely available. Patients are diagnosed with NF1 if they meet two or more of the following seven criteria established by a National Institutes of Health (NIH) consensus panel (Stumpf *et al.*, 1988):

(i)   six or more café-au-lait spots 1.5 cm or larger in postpubertal individuals or 0.5 cm or larger in prepubertal individuals;

(ii)   two or more neurofibromas of any type, or at least one plexiform neurofibroma;

(iii) freckling of the armpits or groin;

(iv) optic glioma;

(v)   two or more Lisch nodules;

(vi) dysplasia of the sphenoid bone, or dysplasia or thinning of long bone cortex;

(vii) first degree relative with NF1.

NF1 diagnosis is often unclear due to phenotypic variability and the fact that affected individuals may not meet diagnostic criteria until their teens or later. In fact, mild cases may go undiagnosed, particularly if the individuals suffer no complications. A number of features which contribute to the variable expressivity of NF1 include macrocephaly, learning disabilities (rarely outright mental retardation), short stature, scoliosis, pseudoarthrosis, seizures, hypertension, pruritus, headaches and malignancy [especially tumours of the central nervous system (Riccardi, 1992)]. About two thirds of NF1 patients have relatively mild symptoms and are able to lead fairly normal lives; however, the average life span is somewhat reduced overall due to the increased occurrence of malignancy and other life-threatening complications (Riccardi, 1992). Malignancies which appear to have an increased incidence in NF1 include neurofibrosarcomas, phaeochromocytomas, neuronal tumours, rhabdomyosarcomas and leukaemias (Bader, 1986; Blatt *et al.*, 1986; Sørensen *et al.*, 1986; Lefkowitz *et al.*, 1990; Shearer *et al.*, 1994).

One of the hallmark features of NF1 is the formation of neurofibromas, benign tumours which arise most commonly along peripheral nerves. They often begin to appear in adolescence, and increase in size and number with age and/or hormonal fluctuations, such as those which occur in pregnancy. Less common, but potentially more serious, are plexiform neurofibromas, which develop along more deeply placed nerves and which may become quite large. While both types of neurofibroma are histologically similar, containing primarily Schwann cells, fibroblasts and mast cells, plexiform tumours have a propensity for progression to malignancy; to neurofibrosarcomas in an estimated 6% or more of NF1 patients (Riccardi, 1992). If present in childhood, plexiform neurofibromas may lead to severe disfigurement and/or functional impairment via overgrowth and subsequent impingement on surrounding tissues.

3.3 Genetics of NF1

NF1 has a very high mutation rate, estimated at 1/10 000 alleles per generation – about 10-fold higher than most genes (Crowe *et al.*, 1956; Sergeyev, 1975). The frequency of this disorder, one of the highest among genetic diseases, is likely to be due to this high mutation rate. In fact, NF1 appears to be the result of new mutations in about half of all patients (Riccardi, 1992). One theory to explain the elevated mutation rate is that the large size of the *NF1* gene makes it a more frequent target for alterations (Cawthon *et al.*, 1990; Viskochil *et al.*, 1990; Wallace *et al.*, 1990); this is consistent with the Duchenne muscular dystrophy (*DMD*) gene, which is similarly very large with a high mutation rate (Monaco *et al.*, 1986). It has been argued, however, that the gene size alone cannot account for this extremely high mutation frequency (Marchuk *et al.*, 1991). Similar to observations in DMD and retinoblastoma, the majority of *NF1* mutations evidently arise in the paternal germline (Jadayel *et al.*, 1990; Stephens *et al.*, 1992). *NF1* is considered to be completely penetrant; however, this issue will ultimately be resolved through DNA studies of the few suspected non-penetrant families (Carey *et al.*, 1979; Riccardi and Lewis, 1988; Riccardi, 1992).

3.3.1 The NF1 gene

The *NF1* gene, at 17q11.2, spans about 350 kilobases (kb) of genomic DNA, has 59 exons (including two that are alternatively spliced) and produces two apparent transcripts (11 and 13 kb, presumably due to different lengths of the 3′ untranslated region) (Viskochil *et al.*, 1993; Marchuk and Collins, 1994). There are three small genes (whose functions are unclear) embedded within intron 27b, encoded on the opposite strand (Cawthon *et al.*, 1991), as well as a pseudogene near the 3′ end (Xu G. *et al.*, 1992). A schematic diagram illustrating the structure of the *NF1* gene is shown in *Figure 3.1*. Two of these embedded genes (*EVI2A* and *EVI2B*) are predominantly lymphoid specific, while the other (*OMGP*) is evidently involved in myelination. It is unknown whether any of these genes are involved in the NF1 phenotype.

3.3.2 Neurofibromin

The *NF1* mRNA encodes a 2818-amino acid hydrophilic protein called neurofibromin. This protein, which appears to associate with microtubules (Gregory *et*

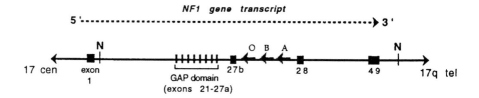

Figure 3.1. Schematic diagram (not to scale) showing the genomic structure of the *NF1* gene and its direction of transcription on chromosome 17q. N, *Not*I restriction sites found in the CpG islands that flank the gene, with the left-hand one being at the 5' end of the *NF1* gene. The entire gene except for the extreme 5' end is contained within the 350-kb *Not*I fragment indicated. Specific exons delineating regions of the gene are shown by black boxes. The three embedded genes in intron 27b, encoded on the opposite strand, are indicated by the shaded arrows: O, *OMGP*; B, *EVI2B*; A, *EVI2A*. The exons encoding the GAP-related domain include 21 through 27a, as indicated by the bracket. No other regions have had functions specifically assigned, although some portions of the 3' half of the gene show protein-level homology to the yeast GAP proteins IRA1 and IRA2.

al., 1993), is enriched in the endoplasmic reticulum of neurons (Nordlund *et al.*, 1993). *NF1* message is present in many tissues (Wallace *et al.*, 1990); however, antibody studies have shown that neurofibromin is in greatest abundance in neurons, Schwann cells and oligodendrocytes (Daston *et al.*, 1992; Golubic *et al.*, 1992; Gutmann and Collins, 1993). The middle eighth of the deduced protein (encoded by exons 21–27b) shows strong homology to GTPase-activating proteins (GAPs) (Buchberg *et al.*, 1990; Xu *et al.*, 1990a), particularly the mammalian p120^{gap} protein and the homologous yeast IRA1 and IRA2 proteins. GAPs are at least partly responsible for keeping RAS, which plays a crucial role in up-regulating cell proliferation, in a GDP-bound, inactive state (Downward, 1992). Several groups have shown that the homologous region of neurofibromin can function as a GAP both *in vitro* and *in vivo* (Ballester *et al.*, 1990; Martin *et al.*, 1990; Xu *et al.*, 1990b). Studies showing a lack of functional neurofibromin in neurofibrosarcomas support the notion that neurofibromin may be an essential negative regulator of RAS in neural crest-derived tissues (DeClue *et al.*, 1991; Basu *et al.*, 1992). Neurofibromin's role(s) in the signal transduction pathway is probably quite complex, since tubulin-bound neurofibromin has diminished GAP activity (Bollag *et al.*, 1993), and alternatively spliced isoforms appear to have somewhat different tissue distributions, levels and GAP activities (see Section 3.3.3).

Studies of melanoma and neuroblastoma cell lines that lack neurofibromin, yet still have normal GAP activity (presumably due to p120^{gap}), suggest that RAS regulation is not the only function for neurofibromin (Johnson *et al.*, 1993). The fact that sequences outside the GAP-related domain show strong evolutionary conservation also indicates that other regions of neurofibromin are likely to have significant functions (Marchuk *et al.*, 1991). Neurofibromin may, therefore, perform separate roles in different tissues and may also have different expression levels and roles in various developmental stages (Daston and Ratner, 1992; Gutmann and Collins, 1993). Gutmann and Collins (1993) have also suggested that, given the widespread expression of the *NF1* message, defects in only certain tissues of NF1 patients could have one or more explanations. Neurofibromin may: (i) normally be required at significantly higher levels and have a more crucial role in affected tissues (i.e. dosage effect); (ii) be post-translationally modified in a tissue-specific manner; and/or (iii) associate with other signal transduction molecules that are expressed specifically in these tissues. In fact, neurofibromin and p120^{gap} appear to be regulated differently (Bollag and McCormick, 1991), with recent evidence that cross-linking of the surface immunoglobulin receptor in B cells induces cellular redistribution of neurofibromin but not p120^{gap} (Boyer *et al.*, 1994).

3.3.3 Alternative splicing of the NF1 gene

As mentioned above, there are at least two alternatively spliced exons. One, exon 23a (near the middle of the gene in the GAP-related domain), encodes a 21-amino acid sequence that is perfectly conserved across many species (Nishi *et al.*, 1991). Tissue specificities or functions for this larger isoform of neurofibromin are not entirely clear, but both forms (normal and larger) are present in most adult tissues. There are a number of studies testing these isoforms to elucidate their roles in development and tumorigenesis (Nishi *et al.*, 1991; Teinturier *et al.*, 1992; Gutmann *et al.*, 1993a; Huynh *et al.*, 1994a); some of these are discussed in Section 3.5.4. The inclusion of the short peptide encoded by exon 23a slightly decreases the GAP activity of the molecule, although this isoform has higher affinity for RAS–GTP (Andersen *et al.*, 1993a; Viskochil *et al.*, 1993). Another alternatively spliced exon, 48a, encodes 18 amino acids very near the carboxy end of the molecule (Cawthon *et al.*, 1990). This isoform is expressed primarily in muscle tissue and thus may be involved in muscle cell development (Gutmann *et al.*, 1993a). There is also a report of an *NF1* cDNA isolated from placenta that only contains the 5′ part of the gene and terminates prior to the GAP-related domain with a novel 3′ untranslated region (Suzuki *et al.*, 1992). This isoform could represent an alternative splice variant and could indicate that significantly different versions of neurofibromin exist.

3.4 Constitutional mutations in the *NF1* gene

The search for *NF1* mutations has proven difficult due to: (i) the large size of the gene; (ii) the presence of the normal allele; (iii) the high mutation rate and subsequent different mutations in most families; (iv) wide variation in the size and type of mutations (from complete deletion of the locus to missense point changes), which requires the use of multiple detection methods; and (v) several homologous loci on other chromosomes which can interfere with interpretation of Southern blot and PCR analyses (Marchuk *et al.*, 1992). In addition, some abnormalities must be further characterized to exclude functionally normal polymorphisms/variants.

An example of mutation detection and characterization is shown in *Figure 3.2*. Published germline *NF1* mutations, numbering about 50 (see *Table 3.1*), include single base substitutions, deletions of all sizes, insertions, alterations leading to splicing errors and two translocations (references listed in *Table 3.1*). All of these mutations, with the exception of five missense mutations and one 6-bp in-frame deletion, are predicted to result in premature truncation of the neurofibromin protein and subsequent disruption of normal function. One of the amino acid changes lies at the highly conserved Lys1423 residue in the GAP-related domain (exon 24); this mutation presumably has an effect on neurofibromin function (a constitutional missense mutation at this codon has been shown to encode a neurofibromin molecule with reduced GAP activity (Li *et al.*, 1992)). It is not obvious whether the other four amino acid changes have a functional effect: these could be rare normal variants, in which case, the NF1 patients who carry these variants must also carry other, yet-to-be identified, constitutional mutations.

As mutations continue to be characterized, questions regarding mutational hot spots and possible genotype – phenotype correlations (which may suggest heretofore unknown functions for neurofibromin) can be addressed. The only *NF1* mutational hot spot yet discerned is a recurrent C → T transition in exon 31, changing an arginine codon to a stop codon. This mutation is, however, relatively rare and no genotype–phenotype correlation has been inferred. One might expect that C → T transitions at CpG dinucleotides such as this would be more common in mutations of paternal origin, due to methylation differences during gametogenesis (Driscoll and Migeon, 1990), but it remains to be seen if this holds true for *NF1*. The involvement of the *NF1* gene in other genetic disorders that show phenotypic overlap is also under study, including investigations of the role of the *NF1* gene in the 'NF-Noonan syndrome' (Stern *et al.*, 1992) and in Watson syndrome (Upadhyaya *et al.*, 1992; Tassabehji *et al.*, 1993).

Overall, due to the difficulty of mutation analysis, direct DNA diagnosis of NF1 is not currently used clinically. However, researchers are developing better methods for this search and it appears that an RNA-based approach may be efficacious, based on several studies demonstrating the presence of mutant *NF1* transcripts (Wallace *et al.*, 1991; Xu, W. *et al.*, 1992a; Colman *et al.*, 1993; Abernathy *et al.*, 1994; Purandare *et al.*, 1994a). Due to identification of a number of polymorphisms within the *NF1* gene (e.g. Jorde *et al.*, 1993), linkage analysis is quite feasible and is a very effective approach for DNA diagnosis in families with a history of NF1. These markers can also be used in direct mutation searches based on loss of heterozygosity analyses (see Section 3.5.2) that can detect large deletions.

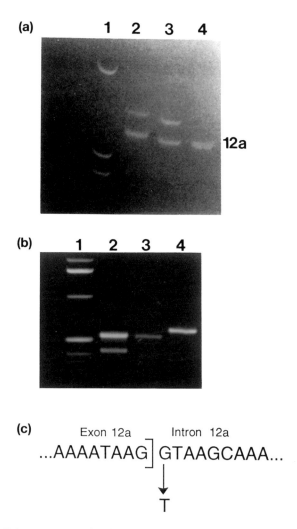

(c)

Exon 12a   Intron 12a

...AAAATAAG⌉GTAAGCAAA...

↓

T

Figure 3.2. Splicing error mutation in the *NF1* gene at exon 12a (unpublished data). (a) Heteroduplex analysis of exon polymerase chain reaction (PCR) products revealed a subtle doublet in the lower band in lane 4, indicating an abnormality in exon 12a in patient UF54 (the top band in lanes 2 and 3 is unrelated). The molecular weight marker is a 1-kb ladder (lane 1) (BRL, Gaithersburg, MD, USA). (b) After sequencing of the cloned DNA PCR product (which revealed a G → T transversion in the splice donor site), reverse transcriptase (RT)-PCR analysis of lymphocyte RNA was performed as described in Colman *et al.* (1993). Primers lay in exons 11 and 13, which amplify a normal cDNA product of 610 bp (as seen in lane 3, control normal RNA). RT-PCR of patient UF54 (lane 2) showed an extra, abnormally small fragment, which was subsequently cloned. Sequencing revealed that this lower band represents a transcript lacking exon 12a, which has been skipped in splicing due to the mutation. Lane 4 shows the normal actin RT-PCR product from UF54 RNA, as a further control. (c) Sequence of the intron/exon boundary showing the G → T point mutation in the first position of the splice donor site.

Table 3.1. Published germline *NF1* mutations

Mutation type	Locus	cDNA Nt#[a]	Predicted result	Reference
A→G transition	Intron 11	1721 (+3 bp)	Loss of exon 11; frameshift	Purandare *et al.* (1994a)
3-bp deletion (2)[b]	Exon 17	2970–2972	Loss of aa991 (Met)	Shen *et al.* (1993a)
G→A transition	Exon 21	3497	Gly→Asp aa1166	Purandare *et al.* (1994a)
1-bp deletion	Exon 24	4152	Frameshift	Abernathy *et al.* (1994)
1-bp deletion	Exon 24	4190	Frameshift	Anglani *et al.* (1993)
A→G transition	Exon 24	4256	Lys→Arg aa1419	Li *et al.* (1992)
A→C transversion	Exon 24	4267	Lys→Gln aa1423	Purandare *et al.* (1994a)
A→G transition (2)	Exon 24	4267	Lys→Glu aa1423	Li *et al.* (1992)
5-bp deletion, T→G	Exon 28	5124–5128; 5129	Frameshift	Stark *et al.* (1991)
2-bp deletion	Exon 28	4868–4869	Frameshift	Xu, W. *et al.* (1992a)
1-bp deletion	Exon 28	5010	Frameshift	Colman *et al.* (1993)
13-bp deletion	Exon 28	5077–5089	Frameshift	Shen and Upadhyaya (1993)
2-bp deletion	Exon 28	5108–5109	Frameshift	Zhong *et al.* (1993)
42-bp insertion	Exon 28	5137	Insertion of 14 aa	Tassabehji *et al.* (1993)
C→T transition	Exon 29	5242	Arg→STOP aa1748	Valero *et al.* (1994)
C→T transition	Exon 29	5260	Glu→STOP aa1754	Valero *et al.* (1994)
1-bp insertion	Exon 29	5662	Frameshift	Upadhyaya *et al.* (1992)
1-bp insertion	Exon 29	5662	Frameshift	Upadhyaya *et al.* (1992)
1-bp insertion	Exon 29	5678	Frameshift	Upadhyaya *et al.* (1992)
4-bp deletion	Exon 30	5679–5682	Frameshift	Hatta *et al.* (1994)
T→G transversion	Intron 30	5749 (+2 bp)	Loss of exon 30; frameshift	Purandare *et al.* (1994a)
1-bp insertion	Exon 31	5816	Frameshift	Zhong *et al.* (1993)
C→T transition	Exon 31	5839	Arg→STOP	Ainsworth *et al.* (1993)
C→T transition	Exon 31	5839	Arg→STOP	Cawthon *et al.* (1990)
C→T transition	Exon 31	5839	Arg→STOP	Estivill *et al.* (1991)
C→T transition	Exon 31	5839	Arg→STOP	Horiuchi *et al.* (1994)
C→T transition	Exon 31	5839	Arg→STOP	Valero *et al.* (1994)
2-bp insertion	Exon 31	5852	Frameshift	Ainsworth *et al.* (1993)
2-bp deletion	Exon 31	5843–5844	Frameshift	Valero *et al.* (1994)
A→G transition	Intron 31	5945 (−5 bp)	4-bp insertion in mRNA; frameshift	Ainsworth *et al.* (1994)
1-bp deletion	Exon 32	5949	Frameshift	Hatta *et al.* (1994)
571-bp deletion	Intron 32–intron 33	6085 (−32 bp) −6364 (+257 bp)	Loss of exon 33; frameshift	Xu, W. *et al.* (1992a)
320-bp Alu insertion	Intron 32	6095 (−44 bp)	Loss of exon 33; frameshift	Wallace *et al.* (1991)

A→G transition	Intron 33	6364 (+4 bp)	Loss of exon 33; frameshift	Hutter et al. (1994)
1-bp insertion	Exon 34	6519	Frameshift	Purandare et al. (1994b)
C→A transversion	Exon 34	6639	Leu→Met aa2143	Upadhyaya et al. (1992)
T→G transversion	Exon 34	6724	Tyr→Asp aa2213	Upadhyaya et al. (1992)
10-bp insertion	Exon 38	6922	Frameshift	Legius et al. (1994a)
6-bp deletion	Exon 39	7096–7101	Loss of aa2366 (Asn) and 2367 (Phe)	Abernathy et al. (1994)
2-bp insertion	Exon 42	7486	Frameshift	Purandare et al. (1994a)
C→T transition	Exon 42	7486	Arg→STOP aa2496	Purandare et al. (1994a)
10-bp deletion	Exon 44	7745–7754	Frameshift	Shen and Upadhyaya (1993)
10-kb intragenic insertion			Disruption of NF1 gene	Upadhyaya et al. (1992)
11-kb deletion			Disruption of NF1 gene	Viskochil et al. (1990)
40-kb deletion			Disruption of NF1 gene	Viskochil et al. (1990)
40-kb intragenic deletion			Disruption of NF1 gene	Upadhyaya et al. (1992)
90-kb intragenic deletion			Disruption of NF1 gene	Upadhyaya et al. (1992)
90-kb deletion			Disruption of NF1 gene	Viskochil et al. (1990)
110-kb insertion			Disruption of NF1 gene	Colley et al. (1991)
200-kb deletion			Disruption of NF1 gene	Colley et al. (1991)
Deletion I27–E37 (2)			Disruption of NF1 gene	Lazaro et al. (1993)
>350-kb deletions (5)			Disruption of NF1 gene	Kayes et al. (1994)
LOH of all loci tested			Disruption of NF1 gene	Colley et al. (1991)
LOH of 17L1A (5' end NF1 gene) (3)			Disruption of NF1 gene	Kamei et al. (1992)
Intragenic translocations			Disruption of NF1 gene	Menon et al. (1989)
Intragenic translocations			Disruption of NF1 gene	Schmidt et al. (1987)

a Nt, nucleotide. Nt#1 is 1st base of ATG start codon.
b Number in brackets indicates number of times the mutation was documented in the same paper.
aa, amino acid.

3.5 The *NF1* gene and neurofibromin in benign and malignant tumours

3.5.1 Tumour suppressor hypothesis

Virtually all *NF1* mutations reported thus far are predicted to disrupt or inactivate neurofibromin. Because its inactivation is associated with excessive proliferation of certain cell types (part of the NF1 phenotype), and part of its function is involved in RAS regulation, the *NF1* gene has been classified as a tumour suppressor gene (Levine, 1993; Seizinger, 1993). Tumour suppressor genes encode proteins that participate in the negative regulation of growth-promoting genes/proteins, and their mutation or 'knock-out' is associated with the development of tumours (Levine, 1993). Both alleles of tumour suppressor genes are inactivated in neoplastic tissues (Knudson, 1971, 1985) and, in many cases, one allele is constitutionally inactivated (inherited) while the other allele is subsequently inactivated ('second hit') via somatic mutation. This has been demonstrated for other inherited cancer syndromes such as retinoblastoma (Dunn *et al.*, 1988) and familial adenomatous polyposis (Ichii *et al.*, 1992). In NF1, since neurofibromin expression presumably acts to down-regulate cell growth, lack of this protein due to knock-out would be expected to result in an increased rate of cell division. It is unclear, however, whether the symptoms of NF1 (benign and/or malignant) are due to such a second inactivation, or alternatively whether heterozygous cells can overproliferate under certain circumstances [via reduced levels of functional neurofibromin (dosage effect)], or due to abnormal activity specific to a mutant molecule ('dominant negative' effect). The clinical variability within families could be explained by a complex genetic/environmental process or by random somatic mutation controlling the progression and severity of the disease. The major lines of evidence supporting the tumour suppressor role of the *NF1* gene are discussed in Sections 3.5.2 and 3.5.3.

3.5.2 NF1 *gene alterations in malignancies*

Loss of heterozygosity (LOH), the somatic deletion of part or all of one homologue in a chromosome pair, is a phenomenon commonly observed at tumour suppressor loci and which is thought to be involved in the aetiology of many tumours. LOH, which is typically assessed by comparing polymorphisms in blood and tumour DNA, has been observed within and spanning the *NF1* gene in a number of malignant tumours in *NF1* patients. For example, Glover *et al.* (1991) found LOH for most or all of chromosome 17 in two out of eight neurofibrosarcomas. A 200-kb deletion involving the 3' half of the *NF1* gene was later found in the remaining allele of one of those tumours missing a 17 homologue, supporting the tumour suppressor hypothesis for the *NF1* gene (Legius *et al.*, 1993). Lothe *et al.* (1993) also reported loss of a 17 homologue in an *NF1* malignant schwannoma.

Another laboratory detected LOH with markers flanking the *NF1* gene (and in some cases, losses on other parts of 17) in a series of seven NF1 phaeochromocytomas and, in all three informative cases, the retained allele contained an inherited *NF1* mutation (Xu, W. *et al.*, 1992b). Western blot analysis subsequently revealed that those seven phaeochromocytomas, as well as several other adrenal gland tumours, lacked neurofibromin (Gutmann *et al.*, 1994). Two out of eight melanoma cell lines showed complete inactivation of the *NF1* gene (one by homozygous deletion, the other by RNA and protein assays); the other six lines showed little or no decrease in the *NF1* message or protein (Andersen *et al.*, 1993b). In another study, five out of six children with NF1 and malignant myeloid disorders showed loss of the normal *NF1* allele in their bone marrow (tumour cells); these deletions did not include 17p (Shannon *et al.*, 1994), ruling out the loss of an entire chromosome 17 homologue [which precludes the involvement of the tumour suppressor *p53* (Levine *et al.*, 1991)]. Although knock-out of the *NF1* gene was reported for neuroblastoma cell lines (The *et al.*, 1993), activated RAS–GTP levels were not greatly increased. Furthermore, the GAP activity in these cells was normal (Johnson *et al.*, 1993), suggesting that neurofibromin must have additional functions related to cell cycle control.

LOH in NF1 tumours is not, however, a universal phenomenon; while Menon *et al.* (1990) found LOH in five out of six neurofibrosarcomas, the common deleted region was 17p, rather than 17q. Glover *et al.* (1991) also reported 17p (but not 17q) LOH in a glioblastoma from an NF1 patient. It is also possible that, at least in some cases, LOH of the *NF1* gene is secondary or coincidental to the loss of other loci. For example, it has been demonstrated that LOH of the *p53* gene locus is a common occurrence in NF1 malignancies (Legius *et al.*, 1994b).

3.5.3 Studies in benign tumours

The studies summarized above indicate that the *NF1* gene is probably involved in cancer development. However, the tumour suppressor mechanism in the development of benign NF1 features has not been adequately resolved. Are only malignant, and not benign, tumours associated with an *NF1* knock-out? Or is it possible that second *NF1* mutations (*NF1* knock-out) are also associated with benign tumours (and other benign NF1 features) and that other genetic events are necessary for transformation to a malignant phenotype? Investigation into these questions might be approached by studying neurofibromas, as gross genetic changes (as indicated by an abnormal karyotype) have not typically been found in these tumours (Rey *et al.*, 1987; Glover *et al.*, 1991).

A report by Skuse *et al.* (1991) purported that neurofibromas are monoclonal in origin and thus genetic analysis, despite minor contamination from normal cells, is possible. Two neurofibroma LOH studies did not reveal deletions which would include the *NF1* gene; however, those studies, conducted before intragenic *NF1* polymorphisms were available, only examined loci outside the *NF1* gene (Skuse *et al.*, 1989, 1991). Preliminary data in our laboratory examining the *NF1* gene itself suggest that somatic *NF1* deletions are present in at least some cutaneous neurofibromas, supporting the tumour suppressor hypothesis for *NF1* gene inactivation in neurofibroma formation (Colman and Wallace, unpublished data). In this scenario, one would expect each tumour from the same individual to have an independent second mutational event, and our preliminary data suggest that this is, in fact, the case. It is commonly believed that neurofibromas are originally derived from Schwann cells (by far the predominant cell type in neurofibromas) and thus many molecular and cell biological studies are examining the specific properties of Schwann cells in NF1 (summarized by Gutmann and Collins, 1995), although this presents a particular challenge in neurofibromas due to their cellular heterogeneity.

3.5.4 NF1 *expression in malignant tumours*

An abnormal neurofibromin level or GAP activity has been found in malignancies in patients with and without NF1. Li *et al.* (1992) described *NF1* missense mutations in the highly conserved Lys1423 codon in a non-NF1 astrocytoma and in a non-NF1 colon cancer sample. An *in vitro* expression assay demonstrated that these mutations cause reduced GAP activity of the resulting mutant neurofibromin GAP domains. Two other studies reported decreased neurofibromin levels and abnormal RAS regulation (decreased GAP activity) in several NF1 neurofibrosarcoma cell lines (Basu *et al.*, 1992; DeClue *et al.*, 1992). In addition, an NF1 neurofibrosarcoma line which had lost an entire 17 homologue showed a dramatic reduction in *NF1* mRNA levels, suggesting that the remaining allele is abnormal (presumably representing the constitutional mutation) (Reynolds *et al.*, 1992).

A number of other studies have also aimed at elucidating the involvement of the alternatively spliced GAP-related domain isoforms of neurofibromin: apparent preferential expression of neurofibromin type II (containing the 21-amino acid insertion) was reported in non-NF1 brain tumours (Suzuki *et al.*, 1991), while Uchida *et al.* (1992) found no differential expression in non-NF1 gastric cancers (although type II predominated in normal stomach mucosa). Thus, there is some evidence that isoform distribution may be a relevant variable in overall neurofibromin function and tumorigenesis. Two groups have also studied the effect of increased neurofibromin expression using transfection studies in NIH 3T3 cells. In one report, overexpression in oncogenic *RAS*-transformed cells virtually eliminated formation of colonies in soft agar, and results from transfection of deletion mutants indicated that a 91-residue region (amino acids 1441–1531) is crucial for RAS binding although, by itself, this peptide had much lower GAP activity (Nur-E-Kamal *et al.*, 1993). Similarly, Johnson *et al.* (1994) found that: (i) cells with increased levels of neurofibromin grew significantly more slowly (although GTP-bound RAS levels did not change); (ii) *RAS* transformation of these cells was greatly inhibited; and (iii) certain melanoma cell lines known to be deficient in neurofibromin showed growth inhibition and differentiation when neurofibromin was induced. Thus, the literature provides a convincing argument that NF1 is a tumour suppressor gene whose protein product has RAS-inactivating ability and can inhibit cell growth by at least one other independent mechanism.

Reports of missense mutations affecting Lys1423 in several tumours (Li *et al.*, 1992) triggered a number of mutagenesis/*in vitro* expression studies aimed at testing several amino acid residues within the GAP-related domain. A Lysl423Met substitution appeared to reduce the thermal stability of the GAP-related domain thereby imparting a variable effect on its GAP activity (Wiesmüller and Wittinghofer, 1992). Nakafuku *et al.* (1993) created two variants (Phe1434Leu and Lys1436Arg) whose resulting GAP domains not only maintained their ability to inactivate normal RAS, but were also able to suppress the oncogenic RAS phenotype in yeast, an ability not demonstrated by normal neurofibromin. These neurofibromin GAP domain variants were able to induce the morphological reversion of *RAS*-transformed NIH 3T3 cells, further suggesting a role for neurofibromin in cell cycle regulation. Poullet *et al.* (1994) tested the effect of Lys1423 substitutions using GAP and yeast complementation assays: mutants at this residue had reduced RAS-GTP affinities. They also confirmed that the 1434 residue is crucial, since 1434 substitutions compensated for Lys1423 mutations and suppressed oncogenic RAS activity. Another study tested a variety of GAP-related domain substitutions (including one at 1423) and found that some mutants conferred reduced abilities to regulate RAS negatively in yeast while others had either no effect or just an intermediate effect (Gutmann *et al.*, 1993b). Together, these studies confirm that at least several residues in the GAP domain of neurofibromin are necessary for normal GAP activity (not unexpected considering the degree of conservation), and that these residues play an important role in complex interactions in the RAS, and perhaps other, cell cycle regulatory pathways.

3.6 NF2 clinical and genetic information

NF2 is an autosomal dominant disorder caused by disruption of a tumour suppressor gene located on 22q12 (Rouleau *et al.*, 1993; Trofatter *et al.*, 1993). NF2 is considered to be 'highly' penetrant; the few apparently normal individuals who are obligate carriers (by virtue of having a parent and child with NF2) have yet to be thoroughly examined (Eldridge, 1990). The new mutation rate for this gene is believed to be 50%, that is, half of affected individuals have no prior family history (Short *et al.*, 1994). NF2 is characterized by the formation of bilateral vestibular schwannomas, an increased predisposition to central nervous system (CNS) tumours, café-au-lait spots or other skin lesions, and lens opacities (Short *et al.*, 1994). NF2 may present in either a mild (type 2A) or severe (type 2B) form.

These are distinguished by the relatively late onset of the mild form, with little skin involvement and fewer tumours, as opposed to the early onset of the severe form, with the formation of large numbers of meningiomas and spinal tumours (Short *et al.*, 1994). There is a difference in natural history which correlates with the parent from whom the disease is inherited; patients with maternally derived NF2 have an earlier presentation and more rapid course than patients with paternally derived NF2 (Short *et al.*, 1994).

The primary clinical feature of NF2 is the formation of CNS tumours, almost always including bilateral vestibular schwannomas (also referred to as acoustic neuromas) (Riccardi, 1992). Other tumours associated with NF2 include meningiomas, astrocytomas of the spinal cord, neurofibromas and ependymomas (Eldridge, 1990; Riccardi, 1992). The NIH diagnostic criteria for NF2 are as follows (Short *et al.*, 1994):

(i) bilateral eighth cranial nerve masses seen by magnetic resonance imaging with gadolinium; or
(ii) a first degree relative with NF2 and either unilateral eighth cranial nerve mass or any of the following: neurofibroma, menigioma, glioma, schwannoma, or posterior capsular cataract or opacity at a young age.

3.7 The *NF2* gene and NF2 protein

The *NF2* gene was initially localized by observation of chromosome 22 LOH in vestibular schwannomas of NF2 patients (Seizinger *et al.*, 1986), which led to very strong suspicions (later verified) that the *NF2* gene is a tumour suppressor gene. Subsequent linkage studies in NF2 families confirmed that the *NF2* gene was located at chromosome 22q12 (Rouleau *et al.*, 1987; Wertelecki *et al.*, 1988). Positional cloning efforts ultimately led to the identification of the *NF2* gene (Rouleau *et al.*, 1993; Trofatter *et al.*, 1993). The gene spans about 100 kb in the genome, has at least 17 exons and encodes transcripts of sizes 7, 4.4 and 2.6 kb (MacCollin *et al.* 1993; Rouleau *et al.*, 1993; Trofatter *et al.*, 1993). Alternative polyadenylation is probably responsible, at least in part, for the multiple forms seen on Northern blots (Trofatter *et al.*, 1993). The cDNA encodes a 595-amino acid protein, which shares strong homology with the cytoskeleton-associated molecules moesin, ezrin, radixin, talin, and erythrocyte protein 4.1. (*Figure 3.3*); some theorize that the proteins from this gene family are involved in linking cytoskeletal components with cell membrane proteins (Sato *et al.*, 1992). The protein has two proposed names, merlin (Trofatter *et al.*, 1993) and schwannomin (Rouleau *et al.*, 1993) (for the sake of simplicity, it will be referred to as merlin here). Of note, merlin is the first member of this gene family identified as a tumour suppressor gene. Addressing the issue of the cellular localization of merlin in the normal vestibular nerve, an anti-merlin antibody study indicated immunoreactivity in the cytoplasm and perinuclear region of Schwann cells,

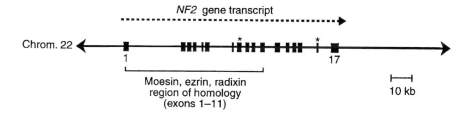

Figure 3.3. Schematic diagram showing the genomic structure of the *NF2* gene on chromosome 22q. The spacing between exons is approximately to scale, although the exon sizes themselves are not. Exons known to be alternatively spliced are indicated by asterisks; other exons (particularly at the 3' end) may also be involved in alternative splicing. Exon '16' represents the 45-bp insertion described in the text. The strongest moesin, ezrin, radixin homology lies in the first 350 amino acids of this protein product, encoded within exons 1–11 as indicated (Trofatter *et al.*, 1993).

ganglion cell bodies and epineural fibroblasts, but not in axons (Sainz *et al.*, 1994).

Based on RNA studies, the *NF2* gene has a very wide tissue distribution (Rouleau *et al.*, 1993; Trofatter *et al.*, 1993), and some alternative splicing has been reported; in the human *NF2* gene, a 45-base-pair (bp) insertion very close to the carboxy terminal results in a frameshift such that the last 16 amino acids of the normal sequence are replaced with a stretch of 11 different amino acids and a new stop codon (Bianchi *et al.*, 1994). This insertion is thought to be a separate, alternatively spliced exon whose inclusion results in a protein that is more hydrophilic at the 3' end (Bianchi *et al.*, 1994; Hara *et al.*, 1994). This alternative splice has also been identified in mouse (Haase *et al.*, 1994; Hara *et al.*, 1994; Huynh *et al.*, 1994b). In some tissues, there also appear to be some differences in quantity and distribution of these two isoforms but the functional implication is not clear (Bianchi *et al.*, 1994; Haase *et al.*, 1994; Hara *et al.*, 1994). Other reported alternative-splice isoforms include: (i) the loss of exon 8 (removing 45 amino acids from the middle of the protein) (Haase *et al.*, 1994); (ii) the previously discussed 45-bp insertion followed by 15 additional bases in the immediate 3' untranslated region (human) (Huynh *et al.*, 1994b); (iii) a 16-bp insertion at the same site as the 45-bp insertion (however the 16 bp are contiguous with the previous exon rather than separated by an intron) with this insertion predicted to result in a 584-amino acid protein (Hara *et al.*, 1994); (iv) the presence of the 16-bp and 45-bp insertions together resulting in a 584-residue protein with a different 3' untranslated region (Huynh *et al.*, 1994b). The significance of finding that most of the currently identified alternative splices fall near the carboxy terminal while the strongest protein homology lies in the amino half of merlin is unclear, although there are some specific differences among the protein family members at the C terminal that may indicate very distinct functions. Researchers are investigating the possibility that these putative isoforms of merlin have specific functional roles in different tissues, possibly related to development or differentiation.

3.8 Germline *NF2* mutations

Constitutional *NF2* mutations were first reported along with the original description of the cloning of the gene and included deletions of various sizes as well as point changes, nearly all of which are predicted to result in a prematurely truncated protein (Rouleau *et al.*, 1993; Trofatter *et al.*, 1993). Subsequent studies, some of which examined tumour tissue as well as constitutional tissue (predominantly blood cells), have discovered a number of other inactivating germline mutations in NF2 patients. Again, these are mostly unique and include deletions (various sizes), insertions (less than 200 bp), point changes and a translocation (Sanson *et al.*, 1993; MacCollin *et al.*, 1993, 1994; Watson *et al.*, 1993; Arai *et al.*, 1994; Bourn *et al.*, 1994; Deprez *et al.*, 1994; Jacoby *et al.*, 1994; Twist *et al.*, 1994). C → T transitions may account for relatively frequent mutations in the *NF2* gene, as eight patients were reported to have a C → T nonsense mutation in exon 2 (Arg57stop) (Rouleau *et al.*, 1993; Bourn *et al.*, 1994; Jacoby *et al.*, 1994; MacCollin *et al.*, 1994). Nonsense C → T transitions have also been reported in two patients each for codon 262 (MacCollin *et al.*, 1994; Twist *et al.*, 1994) and codon 466 (Jacoby *et al.*, 1994; MacCollin *et al.*, 1994), and there are three reports describing a C → T nonsense mutation at codon 341 (Bourn *et al.*, 1994; MacCollin *et al.*, 1994; Twist *et al.*, 1994). The codon 57 nonsense mutation appears to be associated with a severe NF2 phenotype; the one patient having mild features showed signs of somatic mosaicism (Bourn *et al.*, 1994). In contrast to NF1, mutation analysis in NF2 has been much more successful, presumably due to the significantly smaller size of the gene and the recurrence of at least one C → T transition with some frequency. This property has proven to be very useful in the interpretation of tumour mutation data (see Section 3.9), since the failure to find an *NF2* mutation, after a thorough search, strongly suggests that no mutation exists.

3.9 The *NF2* gene is a tumour suppressor

NF2 region LOH (in some cases the entire 22 homologue) has been routinely observed in vestibular schwannomas and meningiomas of NF2 patients, as well as in the same types of tumours of non-NF2 patients (e.g. Seizinger *et al.*, 1986, 1987). This has implicated the *NF2* gene as a tumour suppressor, a hypothesis now proven with the detection of *NF2* gene 'second hits' in such tumours (described below). One study showed a threefold preference for the maternal allele being somatically lost in tumours of NF2 patients carrying new mutations, which suggests that *de novo* germline mutations are predominantly paternal in origin (Fontaine *et al.*, 1991). This is consistent with the situation seen in NF1 and a number of other disorders (Jadayel *et al.*, 1990; Stephens *et al.*, 1992).

A number of laboratories have searched for *NF2* gene mutations in a variety of tumours, especially those of the CNS, to investigate the gene's involvement in the formation of these neoplasms. The first report of somatic *NF2* mutations described nonsense mutations in two NF2 schwannomas and two NF2 meningiomas (Rouleau *et al.*, 1993). Several laboratories have published inactivating *NF2* somatic mutations in vestibular schwannomas and/or non-vestibular schwannomas (those not arising from the eighth cranial nerve). In some cases, the presence of LOH clearly indicates that there is complete inactivation of the *NF2* gene, supporting the tumour suppressor hypothesis. The earlier of these publications, which examined only a portion of the *NF2* gene, detected mutations in 15–20% of tumours (Irving *et al.*, 1994; Twist *et al.*, 1994), while later studies, which encompassed all exons, detected mutations in 50–75% of these tumours (Deprez *et al.*, 1994; Jacoby *et al.*, 1994; Sainz *et al.*, 1994). Further support to one of these later studies was provided by protein-immunostaining experiments, which indicated that the tumour Schwann cells were devoid of merlin (Sainz *et al.*, 1994). In another study, 12 sporadic and three NF2 vestibular schwannomas were examined, and *NF2* mutations were detected in six of the sporadic and two of the hereditary tumours, including four which showed hemizygous or homozygous deletion of part of the *NF2* gene (Bianchi *et al.*, 1994). These latter mutations clearly proved the two-hit hypothesis in vestibular schwannomas. This group also examined other tumour types, including sporadic phaeochromocytomas and melanomas, and found somatic mutations in six melanomas (four frameshift deletions, one in-frame deletion, and one with an in-frame deletion as well as non-conservative missense mutation). No mutations were detected in any of the phaeochromocytomas. Also, the report described examination of 14 primary breast carcinomas with the subsequent detection of a 211-bp frameshift deletion and a missense mutation, although it was not mentioned as to whether both chromosome 22 homologues were present (i.e. whether or not the *NF2* gene was completely inactivated). Another study examined eight ependymomas and 30 astrocytomas (all from non-NF2 patients), both of which occur with increased frequency in NF2 patients and commonly have LOH of the *NF2* region (Rubio *et al.*, 1994). One *NF2* mutation was found in an ependymoma; however, none were detected in astrocytomas, suggesting that the *NF2* gene is not generally involved in astrocytoma development. In analyses of sporadic meningiomas, Deprez *et al.* (1994) found *NF2* mutations in 32% of tumours, while Ruttledge *et al.* (1994) detected specific mutations and/or chromosome 22 LOH in 60% of the tumours (having examined eight *NF2* exons). The accumulated data from the above studies clearly indicate a tumour suppressor role for the *NF2* gene in the neoplasms commonly seen in NF2, as well as probable (if limited) involvement in other tumour types. Based on frequencies, it could be interpreted that NF2 knock-out is the rate-limiting step in NF2-associated tumours while, in the other tumours (such as breast carcinoma, melanoma), loss of merlin function could be one of the steps involved in later stages of tumour development (Bianchi *et al.*, 1994).

3.10 Conclusions and future directions

Since the cloning of the *NF1* gene in 1990, a great deal has been elucidated about the genetics and biology of this disorder, although the current data only scratch the surface of what remains to be learned. There may be additional alternative splices or internal promoters which could add to the variety of forms of neurofibromin and merlin (as has been recently discovered for dystrophin), which might subsequently reveal developmental- or tissue-specific regulation and functions. Direct mutation analysis in NF1 is among the most difficult for any genetic disorder, with most laboratories currently having only a 15–30% success rate in detection. Yet, the accumulation of mutational data will reveal possible functional domains, correlations between clinical features and DNA alterations, differences (if any) between NF1 arising maternally versus paternally, and involvement of the *NF1* gene in other genetic disorders and cancer. Any indication of mutational hot spots or mechanisms will have immediate application in direct DNA diagnosis. Although *NF2* gene mutation analysis is further along, much still remains to be discovered about the structure, interactions, regulation and functions of both full-length neurofibromin and merlin. Accumulated data may also resolve the issue of complete penetrance for both disorders, phenotype–genotype correlations, and will prove or disprove the tumour suppressor hypothesis for benign overgrowths in NF1. Clearly, the data from the last several years essentially prove that neurofibromin and merlin are tumour suppressors and are involved in multiple tumour types. Many resources, including genomic and cDNA probes, polymorphism information, and antibodies, are now available to answer additional questions in NF1 and NF2 research. The production of mice carrying inactivating mutations in the *NF1* gene was recently described, and while the heterozygous animals fail to show classic human NF1 features, they do show late onset predisposition to several tumour types, including phaeochromocytoma and leukaemia (Jacks *et al.*, 1994). Half of the tumours showed somatic loss of the normal *NF1* allele by LOH studies. The homozygous knock-out mice die in embryogenesis. Animal models such as this, for both NF1 and NF2, will be useful tools in understanding disease pathogenesis and in the development and testing of diagnostic procedures and treatments. Investigators in numerous fields are now studying the *NF1* and *NF2* genes and their proteins, particularly with regard to normal development and cancer. Perhaps most importantly, advances in these fields will ultimately result in the development of therapies for *NF1* and *NF2*, which may have application in general cancer therapeutic approaches (Gutmann and Collins, 1995).

Acknowledgements

We would like to acknowledge Dr David Gutmann for helpful suggestions, Lynn Taylor for assistance in preparing the manuscript, and support from the National Institutes of Health (R29 NS31550), Children's Miracle Network and the Hayward Foundation.

References

Abernathy CR, Colman SD, Kousseff B, Wallace MR. (1994) Two NF1 mutations: frameshift in the GAP-related domain, and loss of two codons toward the 3' end of the molecule. *Hum. Mutat.* **3**: 347–352.

Ainsworth PJ, Rodenhiser DI, Costa MT. (1993) Identification and characterization of sporadic and inherited mutations in exon 31 of the neurofibromatosis (NF1) gene. *Hum. Genet.* **91**: 151–156.

Ainsworth P, Rodenhiser D, Stuart A, Jung J. (1994) Characterization of an intron 31 splice junction mutation in the neurofibromatosis type 1 (NF1) gene. *Hum. Mol. Genet.* **3**: 1179–1181.

Andersen LB, Ballester R, Marchuk DA, Chang E, Gutmann DH, Saulino AM, Camonis J, Wigler M, Collins FS. (1993a) A conserved alternative splice in the von Recklinghausen neurofibromatosis (NF1) gene produces two neurofibromin isoforms, both of which have GTPase-activating protein activity. *Mol. Cell Biol.* **13**: 487–495.

Andersen LB, Fountain JW, Gutmann DH, Tarlé SA, Glover TW, Dracopoli NC, Housman DE, Collins FS. (1993b) Mutations in the neurofibromatosis 1 gene in sporadic malignant melanoma lines. *Nature Genet.* **3**: 118–121.

Anglani F, Murgia A, Bedin S, Bresin E, Bernardi F, Clementi M, Tenconi R. (1993) A new disease-causing mutation in the GAP-related domain of the NF1 gene. *Hum. Mol. Genet.* **2**: 1057–1059.

Arai E, Ikeuchi T, Nakamura Y. (1994) Characterization of the translocation breakpoint on chromosome 22q12.2 in a patient with neurofibromatosis type 2 (NF2). *Hum. Mol. Genet.* **3**: 937–939.

Bader JL. (1986) Neurofibromatosis and cancer. *Ann. NY Acad. Sci.* **486**: 57–65.

Ballester R, Marchuk D, Boguski M, Saulino A, Letcher R, Wigler M, Collins FS. (1990) The NF1 locus encodes a protein functionally related to mammalian GAP and yeast IRA proteins. *Cell* **63**: 851–859.

Basu TN, Gutmann DH, Fletcher JA, Glover TW, Collins FS, Downward J. (1992) Aberrant regulation of ras proteins in malignant tumour cells from type 1 neurofibromatosis patients. *Nature* **356**: 713–715.

Bianchi AB, Hara T, Ramesh V, Gao J, Klein-Szanto AJP, Morin F, Menon AG, Trofatter JA, Gusella JF, Seizinger BR, Kley N. (1994) Mutations in transcript isoforms of the neurofibromatosis 2 gene in multiple human tumor types. *Nature Genet.* **6**: 185–192.

Blatt J, Jaffe R, Deutsch M, Adkins JC. (1986) Neurofibromatosis and childhood tumors. *Cancer* **57**: 1225–1229.

Bollag G, McCormick F. (1991) Differential regulation of ras, GAP and neurofibromatosis gene product activities. *Nature* **351**: 576–579.

Bollag G, McCormick F, Clark R. (1993) Characterization of full-length neurofibromin: tubulin inhibits Ras GAP activity. *EMBO J.* **12**: 1923–1927.

Borberg A. (1951) Clinical and genetic investigations into tuberous sclerosis and Recklinghausen's neurofibromatosis. *Acta Psychiatr. Neurol.* **71** (Suppl): 1–239.

Bourn D, Carter SA, Mason S, Evans DGR, Strachan T. (1994) Germline mutations in the neurofibromatosis type 2 tumor suppressor gene. *Hum. Mol. Genet.* **3**: 813–816.

Boyer MJ, Gutmann DH, Collins FS, Bar-sagi D. (1994) Crosslinking of the surface immunoglobulin receptor in B lymphocytes induces a redistribution of neurofibromin but not p120-GAP. *Oncogene* **9**: 349–357.

Buchberg AM, Cleveland LS, Jenkins NA, Copeland NG. (1990) Sequence homology shared by neurofibromatosis type-1 gene and IRA-1 and IRA-2 negative regulators of the RAS cyclic AMP pathway. *Nature* **347**: 291–294.

Carey JC, Laub JM, Hall BD. (1979) Penetrance and variability in neurofibromatosis: a genetic study of 60 families. In: *Birth Defects: Original Article Series*, Vol. XV. Center for Birth Defects Information Service, Dorer, MA, pp. 271–281.

Cawthon RM, Weiss R, Xu G, Viskochil D, Culver M, Stevens J, Robertson M, Dunn D, Gesteland R, O'Connell P, White R. (1990) A major segment of the neurofibromatosis type 1 gene: cDNA sequence, genomic structure, and point mutations. *Cell* **62**: 193–201.

Cawthon RM, Andersen LB, Buchberg AM, Xu G, O'Connell P, Viskochil D, Weiss RB, Wallace MR, Marchuk DA, Culver M, Stevens J, Jenkins NA, Copeland NG, Collins FS, White R. (1991) cDNA sequence and genomic structure of EVI2B, a gene lying within an intron of the neurofibromatosis type 1 gene. *Genomics* **9**: 446–460.

Colley P, Colley A, Thakker N, Donnai D, Super M, Harris R, Strachan T. (1991) Large scale mutations at the NF1 locus in Noonan-NF1 and NF1 patients. *J. Med. Genet.* **28**: 572.

Colman SD, Collins FS, Wallace MR. (1993) Characterization of a single base deletion in neurofibromatosis type 1. *Hum. Mol. Genet.* **2**: 1709–1711.

Crowe FW, Schull WJ, Neel JV. (1956) *A Clinical, Pathological, and Genetic Study of Multiple Neurofibromas*. Charles C. Thomas Publishers, Springfield, IL.

Daston MM, Ratner N. (1992) Neurofibromin, a predominantly neuronal GTPase activating protein in the adult, is ubiquitously expressed during development. *Devel. Dynam.* **195**: 216–226.

Daston MM, Scrable H, Nordlund M, Sturbaum AK, Nissen LM, Ratner N. (1992) The protein of the neurofibromatosis type 1 gene is expressed at highest abundance in neurons, Schwann cells, and oligodendrocytes. *Neuron* **8**: 415–428.

DeClue JE, Stone JC, Blanchard RA, Papageorge AG, Martin P, Zhang K, Lowy DR. (1991) A ras effector domain mutant which is temperature sensitive for cellular transformation: interactions with GTPase-activating protein and NF-1. *Mol. Cell Biol.* **11**: 3132–3138.

DeClue JE, Papageorge AG, Fletcher JA, Diehl SR, Ratner N, Vass WC, Lowy DR. (1992) Abnormal regulation of mammalian p21[ras] contributes to malignant tumor growth in von Recklinghausen (type 1) neurofibromatosis. *Cell* **69**: 265–273.

Deprez RHL, Bianchi AB, Groen NA, Seizinger BR, Hagemeijer A, Drunen E Van, Bootsma D, Koper JW, Avezaat CJJ, Kley N, Zwarthoff EC. (1994) Frequent NF2 gene transcript mutations in sporadic meningiomas and vestibular schwannomas. *Am. J. Hum. Genet.* **54**: 1022–1029.

Downward J. (1992) Regulation of p21^{ras} by GTPase activating proteins and guanine nucleotide exchange proteins. *Curr. Op. Genet. Devel.* **2**: 13–18.

Driscoll DJ, Migeon BR. (1990) Sex difference in methylation of single copy genes in human meiotic germ cells; implications for X chromosome inactivation, parental imprinting, and origin of CpG mutations. *Somat. Cell Mol. Genet.* **16**: 267–282.

Dunn JM, Phillips RA, Becker AJ, Gallie BL. (1988) Identification of germline and somatic mutations affecting the retinoblastoma gene. *Science* **241**: 1797–1800.

Eldridge R. (1990) Neurofibromatosis type 2. In: *Neurofibromatosis: a Handbook for Patients, Families, and Health-care Professionals* (eds AE Rubenstein and B Korf). Thieme Medical Publishers, New York, pp. 29–39.

Estivill X, Lazaro C, Casals T, Ravella A. (1991) Recurrence of a nonsense mutation in the NF1 gene causing classical neurofibromatosis type 1. *Hum. Genet.* **88**: 185–188.

Fontaine B, Sanson M, Delattre O, Menon A, Rouleau GA, Seizinger BR, Jewell AF, Hanson MP, Aurias A, Martuza RL, Gusella JF. (1991) Parental origin of chromosome 22 loss in sporadic and NF2 neuromas. *Genomics* **10**: 280–283.

Glover TW, Stein CK, Legius E, Andersen LB, Brereton A, Johnson S. (1991) Molecular and cytogenetic analysis of tumors in von Recklinghausen neurofibromatosis. *Genes Chrom. Cancer* **3**: 62–70.

Golubic M, Roudebush M, Dorbrowolski S, Wolfman A, Stacey DW. (1992) Catalytic properties, tissue and intracellular distribution of neurofibromin. *Oncogene* **7**: 2151–2159.

Gregory PE, Gutmann DH, Mitchell A, Park S, Boguski M, Jacks T, Wood DL, Jove R, Collins FS. (1993) Neurofibromatosis type 1 gene product (neurofibromin) associates with microtubules. *Somat. Cell Mol. Genet.* **19**: 265–274.

Gutmann DH, Collins FS. (1993) The neurofibromatosis type 1 gene and its protein product, neurofibromin. *Neuron* **88**: 9658–9662.

Gutmann DH, Collins FS. (1995) von Recklinghausen neurofibromatosis. In: *The Metabolic and Molecular Basis for Inherited Disease*, 7th Edn. McGraw Hill, New York, pp. 677–696.

Gutmann DH, Andersen LB, Cole JL, Swaroop M, Collins FS. (1993a) An alternatively-spliced mRNA in the carboxy terminus of the neurofibromatosis type 1 (NF1) gene is expressed in muscle. *Hum. Mol. Genet.* **2**: 989–992.

Gutmann DH, Boguski M, Marchuk D, Wigler M, Collins FS, Ballester R. (1993b) Analysis of neurofibromatosis type 1 (NF1) GAP-related domain by site-directed mutagenesis. *Oncogene* **8**: 761–769.

Gutmann DH, Cole JL, Stone WJ, Ponder BAJ, Collins FS. (1994) Loss of neurofibromin in adrenal gland tumors from patients with neurofibromatosis type 1. *Genes Chrom. Cancer* **10**: 55–58.

Haase VH, Trofatter JA, MacCollin M, Tarttelin E, Gusella JF, Ramesh V. (1994) The murine NF2 homologue encodes a highly conserved merlin protein with alternative forms. *Hum. Mol. Genet.* **3**: 407–411.

Hara T, Bianchi AB, Seizinger BR, Kley N. (1994) Molecular cloning and characterization of alternatively spliced transcripts of the mouse neurofibromatosis 2 gene. *Cancer Res.* **54**: 330–335.

Hatta N, Horiuchi T, Fujita S. (1994) Analysis of NF1 gene mutations in neurofibromatosis type 1 patients in Japan. *Biochem. Biophys. Res. Commun.* **199**: 207–212.

Horiuchi T, Hatta N, Matsumoto M, Ohtsuka H, Collins FS, Kobayashi Y, Fujita S. (1994) Nonsense mutations at Arg-1947 in two cases of familial neurofibromatosis type 1 in Japanese. *Hum. Genet.* **93**: 81–83.

Huson SM, Hughes RAC (eds). (1994) *The Neurofibromatoses: a Pathogenetic and Clinical Overview.* Chapman and Hall Medical Press, London.

Hutter P, Antonarakis SE, Delozier-Blanchet CD, Morris MA. (1994) Exon skipping associated with A to G transition at +4 of the IVS33 splice donor site of the neurofibromatosis type 1 (NF1) gene. *Hum. Mol. Genet.* **3**: 663–665.

Huynh DP, Nechiporuk T, Pulst SM. (1994a) Differential expression and tissue distribution of type I and type II neurofibromins during mouse fetal development. *Devel. Biol.* **161**: 538–551.

Huynh DP, Nechiporuk T, Pulst SM. (1994b) Alternative transcripts in the mouse neurofibromatosis type 2 (*NF2*) gene are conserved and code for schwannomins with distinct C-terminal domains. *Hum. Mol. Genet.* **3**: 1075–1079.

Ichii S, Horii A, Nakatsura S, Furuyama J, Utsunomiya J, Nakamura Y. (1992) Inactivation of both APC alleles in an early stage of colon adenomas in a patient with familial adenomatous polyposis (FAP). *Hum. Mol. Genet.* **1**: 387–390.

Irving RM, Moffat DA, Hardy DG, Barton DE, Xuereb JH, Maher ER. (1994) Somatic NF2 gene mutations in familial and non-familial vestibular schwannoma. *Hum. Mol. Genet.* **3**: 347–350.

Jacks T, Shih ST, Schmitt, EM, Bronson RT, Bernards A, Weinberg RA. (1994) Tumour predisposition in mice heterozygous for a targeted mutation in NF1. *Nature Genet.* **7**: 353–361.

Jacoby LB, MacCollin M, Louis DN, Mohney T, Rubio MP, Pulaski K, Trofatter JA, Kley N, Seizinger B, Ramesh V, Gusella JF. (1994) Exon scanning for mutation of the NF2 gene in schwannomas. *Hum. Mol. Genet.* **3**: 413–419.

Jadayel D, Fain P, Upadhyaya M, Ponder MA, Huson SM, Carey J, Fryer A, Mathew CGP, Barker DF, Ponder BAJ. (1990) Paternal origin of new mutations in von Recklinghausen neurofibromatosis. *Nature* **343**: 558–559.

Johnson MR, Look AT, DeClue JE, Valentine MB, Lowy DR. (1993) Inactivation of the NF1 gene in human melanoma and neuroblastoma cell lines without impaired regulation of GTP-Ras. *Proc. Natl Acad. Sci. USA* **90**: 5539–5543.

Johnson MR, DeClue JE, Felzmann S, Vass WC, Xu G, White R, Lowy DR. (1994) Neurofibromin can inhibit ras-dependent growth by a mechanism independent of its GTPase-accelerating function. *Mol. Cell Biol.* **14**: 641–645.

Jorde LB, Watkins WS, Viskochil D, O'Connell P, Ward K. (1993) Linkage disequilibrium in the neurofibromatosis 1 (NF1) region: implications for gene mapping. *Am. J. Hum. Genet.* **53**: 1038–1050.

Kamei T, Fukushima Y, Shibaba A, Hayashi Y, Tachibana N, Takeda I, Niikawa N, Collins FS, Takahashi K, Masumara S. (1992) DNA deletion in patients with von Recklinghausen neurofibromatosis. *Clin. Genet.* **42**: 53–54.

Kayes LM, Burke W, Riccardi VM, Bennett R, Ehrlich P, Rubenstein A, Stephens K. (1994) Deletions spanning the neurofibromatosis 1 gene: identification and phenotype of five patients. *Am. J. Hum. Genet.* **54**: 424–436.

Knudson AG Jr. (1971) Mutation and cancer: statistical study of retinoblastoma. *Proc. Natl Acad. Sci. USA* **68**: 820–823.

Knudson AG Jr. (1985) Hereditary cancer, oncogenes, and antioncogenes. *Cancer Res.* **45**: 1437–1443.

Lazaro C, Gaona A, Ravella A, Volpini V, Casals T, Fuentes J-J, Estivill X. (1993) Novel alleles, hemizygosity and deletions at an Alu-repeat within the neurofibromatosis type 1 (NF1) gene. *Hum. Mol. Genet.* **2**: 725–730.

Lefkowitz IB, Obringer AC, Meadows AT. (1990) Neurofibromatosis and cancer: incidence and management. In: *Neurofibromatosis: a Handbook for Patients, Families, and Healthcare Professionals* (eds AE Rubenstein and BR Korf). Thieme Medical Publishers, New York, pp. 99–111.

Legius E, Marchuk DA, Collins FS, Glover TW. (1993) Somatic deletion of the neurofibromatosis type 1 gene in a neurofibrosarcoma supports a tumour suppressor gene hypothesis. *Nature Genet.* **3**: 122–126.

Legius E, Hall BK, Wallace MR, Collins FS, Glover TW. (1994a) Ten base pair duplication in exon 38 of the NF1 gene. *Hum. Mol. Genet.* **3**: 829–830.

Legius E, Dierick H, Wu R, Hall BK, Marynen P, Cassiman J-J, Glover TW. (1994b) TP53 mutations are frequent in malignant NF1 tumors. *Genes Chrom. Cancer* **10**: 250–255.

Levine AJ. (1993) The tumor suppressor genes. *Annu. Rev. Biochem.* **62**: 623–651.

Levine AJ, Momand J, Finley CA. (1991) The p53 tumor suppressor gene. *Nature* **351**: 453–456.

Li Y, Bollag G, Clark R, Stevens J, Conroy L, Fults D, Ward K, Friedman E, Samowitz W, Robertson M, Bradley P, McCormick F, White R, Cawthon R. (1992) Somatic mutations in the neurofibromatosis 1 gene in human tumors. *Cell* **69**: 275–281.

Lothe RA, Saeter G, Danielsen HE, Stenwig AE, Hoyheim B, O'Connell P, Borresen AL. (1993) Genetic alterations in a malignant schwannoma from a patient with neurofibromatosis (NF1). *Pathol. Res. Pract.* **189**: 465–471.

MacCollin M, Mohney T, Trofatter J, Wertelecki W, Ramesh V, Gusella J. (1993) DNA diagnosis of neurofibromatosis 2: altered coding sequence of the merlin tumor suppressor in an extended pedigree. *JAMA* **270**: 2316–2320.

MacCollin M, Ramesh V, Jacoby LB, Louis DN, Rubio MP, Pulaski K, Trofatter JA, Short MP, Bove C, Eldridge R, Parry DM, Gusella JF. (1994) Mutational analysis of patients with neurofibromatosis 2. *Am. J. Hum. Genet.* **55**: 314–320.

Marchuk DA, Collins FS. (1994) Molecular genetics of neurofibromatosis 1. In: *The Neurofibromatoses: a Pathogenetic and Clinical Overview* (eds SM Huson and RAC Hughes). Chapman and Hall Medical Press, London, pp. 23–49.

Marchuk DA, Saulino AM, Tavakkol R, Swaroop M, Wallace MR, Andersen LB, Mitchell AL, Gutmann DH, Boguski M, Collins FS. (1991) cDNA cloning of the type 1 neurofibromatosis gene: complete sequence of the NF1 gene product. *Genomics* **11**: 931–940.

Marchuk DA, Tavakkol R, Wallace MR, Brownstein BH, Taillon-Miller P, Fong C, Legius E, Andersen LB, Glover TW, Collins FS. (1992) A yeast artificial chromosome contig encompassing the type 1 neurofibromatosis gene. *Genomics* **13**: 672–680.

Martin GA, Viskochil D, Bollag G, McCabe PC, Crosier WJ, Halbruck H, Conroy L, Clark R, O'Connell P, Cawthon RM, Innis MA, McCormick F. (1990) The GAP-related domain of the NF1 gene interacts with ras p21. *Cell* **63**: 843–849.

Menon AG, Ledbetter DH, Rich DC, Seizinger BR, Rouleau GA, Michels VV, Schmidt MA, Dewald G, DallaTorre CM, Haines JL, Gusella JF. (1989) Characterization of a translocation within the von Recklinghausen neurofibromatosis region of chromosome 17. *Genomics* 5: 245–249.

Menon AG, Anderson KM, Riccardi VM, Chung RY, Whaley JM, Yandell DW, Farmer GE, Freiman RN, Lee JK, Li FP, Barker DF, Ledbetter DH, Kleider A, Martuza RL, Gusella JF, Seizinger BR. (1990) Chromosome 17p deletions and p53 gene mutations associated with the formation of malignant neurofibrosarcomas in von Recklinghausen neurofibromatosis. *Proc. Natl Acad. Sci. USA* 87: 5435–5439.

Monaco AP, Neve RL, Colletti-Feener C, Bertelson CJ, Kurnit DM, Kunkel LM. (1986) Isolation of candidate cDNAs for portions of the Duchenne muscular dystrophy gene. *Nature* 323: 646–650.

Mulvihill JJ, Parry DM, Sherman JL, Pikus A, Kaiser-Kupfer MI, Eldridge R. (1990) NIH conference: neurofibromatosis 1 (Recklinghausen disease) and neurofibromatosis 2 (bilateral acoustic neurofibromatosis). An update. *Ann. Intern. Med.* 113: 39–52.

Nakafuku M, Nagamine M, Ohtoshi A, Tanaka K, Toh-e A, Kaziro Y. (1993) Suppression of oncogenic ras by mutant neurofibromatosis type 1 genes with single amino acid substitutions. *Proc. Natl Acad. Sci. USA* 90: 6706–6710.

Nishi T, Lee PSY, Oka K, Levin VA, Tanase S, Morino Y, Saya H. (1991) Differential expression of two types of the neurofibromatosis type 1 (NF1) gene transcripts related to neuronal differentiation. *Oncogene* 6: 1555–1559.

Nordlund M, Gu X, Shipley MT, Ratner N. (1993) Neurofibromin is enriched in the endoplasmic reticulum of CNS neurons. *J. Neurosci.* 13: 1588–1600.

Nur-E-Kamal MSA, Varga M, Maruta H. (1993) The GTPase-activating NF1 fragment of 91 amino acids reverses v-Ha-Ras-induced malignant phenotype. *J. Biol. Chem.* 268: 22331–22337.

Poullet P, Lin B, Esson K, Tamanoi F. (1994) Functional significance of lysine 1423 of neurofibromin and characterization of a second site suppressor which rescues mutations at this residue and suppresses RAS2Val-19-activated phenotypes. *Mol. Cell Biol.* 14: 815–821.

Purandare SM, Lanyon WG, Connor JM. (1994a) Characterization of inherited and sporadic mutations in neurofibromatosis type-1. *Hum. Mol. Genet.* 3: 1109–1115.

Purandare SM, Davidson HR, Lanyon WG, Connor JM. (1994b) A novel insertional mutation of a single base in exon 34 of the neurofibromatosis 1 gene. *Hum. Mutat.* 3: 76–78.

Rey JA, Bello MJ, de Campos JM, Benitez J, Sarasa JL, Boixados JR, Sanchez Cascos A. (1987) Cytogenetic clones in a recurrent neurofibroma. *Cancer Genet. Cytogenet.* 26: 157–163.

Reynolds, JE, Fletcher JA, Lytle CH, Nie L, Morton CC, Diehl SR. (1992) Molecular characterization of a 17q11.2 translocation in a malignant schwannoma cell line. *Hum. Genet.* 90: 450–456.

Riccardi VM. (1992) *Neurofibromatosis. Phenotype, Natural History and Pathogenesis.* The Johns Hopkins University Press, Baltimore, Maryland.

Riccardi VM, Lewis RA. (1988) Penetrance of von Recklinghausen neurofibromatosis: a distinction between predecessors and descendants. *Am. J. Hum. Genet.* 42: 284–289.

Rouleau GA, Wertelecki W, Haines JL, Hobbs WJ, Trofatter JA, Seizinger BR, Martuza RL, Superneau DW, Conneally PM, Gusella JF. (1987) Genetic linkage of bilateral acoustic neurofibromatosis to a DNA marker on chromosome 22. *Nature* **329**. 246–248.

Rouleau GA, Merel P, Lutchman M, Sanson M, Zucman J, Marineau C, Hoang-Xuan K, Demczuk S, Desmaze C, Plougastel B, Pulst SM, Lenoir G, Bijlsma E, Fashold R, Dumanski J, Jong P Dde Parry D, Eldrige R, Aurias A, Delattre O, Thomas G. (1993) Alteration in a new gene encoding a putative membrane-organizing protein causes neuro-fibromatosis type 2. *Nature* **363**: 515–521.

Rubenstein AE, Korf BR (eds) (1990) *Neurofibromatosis: a Handbook for Patients, Families, and Health-care Professionals.* Thieme Medical Publishers, New York.

Rubenstein AE, Bunge RP, Housman DE (eds) (1986) Neurofibromatosis. *Ann. NY Acad. Sci.* **486**: 1–415.

Rubio MP, Correa KM, Ramesh V, MacCollin MM, Jacoby LB, Von Deimling A, Gusella JF, Louis DN. (1994) Analysis of the neurofibromatosis 2 gene in human ependymomas and astrocytomas. *Cancer Res.* **54**: 45–47.

Ruttledge MH, Sarrazin J, Rangaratnam S, Phelan CM, Twist E, Merel P, Delattre O, Thomas G, Nordenskjold M, Collins VP, Dumanski JP, Rouleau GA. (1994) Evidence for the complete inactivation of the NF2 gene in the majority of sporadic meningiomas. *Nature Genet.* **6**: 180–184.

Sainz J, Huynh DP, Figueroa K, Ragge NK, Baser ME, Pulst SM. (1994) Mutations of the neurofibromatosis type 2 gene and lack of the gene product in vestibular schwannomas. *Hum. Mol. Genet.* **3**: 885–891.

Sanson M, Marineau C, Desmaze C, Lutchman M, Ruttledge M, Baron C, Narod S, Delattre O, Lenoir G, Thomas G, Aurias A, Rouleau GA. (1993) Germline deletion in a neurofibromatosis type 2 kindred inactivates the NF2 gene and a candidate meningioma locus. *Hum. Mol. Genet.* **2**: 1215–1220.

Sato N, Funayama N, Nagafuchi A, Yonemura S, Tsukita S, Tsukita S. (1992) A gene family consisting of ezrin, radixin and moesin. Its specific localization at actin filament/plasma membrane association sites. *J. Cell Sci.* **103**: 131–143.

Schmidt MA, Michels VV, Dewald GW. (1987) Cases of neurofibromatosis with rearrangements of chromosome 17 involving band 17q11.2. *Am. J. Med. Genet.* **28**: 771–777.

Seizinger B. (1993) NF1: a prevalent cause of tumorigenesis in human cancers? *Nature Genet.* **3**: 97–99.

Seizinger BR, Martuza RL, Gusella JF. (1986) Loss of genes on chromosome 22 in tumorigenesis of human acoustic neuroma. *Nature* **322**: 644–647.

Seizinger BR, Rouleau G, Ozelius LJ, Lane AH, St. George-Hyslop P, Huson S, Gusella JF, Martuza RL. (1987) Common pathogenetic mechanism for three tumor types in bilateral acoustic neurofibromatosis. *Science* **236**: 317–319.

Sergeyev AS. (1975) On the mutation rate of neurofibromatosis. *Hum Genet.* **28**: 129–138.

Shannon KM, O'Connell P, Martin GA, Paderanga D, Olson K, Dinndorf P, McCormick F. (1994) Loss of the normal NF1 allele from the bone marrow of children with type 1 neurofibromatosis and malignant myeloid disorders. *N. Engl. J. Med.* **330**: 597–601.

Shearer P, Parham D, Kovnar E, Kun L, Rao B, Lobe T, Pratt C. (1994) Neurofibromatosis type 1 and malignancy: review of 32 pediatric cases treated at a single institution. *Med. Pediatr. Oncol.* **22**: 78–83.

Shen MH, Upadhyaya M. (1993) A *de novo* nonsense mutation in exon 28 of the neurofibromatosis type 1 (NF1) gene. *Hum. Genet.* **92**: 410–412.

Shen MH, Harper PS, Upadhyaya M. (1993) Neurofibromatosis type 1 (NF1): the search for mutations by PCR-heteroduplex analysis on Hydrolink gels. *Hum. Mol. Genet.* **2**: 1861–1864.

Short PM, Martuza RL, Huson SM. (1994) In: *The Neurofibromatoses* (eds SM Huson and RAC Hughes). Chapman and Hall Medical Press, London, pp. 414–444.

Skuse GR, Kosciolek BA, Rowley PT. (1989) Molecular genetic analysis of tumors in von Recklinghausen neurofibromatosis: loss of heterozygosity for chromosome 17. *Genes Chrom. Cancer* **1**: 36–41.

Skuse GR, Kosciolek BA, Rowley PT. (1991) The neurofibroma in von Recklinghausen neurofibromatosis has a unicellular origin. *Am. J. Hum. Genet.* **49**: 600–607.

Sørensen SA, Mulvihill JJ, Nielsen A. (1986) Long-term follow-up of von Recklinghausen neurofibromatosis. Survival and malignant neoplasms. *N. Engl. J. Med.* **314**: 1010–1015.

Stark M, Assum G, Krone W. (1991) A small deletion and an adjacent base exchange in a potential stem-loop region of the neurofibromatosis 1 gene. *Hum. Genet.* **87**: 685–687.

Stephens K, Kayes L, Riccardi VM, Rising M, Sybert VP, Pagon RA. (1992) Preferential mutation of the neurofibromatosis type 1 gene in paternally derived chromosomes. *Hum. Genet.* **88**: 279–282.

Stern HG, Saal HM, Lee JS, Fain PR, Goldgar DE, Rosenbaum KN, Barker DF. (1992) Clinical variability of type 1 neurofibromatosis: is there a neurofibromatosis-Noonan syndrome? *J. Med. Genet.* **29**: 184–187.

Stumpf DA, Alksne JF, Annegers JF, Brown SS, Conneally PM, Housman D, Leppert M, Miller JP, Moss ML, Pileggi AJ, Rapin I, Strohman RC, Swanson LW, Zimmerman A. (1988) Neurofibromatosis. *Arch. Neurol.* **45**: 575–578.

Suzuki H, Takahashi K, Kubota Y, Shibahara S. (1992) Molecular cloning of a cDNA coding for neurofibromatosis type 1 protein isoform lacking the domain related to ras-GTPase-activating protein. *Biochem. Biophys. Res. Commun.* **187**: 984–990.

Suzuki Y, Suzuki H, Takamasa K, Yoshimoto T, Shibahara S. (1991) Brain tumors predominantly express the neurofibromatosis type 1 gene transcripts containing the 63 base insert in the region coding for GTPase activating protein-related domain. *Biochem. Biophys. Res. Commun.* **181**: 955–961.

Tassabehji M, Strachan T, Sharland M, Colley A, Donnai D, Harris R, Thakker N. (1993) Tandem duplication within a neurofibromatosis type 1 (NF1) gene exon in a family with features of Watson syndrome and Noonan syndrome. *Am. J. Hum. Genet.* **53**: 90–95.

Teinturier C, Danglot G, Slim R, Pruliere D, Launay JM, Bernhiem A. (1992) The neurofibromatosis 1 gene transcripts expressed in peripheral nerve and neurofibromas bear the additional exon located in the GAP domain. *Biochem. Biophys. Res. Commun.* **188**: 851–857.

The I, Murthy AE, Hannigan GE, Jacoby LB, Menon AG, Gusella JF, Bernards A. (1993) Neuro-fibromatosis type 1 gene mutations in neuroblastoma. *Nature Genet.* **3**: 62–66.

Trofatter JA, MacCollin MM, Rutter JL, Murrell JR, Duyao MP, Parry DM, Eldridge R, Kley N, Menon AG, Pulaski K, Haase VH, Ambrose CM, Munroe D, Bove C, Haines JL, Martuz RL, MacDonald ME, Seizinger BR, Short MP,

Buckler AJ, Gusella JF. (1993) A novel moesin-, ezrin-, radixin-like gene is a candidate for the neurofibromatosis 2 tumor suppressor. *Cell* **72**: 791–800.

Twist EC, Ruttledge MH, Rousseau M, Sanson M, Papi L, Merel P, Delattre O, Thomas G, Rouleau GA. (1994) The neurofibromatosis type 2 gene is inactivated in schwannomas. *Hum. Mol. Genet.* **3**: 147–151.

Uchida T, Matozaki T, Suzuki T, Matsuda K, Wada K, Nakano O, Konda Y, Nishisaki H, Nagao M, Sakamoto C, Kasuga M. (1992) Expression of two types of neurofibromatosis type 1 gene transcripts in gastric cancers and comparison of GAP activities. *Biochem. Biophys. Res. Commun.* **1**: 332–339.

Upadhyaya M, Cheryson A, Broadhead W, Fryer A, Shaw DJ, Huson S, Wallace MR, Andersen LB, Marchuk DA, Viskochil D, Black D, O'Connell P, Collins FS, Harper PS. (1990) A 90 kb DNA deletion associated with neurofibromatosis type 1. *J. Med. Genet.* **27**: 738–741.

Upadhyaya M, Shen M, Cherryson A, Farnham J, Maynard J, Huson SM, Harper PS. (1992) Analysis of mutations at the neurofibromatosis 1 (NF1) locus. *Hum. Mol. Genet.* **1**: 735–740.

Valero MC, Velasco E, Moreno F, Hernandez-Chico C. (1994) Characterization of four mutations in the neurofibromatosis type 1 gene by denaturing gradient gel electrophoresis (DGGE). *Hum. Mol. Genet.* **3**: 639–641.

Viskochil D, Buchberg AM, Xu G, Cawthon RM, Stevens J, Wolff RK, Culver M, Carey JC, Copeland NG, Jenkins NA, White R, O'Connell P. (1990) Deletions and a translocation interrupt a cloned gene at the neurofibromatosis type 1 locus. *Cell* **62**: 187–192.

Viskochil D, White R, Cawthon R. (1993) The neurofibromatosis type 1 gene. *Annu. Rev. Neurosci.* **16**: 183–205.

Wallace MR, Collins FS. (1991) Molecular genetics of von Recklinghausen neurofibromatosis. *Adv. Hum. Genet.* **20**: 267–307.

Wallace MR, Marchuk DA, Andersen LB, Letcher R, Odeh HM, Saulino AM, Fountain JW, Brereton A, Nicholson J, Mitchell AL, Brownstein BH, Collins FS. (1990) Type 1 neurofibromatosis gene: identification of a large transcript disrupted in three NF1 patients. *Science* **249**: 181–186.

Wallace MR, Andersen LB, Saulino AM, Gregory PE, Glover TW, Collins FS. (1991) A *de novo* Alu insertion results in neurofibromatosis type 1. *Nature* **353**: 864–866.

Warkany J. (1981) Friedrich Daniel von Recklinghausen and his times. *Adv. Neurol.* **29**: 251–257.

Watson CJ, Gaunt L, Evans G, Patl K, Harris R, Strachan T. (1993) A disease-associated germline deletion maps the type 2 neurofibromatosis (NF2) gene between the Ewing sarcoma region and the leukemia inhibitory factor locus. *Hum. Mol. Genet.* **2**: 701–704.

Wertelecki W, Rouleau GA, Superneau DW, Forehand LW, Williams JP, Haines JL, Gusella JF. (1988) Neurofibromatosis 2: clinical and DNA linkage study of a large kindred. *N. Engl. J. Med.* **319**: 278–283.

Wiesmüller L, Wittinghofer A. (1992) Expression of the GTPase activating domain of the neurofibromatosis type 1 (NF1) gene in *E. coli* and role of the conserved lysine residue. *J. Biol. Chem.* **267**: 10207–10210.

Xu G, O'Connell P, Viskochil D, Cawthon R, Robertson M, Culver M, Dunn D, Stevens J, Gesteland R, White R, Weiss R. (1990a) The neurofibromatosis type 1 gene encodes a protein related to GAP. *Cell* **62**: 599–608.

Xu G, Lin B, Tanaka K, Dunn D, Wood D, Gesteland R, White R, Weiss R, Tamanoi F. (1990b) The catalytic domain of the neurofibromatosis type 1 gene product stimulates ras GTPase and complements IRA mutants of *S. cerevisiae. Cell* **63**: 835–841.

Xu G, O'Connell P, Stevens J, White R. (1992) Characterization of human adenylate kinase 3 (AK3) cDNA and mapping of the AK3 pseudogene to an intron of the NF1 gene. *Genomics* **13**: 537–542.

Xu W, Yu Q, Ponder M, Wallace M, Xu G, Ponder B. (1992a) Molecular analysis of neurofibromatosis type 1 mutations. *Hum. Mutat.* **1**: 474–477.

Xu W, Mulligan LM, Ponder MA, Liu L, Smith BA, Mathew CGP, Ponder BAJ. (1992b) Loss of NF1 alleles in phaeochromocytomas from patients with type 1 neurofibromatosis. *Genes Chrom. Cancer* **4**: 337–342.

Zhong J, Spiegel R, Boltshauser E, Schmid W. (1993) Two novel mutations: 5108delAG and 5816insG in the NF1 gene detected by SSCP analysis. *Hum. Mol. Genet.* **2**: 1491–1492.

Multiple endocrine neoplasia types 1 and 2

Catharina Larsson, Magnus Nordenskjöld and Jan Zedenius

Multiple endocrine neoplasia (MEN) is a term that encompasses several 'syndromes' with little or no aetiological relationship to each other. Included in this group are, literally, all cases which involve more than one endocrine gland as well as cases of sporadic tumours which secrete a growth factor (e.g. adrenocorticotrophic hormone, ACTH) that causes a secondary hyperplasia in another gland (e.g. the adrenal cortex). The terms MEN types 1, 2A and 2B have a narrower scope. Earlier terms for MEN type 1 (MEN1) have included multiple endocrine adenoma, multiple endocrine adenomatosis, multiple endocrine adenopathy, pluriglandular syndrome, and Wermer syndrome. In MEN1, the parathyroid glands, anterior pituitary, duodenum and endocrine pancreas are most commonly affected. MEN type 2A (MEN2A) refers to malignancy of the thyroid (medullary thyroid carcinoma, MTC), hyperparathyroidism (HPT) and phaeochromocytoma, whereas the type 2B form (MEN2B) includes MTC, phaeochromocytoma and neurogangliomatosis. Although both hereditary, the aetiological mechanisms are quite different for MEN1 and 2. With the demonstration that germline mutations in the *RET* gene are responsible for MEN2A, MEN2B and isolated familial MTC (Donis-Keller *et al.*, 1993; Mulligan *et al.*, 1993a; Hofstra *et al.*, 1994), it is now believed that tumorigenesis in this syndrome results from an activating mutation which produces a protein product with oncogenic features. Our present knowledge of MEN1 fits the hypothesis that the *MEN1* gene is a tumour suppressor gene located on chromosome 11, and that tumorigenesis involves elimination of the normal allele at the MEN1 locus (Larsson *et al.*, 1988).

4.1 Multiple endocrine neoplasia type 1

4.1.1 Clinical features

The first documented description of MEN1 is generally credited to Erdheim (1903), but the familial nature of the disease was first reported in 1939 by Rossier and Dressler. Over the years, advances in the knowledge of the varied expressions of MEN1 paralleled the development of radio immunoassays for the abnormally high levels of circulating peptides in affected individuals. The clinical picture is largely dependent on the glands involved and whether the lesions hypersecrete symptom-causing hormones or not. Autopsy findings suggest that with increasing age of the patient all three tissues will be affected. However, biochemically, 90–97% of the patients exhibit primary HPT (pHPT), 30–80% have pancreatic tumours and 15–50% pituitary tumours (Brandi et al., 1987).

Hypercalcaemia, as a result of an excess of parathyroid hormone (PTH) secretion by the parathyroids, is the most common biochemical abnormality in MEN1. Hypersecretion of pancreatic polypeptides, e.g. gastrin, insulin, pancreatic polypeptide, results in either distinct syndromes (Zollinger–Ellison with gastrin hypersecretion, hypoglycaemia with insulin hypersecretion etc.) or can be detected by specific assays only. Anterior pituitary hyperfunction most commonly manifests itself as hyperprolactinemia with its specific clinical symptoms or, less commonly, as acromegaly (growth hormone, GH, hypersecretion), hyperthyroidism (thyroid-stimulating hormone hypersecretion) or Cushing's syndrome (ACTH hypersecretion). MEN1 patients also have other tissues affected more often than is seen in the general population. Examples of such lesions are asymptomatic adrenocortical hyperplasia or tumours, carcinoid tumours, thyroid disease, lipomas and pinealomas (Brandi et al., 1987).

4.1.2 Incidence

The true incidence of MEN1 is unknown, since there are no long-term population-based studies. The disease prevalence is estimated to be in the range of 0.02–0.175 per 1000 (Brandi et al., 1987). However, the geographical variation is considerable. For example, over 150 recently identified affected individuals from 2300 family members spanning eight generations have been tracked back to a woman who came to Tasmania in the middle of the last century. Her description recorded by the ship's surgeon is consistent with the diagnosis of MEN1 (Shepherd, 1991). Thus, there are no new mutations in this cohort of patients which illustrates the importance of obtaining a detailed family history.

Among unselected patients with HPT, in up to 18% the condition is associated with the MEN1 syndrome. Similarly, up to 54% of patients with Zollinger–Ellison's syndrome, i.e. gastrinoma, 4% of those with insulinoma and 3% of those with a pituitary tumour subjected to surgery display a concomitant MEN1 trait (Brandi et al., 1987). These figures do, however, differ largely between hospitals.

4.1.3 Suggested aetiological mechanisms

Chromosomal instability. The term chromosomal breakage syndrome has been applied to a number of autosomal recessive disorders that predispose to various malignancies. Such chromosomal instability was first described in Bloom's syndrome and then in Fanconi's anaemia, ataxia teleangiectasia and xeroderma pigmentosum. In 1983, similar findings were reported for MEN1 patients. Cultured lymphocytes from nine patients representing six MEN1 families showed an increased frequency of gaps and chromatid-type abnormalities but a normal range of sister chromatid exchanges. Lymphocytes from several MEN1 patients have also shown an increased incidence of chromosomal breakage when compared with normal controls. Recently, Scappaticci *et al.* (1991) showed that cultured lymphocytes from three MEN1 patients had a high rate of chromosomal instability which consisted of numerical as well as structural abnormalities such as, dicentric chromosomes, ring chromosomes, acentric fragments and double minutes. In addition, karyotyping of fibroblasts from the same patients showed increased frequencies of chromosome breaks and hyperploidy. Surprisingly, although more than 60 chromosomally aberrant cells were analysed in this study, no rearrangements were found that involved chromosome 11, to which the *MEN1* gene has been assigned (Larsson *et al.*, 1988).

Activation of an oncogene.   Oncogenes, initially identified as transforming components of retroviruses, have been associated with tumour formation in several animal and human malignancies. Human non-transforming counterparts, the proto-oncogenes, were then discovered, and are generally believed to be involved in cell proliferation and differentiation. As a result of specific genetic alterations they may gain the capacity to transform a cell into a neoplastic state. These genes act dominantly on the cellular level, since only one of the two gene copies has to be activated to transform a cell. The effects are mediated either through overexpression of the normal protein, or expression of an aberrant protein. These phenomena may be caused by gene amplification, translocation and/or deletion of chromosomal regions, or point mutations within the actual gene. The oncogenes are classified according to the subcellular localization and biochemical function of their products. Some, for example, are found in the nucleus, recognized as transcription factors, while others are found on the cell surface, i.e. growth factors bound to specific membrane receptors.

Two different observations are compatible with abnormal oncogene expression in non-neoplastic tissues from MEN1 patients. Brandi *et al.* (1986) described parathyroid mitogenic activity in plasma from MEN1 patients, and the mitogenic factor appeared to be an unknown protein that was not secreted by the neoplastic parathyroid glands themselves. It did not, however, have a similar mitogenic effect on pancreatic or pituitary cells, despite the frequent presence of tumours affecting these organs in MEN1 patients. The recent finding of double minutes in lymphocyte cultures from MEN1 patients could reflect the same underlying mechanism (Scappaticci *et al.*, 1991), considering that in malignancies the cytogenetic phenomenon of double minutes is usually associated with amplification of cellular oncogenes.

Even if illegitimate expression of an oncogene does not account for the inherited mutation in MEN1, it may still be operative in tumour development. Indeed, mutations in the GSα gene, *GSP*, have been found in a subset of GH-secreting pituitary tumours with constitutive activation of adenylyl cyclase and GH hypersecretion (Lyons *et al.*, 1990). Cyclic AMP, which is an intracellular second messenger for several trophic hormones and has the ability to stimulate cell growth, is thus autonomously synthesized. In addition, DNA transfection experiments performed in NIH-3T3 cells determined the presence of transforming DNA sequences in prolactinomas (Gonsky *et al.*, 1991).

Inactivation of a tumour suppressor gene. The first experimental evidence for the existence of tumour suppressor genes came from studies of somatic cell hybrids, when it was shown that malignancy could be suppressed by the fusion of malignant and non-malignant cells (Harris, 1969). The idea that heritable tumours result from mutations causing a cell to become homozygous for a cancer-causing gene was proposed by De Mars in 1970 and, 1 year later, Alfred Knudson introduced the two-mutational model of tumorigenesis (Knudson, 1971). This model was first suggested and proven for retinoblastoma (RB), which has since served as a prototype for identification of similar mechanisms in other types of tumours. The theory proposes that the inherited defective gene would make the individual a heterozygous carrier predisposed to tumour development. The tumour develops when a second mutational event has occurred, so that the remaining 'normal' gene or gene function is eliminated. Several types of second mutational events are possible, e.g. loss of a chromosome, somatic recombination, chromosomal deletion or point mutation. Since the tumours occur by complete elimination of the normal function, these genes are referred to as tumour suppressor genes. The synonym, 'recessive cancer genes', refers to the fact that at the cellular level their action is recessive, although the traits are dominantly transmitted.

The two-mutational model for tumorigenesis implies that it would be possible to determine on which chromosome the gene for the heritable form of a neoplasia is situated, simply by detecting chromosomal rearrangements which might reflect the 'second mutational event'. Such information may be available from karyotypes of cultured tumour cells, or by genotype comparison of constitutional and primary tumour tissue. The latter type of analysis involves the detection of loss of heterozygosity (LOH) by the use of restriction fragment length polymorphism (RFLP) markers.

If MEN1-associated tumours result from unmasking of a recessive mutation according to the two-mutational model, chromosomal rearrangements in such tumours might indicate on which chromosome the *MEN1* gene is situated. Therefore, constitutional and tumour genotypes were compared at different polymorphic loci in two brothers with neuro-endocrine pancreatic tumours, who had inherited the disease from their mother. Using informative markers from 18 different chromosomes, LOH was only detected on chromosome 11, both tumours having lost one of the constitutional alleles at all informative loci on this chromosome (Larsson *et al.*, 1988).

The significance of these findings was further supported when the parental origin of the lost chromosome was determined. In both cases, the lost alleles were always derived from the unaffected father (*Figure 4.1*). The chromosomal localization of the *MEN1* gene was then determined by linkage analysis in MEN1 families. This method exploits the tendency for genes that lie close together on a chromosome (linked genes) to co-segregate during cell division in gonads, i.e. meiosis. DNA markers detecting RFLPs at the muscle glycogen phosphorylase (*PYGM*) locus, located on 11q13, were found to co-segregate with the disease predisposition. This chromosomal localization was then confirmed in several families and, despite extensive linkage analysis of multiple families, no meiotic recombinations between *MEN1* and *PYGM* have been detected so far (Fujimori *et al.*, 1992; Iwasaki *et al.*, 1992; Larsson *et al.*, 1992, 1995; Thakker *et al.*, 1993a; Teh *et al.*, 1995). This indicates that the *MEN1* gene is located close to the *PYGM* locus and that the disease is genetically homogenous. Isolation of additional polymorphic markers have allowed more precise localization of the *MEN1* gene to within a 900-kb region, i.e. approximately 1% of the entire chromosome 11 (Weber *et al.*, 1994).

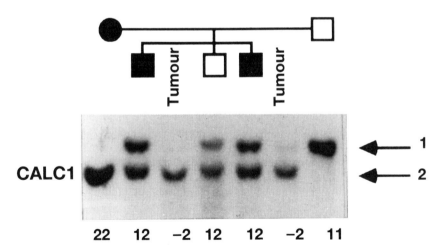

Figure 4.1. Loss of the wild-type allele in MEN1-associated tumours. The pedigree shows an MEN1 family with three affected members (black symbols). The autoradiogram below each family member shows the alleles for the Taq1 RFLP at the *CALCA* locus on chromosome 11. The genotypes with the alleles (1 and 2) for each family member are given below the autoradiogram. The affected mother is homozygous (2, 2) for this marker and the father is homozygous for the alternative allele (1, 1). Hence, all three children are heterozygous (1, 2). Tumour tissue (insulinoma) from both affected sons shows loss of the paternal allele 1.

Taken together these findings strongly suggested that the *MEN1* gene is a tumour suppressor gene on 11q13 whose inactivation is involved in tumour development in MEN1. Alternatively, the predisposing genetic defect in MEN1 could also be a dominantly acting mutation. LOH, serving to eliminate other loci on chromosome 11 as a secondary event, would also selectively involve the chromosome derived from the unaffected parent in this situation. However, several observations make this latter model unlikely: LOH involving the 11q13 region is seen in the majority of parathyroid and pancreatic tumours from MEN1 patients (Friedman *et al.*, 1989; Thakker *et al.*, 1989; Byström *et al.*, 1990). Deletion mapping studies of familial parathyroid and pancreatic tumours have revealed a minimal region of overlapping deletions around the *PYGM* locus at 11q13. Loss of the wild-type chromosome has been found in pancreatic, parathyroid and pituitary tumours (Larsson *et al.*, 1988; Thakker *et al.*, 1989; Byström *et al.*, 1990; Yoshimoto *et al.*, 1991). Two pancreatic tumours from two brothers with MEN1 both showed loss of the chromosome 11 alleles derived from the unaffected parent (Larsson *et al.*, 1988). DNA from microadenomatous 'hyperplastic' pancreatic tissue of one of the brothers was also studied. This tissue was classified histopathologically as benign 'hyperplasia' and showed the same type of chromosomal rearrangement as the insulinoma (Larsson *et al.*, 1988). This is in agreement with the idea that 11q13 deletions are primary events in tumour development and, hence, the secondary mutational events involved in tumour progression still need to be identified.

MEN1 patients mainly develop neoplasias of the parathyroid glands, the neuro-endocrine pancreas and duodenum, and the anterior pituitary gland. Autopsy findings suggest that all three tissues will be affected with increasing age. MEN1 patients, however, also develop tumours in other organs more frequently than in the general population. Examples of such lesions are adrenocortical tumours, carcinoids, lipomas and pinealomas. These associated tumours may reflect a pleiotropic effect of the underlying inherited *MEN1* mutation but may also occur secondary to the MEN1 phenotype. The most common of these associated tumours is benign enlargement of the adrenal cortex, which has been found in about one third of MEN1 necropsy cases. As for the pituitary tumours, they are rarely operated on and genetic analysis must rely on a small number of cases. Biochemical evaluation of 12 such cases failed to demonstrate any disturbances in the hypothalamic–pituitary–adrenal axis (Skogseid *et al.*, 1992). However, pancreatic tumours were over-represented and, in fact, present in all the MEN1 patients with adrenal enlargement (Skogseid *et al.*, 1992). Benign adrenal enlargements from five of these MEN1 patients showed no LOH for markers flanking the *MEN1* region in agreement with results from an analysis of eight sporadic adrenocortical adenomas (Yano *et al.*, 1989). However, one case of MEN1-associated adrenocortical carcinoma displayed a complete loss of the normal chromosome 11 (Skogseid *et al.*, 1992). Although there may be several possible explanations for the lack of 11q13 deletions in the benign adrenocortical tumours, a conclusive answer cannot be offered until the *MEN1* gene has been cloned.

If the *MEN1* gene is a tumour suppressor gene, its expression is expected to be altered or lost in the tumours. Furthermore, introduction of a cloned *MEN1* gene into a tumour cell line should result in reversion of the neoplastic phenotype. Similarly, knock-out experiments using transgenic animals should result in the occurrence of multiple endocrine tumours. This model would be strongly supported by the identification of tumours showing single mutations in both copies of the gene. However, such studies await isolation of the *MEN1* gene. Examination of a unique pedigree where the two unrelated parents and all three siblings are affected by the disease represented an alternative approach to this question (Brandi *et al.*, 1993). When the three siblings, their parents and relatives were genotyped for markers tightly linked to the *MEN1* locus, two of the siblings were found to be homozygous and one heterozygous for *MEN1*. As compared to the heterozygotes, no phenotypic abnormalities were observed in the two homozygotes with regard to the MEN1 syndrome. However, homozygosity for *MEN1* appears to affect fertility. In MEN1 tumours, the putative second mutation is usually a deletion involving most of chromosome 11. Therefore, the two mutations in the two homozygotes may be insufficient to eliminate the function of the *MEN1* gene entirely. Alternatively, loss of the *MEN1* gene function may be insufficient for tumour development. It could also be speculated that elimination of the *MEN1* gene function results in a benign growth, i.e. non-secreting microadenomas which may not be clinically recognizable.

4.1.4 Involvement of the MEN1 gene in sporadic neoplasms

During the last few years, some evidence has indeed accumulated for a two-hit mechanism of tumour development in MEN1. By analogy with findings in retinoblastoma, sporadic counterparts of MEN1-associated tumours are expected to occur by inactivation of the *MEN1* gene. The extent and level of the involvement of the *MEN1* gene in tumorigenesis of sporadic parathyroid, pituitary, and pancreatic–duodenal tumours might also be clarified. Hence, the sporadic tumour forms would also show allelic losses for the 11q13 region.

MEN1 patients characteristically develop multiple primary tumours, equally distributed among men and women, at 20–30 years of age, while the sporadic counterparts mainly show single tumours with a later onset. Morphologically, the lesions are indistinguishable, suggesting that tumorigenesis involves related mechanisms (Sandelin *et al.*, 1992). Indeed, sporadic parathyroid, pancreatic and pituitary tumours, as well as adrenocortical carcinomas, show deletions of chromosome 11 but this has a tendency to be less frequent than in the familial tumour forms.

In the few analysed cases of carcinoid tumours, no allelic losses could be demonstrated for 11q markers. We are not aware of any allelic loss studies having been performed on lipomas and/or pinealomas from MEN1 patients so no firm conclusions about the involvement of the *MEN1* gene in these tumours can be made. Interestingly, an interstitial deletion of 11q in one sporadic pinealoblastoma was noted karyotypically (Streekantaiah *et al.*, 1989). A variety of thyroid neoplasia has been documented in patients with MEN1 and, for the most part, these are not considered to be part of the MEN1 syndrome complex. However, allelic losses involving the long arm of chromosome 11, and including the *MEN1* gene locus, were found in several follicular thyroid tumours (Matsuo *et al.*, 1991). This may be indirect evidence that the *MEN1* gene may also be involved in thyroid tumorigenesis, and that the above mentioned association is truly a genetic one.

Sporadic pHPT is a frequent disorder, mainly affecting post-menopausal women, for reasons not yet known. Parathyroid carcinoma is seen in both sporadic and in familial forms of isolated pHPT but has not been reported in classical MEN1 families. The recent finding that familial isolated pHPT is not associated with chromosome 11 deletions but, instead, shows deletions at the *RB* gene locus on 13q supports the hypothesis that pHPT constitutes a genetically heterogeneous entity (Wassif *et al.*, 1993; Cryns *et al.*, 1994).

4.1.5 MEN1-related diseases

Molecular genetic studies of the rare familial tumour cases have demonstrated a relationship between the mechanisms involved in the tumorigenesis of familial and sporadic tumour forms. The sporadic forms can occur as a result of mutations in the same genes as in the familial forms. Familial MEN1 is characterized by hyperfunction of the parathyroid glands, the neuroendocrine pancreas–duodenum and the anterior pituitary. Genetic linkage analysis suggests the absence of genetic heterogeneity for all the typical MEN1 families of different races tested so far (Larsson *et al.*, 1995). In addition to the classical form, various MEN1-related syndromes have been described.

$MEN1_{BURIN}$ was initially described in the Burin peninsula with associated HPT, prolactinoma and carcinoids (Petty *et al.*, 1994). Unlike typical MEN1, $MEN1_{BURIN}$ manifests pancreatic endocrine tumours at a lower frequency. Linkage analysis with chromosome 11 markers close to the *MEN1* locus was performed in five Newfoundland and Pacific North-West kindreds. All families tested were linked to the same *PYGM* allele and to flanking polymorphic markers, suggesting a founder effect in this cohort and that $MEN1_{BURIN}$ reflects intralocus genetic heterogeneity of the *MEN1* gene. No $MEN1_{BURIN}$-like families were reported in other countries, although some families showed a high incidence of pituitary tumours as compared to epidemiological data for classical MEN1.

Familial isolated HPT (FIHP) is a rare inherited syndrome characterized by hypercalcaemia, high PTH levels and isolated parathyroid tumours (Goldsmith *et al.*, 1976). More than 30 kindreds, mostly small, have been reported to display pHPT only, with no evidence of other associated endocrinopathies to suggest MEN1. In most cases, the inheritance is autosomal dominant, but a single kindred with autosomal recessive transmission has been described. Some families reported initially to have FIHP were subsequently diagnosed as having MEN1. The relationship between this syndrome and MEN1 is still unclear, and it has been suggested that FIHP represents a genetic variant of MEN1. Diagnosis of FIHP remains a challenge, and has to be clearly proven by excluding endocrine lesions in other tissues using recently developed radiological and immunoassay techniques. Recent studies in such a family indicate that some variants of FIHP occur as a genetically and clinically distinct entity with an increased risk of malignant transformation of the parathyroid tumours (Wassif *et al.*, 1993). From these results, it can neither be excluded that FIHP is a genetically heterogeneous disease nor that some of the families in fact represent true variants of MEN1. Among the rare families that have been tested with markers from the *MEN1* region, some are linked while others are not.

Familial pituitary tumours are generally expressed as acromegaly (Yuasa *et al.*, 1990). This syndrome is rare and characterized by clinically expressed GH-secreting adenoma requiring surgery in most cases. Various hypotheses have been presented regarding the genetic origin of familial acromegaly, which might be related to specific mutations of the *MEN1* gene. Several reports suggest that pituitary adenoma and familial acromegaly might result from point mutations in activating subunits of G-proteins resulting in constitutional activation of G-coupled receptors, as described in the McCune–Albright syndrome (Levine, 1991; Thakker *et al.*, 1993b). Recent demonstration of the occurrence of McCune–Albright disease in a typical MEN1 kindred favours similarities between the *MEN1* gene and the G-protein-coupled transduction system on cell membranes (O'Halloran and Shalet, 1994). Linkage data using polymorphic chromosome 11 markers are limited because of the low number of affected cases in such pedigrees (Larsson *et al.*, 1995).

4.2 Multiple endocrine neoplasia type 2

4.2.1 Clinical features

Sipple's syndrome (MEN2A or MEN2).   Sipple's syndrome is an autosomal dominantly inherited disease with a high degree of penetrance and variable expressivity. It is characterized by the association of MTC with phaeochromocytoma and, in some cases, HPT.

MTC originates from the thyroid parafollicular C-cells, which secrete calcitonin. The familial form constitutes about 25% of all cases of MTC. It is seen in almost all hereditary MEN2A patients and usually differs from the sporadic counterparts in that the tumours occur bilaterally and are preceded by general C-cell hyperplasia (Wolfe *et al.*, 1973).

Phaeochromocytoma is seen in half of all MEN2A patients. Like MTC it occurs bilaterally and multicentrically and is preceded by diffuse and nodular hyperplasia (Carney *et al.*, 1976). Extra-adrenal tumours, especially in accessory adrenal glands, have been described in nearly 50% of cases (Lips *et al.*, 1981). The delay in onset between MTC and phaeochromocytoma can be more than 25 years.

Froboese's syndrome (MEN2B or MEN3).   This syndrome is characterized by the association of MTC and phaeochromocytoma with a distinct phenotype, mucosal neuromas, intestinal ganglioneuromatosis and neurological disturbances. A few hundred patients with this condition have been reported so far, and about half of the cases appear to be new mutations (Khairi *et al.*, 1975), i.e. there is no family history of the disease.

Familial medullary thyroid carcinoma (FMTC).   This disease, in which C-cell tumour of the thyroid is the only lesion, is less common than MEN2A. FMTC tends to present later in life and appears to be less aggressive than MEN2A (Farndon *et al.*, 1986). FMTC is only recognizable in large families where there is no sign of the associated MEN2A lesions. In fact, small families with MTC only are more likely to be MEN2A families where phaeochromocytoma or HPT has not yet been diagnosed.

4.2.2 Localization of the MEN2A gene

Cytogenetic analysis.   Cytogenetic analysis has suggested that MEN2A patients carry an interstitial deletion of chromosome band 20p12 (Babu *et al.*, 1984), but these findings could not be confirmed by repeated efforts in other laboratories (Wurster-Hill *et al.*, 1988). Interestingly, an extra ring chromosome in 3% of the cells from an affected individual was found. The precise identification of the origin of this marker chromosome was not possible, but it showed similarities with the centromeric regions of chromosomes 10 and 20 (Temperani *et al.*, 1989).

Linkage analysis.   The *MEN2A* locus was assigned to chromosome 10 by linkage analysis. The long search began with studies using protein marker systems, followed by RFLP analysis. A weak linkage to human leucocyte antigen (HLA) was excluded (Simpson and Falk, 1982). The reporting of an interstitial deletion at 20p12 suggested that the *MEN2* gene could be within this region, but analysis with RFLP markers excluded this possibility (Goodfellow *et al.*, 1985; Farrer *et al.*, 1987). Positive but non-significant lodscores (log of odds for linkage) were also obtained for the haptoglobin and group-specific component loci on chromosomes 16 and 4, respectively (Kruger *et al.*, 1986). Such findings can easily be the result of random events but may also indicate the true location of the disease locus. However, RFLP analysis excluded these 'hints' of linkage (Kidd *et al.*, 1986). A report of a family in which MEN2A was associated with myotonic dystrophy focused interest on chromosome 19, but this localization was also excluded (Carter *et al.*, 1987).

At the Second International Workshop on MEN2, all published data were combined with unpublished data from the laboratories of Kidd, Ponder and Simpson. This analysis showed that approximately 32% of the genome could be excluded as possible sites for the disease locus (Simpson and Kidd, 1987). A preliminary report showed a positive lodscore of 1.7 in one family (Simpson and Kidd, 1987) with an anonymous marker derived from a flow-sorted library designed to isolate chromosomes 9 and 22. This marker, however, was subsequently localized to chromosome region 10q21.1 by *in situ* hybridization (McDermid *et al.*, 1987) and is now known as D10S5.

Finally, odds in favour of linkage between the disease and two other chromosome 10 markers were obtained (Mathew *et al.*, 1987; Simpson *et al.*, 1987). Analysis of additional families gave a significant lodscore for the D10S5 marker (Simpson *et al.*, 1987). The interstitial retinol-binding protein gene, towards the centromeric part of chromosome 10, was also linked to the putative *MEN2A* locus (Mathew *et al.*, 1987; Simpson *et al.*, 1987).

Extensive family studies of MEN2A, FMTC and MEN2B kindreds from all over the world confirmed the location of the disease gene in the pericentromeric region of chromosome 10. Since these studies all pointed in one direction, it was likely that the three clinical entities reflected related genetic lesions at one locus. As a result of extensive family studies with numerous markers, the disease gene was localized to a small region on chromosome 10 (Gardner *et al.*, 1993; Mole *et al.*, 1993).

4.2.3 Identification of the RET proto-oncogene as the MEN2 gene

The region on chromosome 10 predicted by family studies to harbour the *MEN2* gene also contained the *RET* oncogene. This was an obvious candidate gene for MEN2 since it is expressed in the neuroectodermal tissues affected in MEN2, and also because it had transforming capacity in the case of the chimeric gene product *RET/PTC* detected in a subset of human papillary thyroid carcinomas (Grieco *et al.*, 1990). Detailed mutation analysis of the *RET* gene in MEN2 patients revealed specific mutations in gene carriers but not in their unaffected relatives (Donis-Keller *et al.*, 1993; Mulligan *et al.*, 1993a; Hofstra *et al.*, 1994).

The RET protein is a member of the family of receptor tyrosine kinases. The protein spans the cell membrane and is involved in the transduction signals for cell proliferation and differentiation (Ulrich and Schlessinger, 1990). The extracellular cadherin-like 'ligand-binding' site recognizes an as yet unidentified signal, and the transmembrane region connects this with the intracellular tyrosine kinase domain (*Figure 4.2*). Activation results in phosphorylation of tyrosine residues of the RET protein itself, which forms part of an onward signalling cascade.

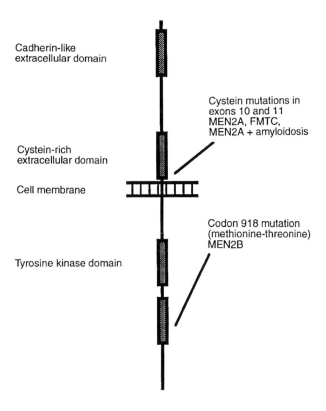

Cadherin-like
extracellular domain

Cystein mutations in
exons 10 and 11
MEN2A, FMTC,
MEN2A + amyloidosis

Cystein-rich
extracellular domain

Cell membrane

Codon 918 mutation
(methionine-threonine)
MEN2B

Tyrosine kinase domain

Figure 4.2. Schematic representation of the *RET* proto-oncogene.

In order to determine whether the *RET* gene is involved in MEN2, the entire gene was scanned for small mutations. These screening efforts revealed that MEN2A and FMTC patients carry heterozygous base pair substitutions altering one out of five cystein-coding codons within exon 10 or 11 of the *RET* gene (Donis-Keller *et al.*, 1993; Mulligan *et al.*, 1993a). The most common mutations involve alterations of the Cys634 amino acid which have been observed in more than 75% of the 170 or so unrelated MEN2A/FMTC families analysed (Ponder, 1994; Donis-Keller, 1995; Gagel *et al.*, 1995). The amino acids Cys609, Cys611, Cys618 and Cys620 have also been shown to be mutated in MEN2A. Furthermore, there appears to be a relationship between the type of mutation and the phenotype. Hence, in FMTC, Cys634 mutations are not as abundant as in MEN2A and, in MEN2A with parathyroid lesions, Cys634→Arg is much more common than any other mutation (Mulligan *et al.*, 1994). In contrast, the cases of MEN2B have a different mutation which, in almost every case, is a Met→Thr, as a result of a change in codon 918. This amino acid substitution occurs in the heart of the catalytic domain of the thyrosine kinase region

(Carlson *et al.*, 1994; Eng *et al.*, 1994; Hofstra *et al.*, 1994). Analysis of sporadic MTC tumours also revealed frequent activating mutations in the *RET* gene (Hofstra *et al.*, 1994). Preliminary data suggest that mutations in codon 918 in MTC are associated with a more aggressive course of the disease than for tumours not carrying such mutations (Zedenius *et al.*, 1994).

Taking these results together, three clinically distinguishable inherited syndromes, MEN2A, MEN2B and FMTC, are all caused by mutations in a single gene. In addition, the related sporadic form of these syndromes, MTC, also shows mutations in this gene.

4.2.4 Somatic mutations in MEN2-related tumours

It was long believed that the *MEN2* gene belonged to the tumour suppressor category of genes, as did genes responsible for several other inherited cancer syndromes. It was a natural strategy in the search for the *MEN2* gene, therefore, to study LOH in tumour tissues from MEN2 patients, as an alternative to linkage analysis. This strategy was based on the two-mutational model of tumorigenesis (see Section 4.1.3). Indeed, the necessity for two mutational events had been proposed from epidemiological studies of MEN2 patients and their sporadic counterparts (Jackson *et al.*, 1979). However, several studies failed to detect significant LOH at the *MEN2* locus in the centromeric region of 10q. Instead, other regions are frequently somatically altered in MEN2-related tumours (Mulligan *et al.*, 1993b). It is tempting to speculate that the seemingly activating germline mutation of the *RET* gene is mainly responsible for the hyperplasia of cells expressing the gene, while the neoplastic transformation requires additional events, such as the inactivation (or activation) of another, as yet unidentified, gene. On the one hand, the fact that somatic *RET* mutations in sporadic MTCs almost never affect the codons involved in MEN2A and FMTC could support this theory. On the other hand, the fact that somatic heterozygous mutations of codon 918 in the *RET* gene (the MEN2B-responsible mutation) are found in many sporadic MTCs may suggest that this alteration, in itself, functions as a trigger for tumour development. Further search for somatic alterations, within the *RET* gene but also at other loci, will elucidate the mechanisms for development of MEN2-related tumours. Finally, transgenic animal models will probably be of crucial importance for the understanding of why MEN2-related tumours occur, and hopefully some constitutional or somatic mutations will be correlated with the biological behaviour of the tumours.

4.2.5 Animal models for RET mutations

By targeted mutagenesis of the *RET* gene, Pachnis and co-workers (Pachnis *et al.*, 1993; Schuchard *et al.*, 1994) constructed transgenic mice with a 0.8-kb deletion of the *RET* gene. The heterozygous animals were then intercrossed to obtain homozygous knock-out mice. The heterozygous mice showed no phenotype, whilst the homozygously deleted animals all died within 16 to 24 h of birth. On autopsy, the mice were found to have either complete renal agenesis or only rudimentary kidneys. The second severe abnormality was failure of the milk to progress from the stomach to the small intestine. This suggested an abnormal development of the enteric nervous system. Histological examination of the gut confirmed the absence of ganglia in the entire bowel. These observations suggested that inactivating mutations of the *RET* gene can cause aganglionosis (Schuchard *et al.*, 1994). No relevant animal model has yet been constructed, however, for the activating *RET* mutations observed in MEN2.

4.2.6 RET *mutations in Hirschsprung's disease*

Independent of these knock-out experiments, family studies in large pedigrees segregating Hirschsprung's disease as an autosomal dominant trait implicated a locus in the same 10q12 region containing *RET*. This disorder is caused by varying degrees of absence of enteric ganglia cells and corresponds in humans to the phenotype observed in the homozygous knock-out mice. The observation of a patient with an interstitial chromosome 10 deletion and Hirschsprung's disease (Martucciello *et al.*, 1992) prompted two groups to perform linkage analysis on chromosome 10. Evidence in favour of positive linkage to the pericentromeric region of chromosome 10 was obtained in a subset of Hirschsprung families (Angrist *et al.*, 1993; Luo *et al.*, 1993; Lyonnet *et al.*, 1993). When the *RET* gene was screened for mutations in patients with Hirschsprung's disease, different types of inactivating mutations in heterozygous form were seen in a subset of the patients (Edery *et al.*, 1994; Romeo *et al.*, 1994). These mutations were either deletions of part of, or the entire, *RET* gene or nonsense mutations within the gene. No hot spot for these mutations has yet been observed. In addition to these autosomal dominant mutations, family studies have shown that Hirschsprung's disease may also be inherited as an autosomal recessive, and a region on chromosome 13 has been implicated by family studies (Puffenberger *et al.*, 1994).

Taken together, activating *RET* gene mutations have been shown to cause three different clinically distinguishable variants of MEN: MEN2A, FMTC and MEN2B, while at least a subfraction of Hirschsprung's disease cases are caused by inactivating mutations in the same gene (*Figure 4.2*). Thus, these relationships illustrate genetic homogeneity for MEN2 and genetic heterogeneity for Hirschsprung's disease and, furthermore, that one gene can be responsible for several different clinical syndromes.

4.2.7 Clinical management of MEN2 kindreds

Biochemical screening for C-cell hyperplasia and phaeochromocytoma is an effective method of identifying early stages of MEN2. Family registers have been established in many countries which has helped to reduce morbidity and mortality for the disease (Gagel *et al.*, 1988). The screening procedure, however, is not only unpleasant for the patient but also costly, especially since the programme has to be repeated many times. The screening ought to be negative in non-carriers but there is always uncertainty as to when to stop screening, due to the relative risk of late presenting disease. Furthermore, the screening procedure sometimes leads to false positive results, and we now know that, in some instances, this has lead to unnecessary thyroidectomy. In some of these cases elevation of calcitonin levels appears to result from causes other than C-cell hyperplasia (Lips *et al.*, 1994; Gagel *et al.*, 1995).

Using DNA testing in familial cases, it is now possible to exclude the non-carriers from screening and also to perform thyroidectomy on gene carriers before they have developed MTC (Lips *et al.*, 1994). This is important since, in a few cases, metastatic MTC has already developed in small children. Family mutational screenings are relatively simple to perform but rely on the identification of the individual mutation in each family by DNA sequencing. *Figure 4.3* illustrates the different MEN2A-associated *RET* mutations identified in Swedish cases. For each individual mutation, a simple restriction enzyme-based test is developed wherever possible in order to screen the at-risk family members. By amplification of the mutated exon, and digestion with the appropriate restriction enzyme, the risk of MEN2A can be confirmed or excluded using a simple blood sample.

At the Fifth International Workshop on Multiple Endocrine Neoplasia in the summer of 1994, a consensus was reached concerning procedures for DNA testing for MEN2. It is important to determine the individual mutation segregating in each MEN2 kindred. One affected individual should be subject to either DNA sequencing of exon 10 and/or 11 of the *RET* gene or a restriction endonuclease-based test which will identify the mutation. Each at-risk individual should then be offered a DNA test, which may be based on sequencing or on a restriction digest. If an at-risk individual tests normal (non-carrier), the test should be repeated on a newly drawn blood sample to avoid sample mix up. After two negative DNA tests, the individual and her/his offspring can be reassured and withdrawn from the screening programme.

In patients (twice) tested positive for a MEN2-associated mutation, total thyroidectomy should be performed. The time for this operation is subject to some controversy. Metastatic MTC has been reported at as early as 6 years of age (Graham *et al.*, 1987). The consensus of opinion at the Workshop, therefore, was that since the mortality risk of operating on children is small, and considering the risk of metastatic disease, total thyroidectomy should be performed upon children at the age of 5–6 years. It should be emphasized that prophylactic thyroidectomies should be performed at surgical centres with properly documented experience. This procedure should reduce the mortality in at least 95% of gene carriers for MEN2. The patients should then continue to be screened for early detection of phaeochromocytoma.

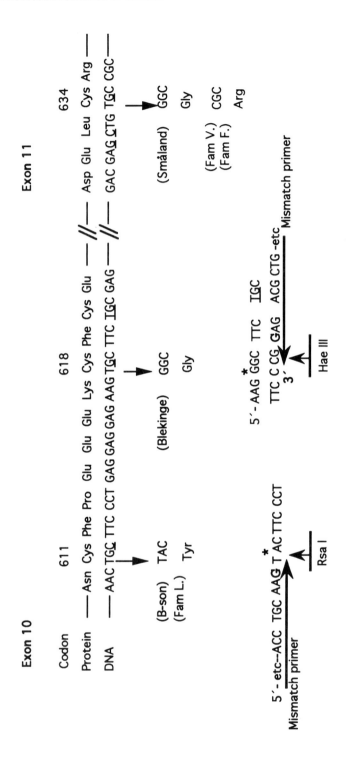

Figure 4.3. Mutations in Swedish MEN2A cases. The normal peptide (top) and DNA sequence for part of exons 10 and 11 are shown. Mutations in codon 611, 618 and 634 are shown. The exon 634 mutations introduce a new *Hae*III site (Cys634 → Gly) or a new *Cfo*I site (Cys634 → Arg). The exon 611 and 618 mutations are identified by amplification with a modified primer which will introduce a new restriction site in the presence of the mutation. The altered base is marked by an asterisk.

Acknowledgements

The authors wish to thank the Swedish Cancer Foundation, the Margaret and Axel Axson Johnson Foundation, the Medical Research Council, the Magnus Bergwall Foundation and the Lars Hierta Foundation for their support.

References

Angrist M, Kauffman ER, Slaugenhaupt SA, Matise TC, Puffenberger EG, Washington SS, Welssenbach J, Lipson A, Cass DT, Reyna T, Weeks DE, Sleber W, Chakravarti A. (1993) A gene for Hirschsprungs disease (megacolon) in the pericentromeric region of human chromosome 10. *Nature Genet.* **4**: 351–356.

Babu VR, Van Dyke DL, Jackson CE. (1984) Chromosome 20 deletions in human multiple endocrine neoplasia types 2A and 2B: a double blind study. *Proc. Natl Acad. Sci. USA* **81**: 2525–2528.

Brandi ML, Aurbach GD, Fitzpatrick LA, Quarto R, Spiegel AM, Blitziotes MM, Norton JA, Doppman JL, Marx SJ. (1986) Parathyroid mitogenic activity in plasma from patients with familial multiple endocrine neoplasia type I. *N. Engl. J. Med.* **314**: 1287–1293.

Brandi ML, Marx SJ, Aurbach GD, Fitzpatrick LA. (1987) Familial multiple endocrine neoplasia type I: a new look at pathophysiology. *Endocrinol. Rev.* **8**: 391–405.

Brandi M-L, Weber G, Svensson A, Falchetti A, Tonelli F, Castello R, Furlani L, Scappaticci S, Fraccaro M, Larsson C. (1993) Homozygotes for the autosomal dominant neoplasia syndrome MEN1. *Am. J. Hum. Genet.* **53**: 1167–1172.

Byström C, Larsson C, Blomberg C, Sandelin K, Falkmer U, Skogseid B, Öberg K, Werner S, Nordenskjöld M. (1990) Localisation of the MEN I gene to a small region within chromosome band 11q13 by deletion mapping in tumors. *Proc. Natl Acad. Sci. USA* **87**: 1968–1972.

Carlson KM, Dou S, Chi D, Scavarda N, Toshima K, Jackson CE, Wells SA Jr, Goodfellow PJ, Donis–Keller H. (1994) A single missense mutation in the tyrosine kinase catalytic domain of the RET proto-oncogene is associated with multiple endocrine neoplasia type 2B. *Proc. Natl Acad. Sci. USA* **91**: 1579–1583.

Carney JA, Sizemore GW, Sheps SG. (1976) Adrenal medullary disease in multiple endocrine neoplasia type 2. Pheochromocytoma and its precursors. *Am. J. Clin. Pathol.* **83**: 177–196.

Carter C, Easton DF, Mathew CGP, Welander G, Telenius H, Telenius-Berg M, Ponder BAJ. (1987) Exclusion of linkage of loci on chromosome 19 with multiple endocrine neoplasia, type 2. *Cytogenet. Cell Genet.* **45**: 33–37.

Cryns VL, Thor A, Hong–Ji X, Shi–Xue H, Wierman ME, Vickery AL, Benedict WF, Arnold A. (1994) Loss of the retinoblastoma tumor-suppressor gene in parathyroid carcinoma. *N. Engl. J. Med.* **330**: 757–761.

Donis Keller H. (1995) The RET proto-oncogene and cancer. *Int. J. Med.*, in press.

Donis Keller H, Chi D, Carlson KM, Toshima K, Lairmore TC, Howe JR, Moley JF, Goodfellow P, Wells SA Jr. (1993) Mutations in the RET proto-oncogene are associated with MEN2A and FMTC. *Hum. Mol. Genet.* **2**: 851–856.

Edery P, Lyonnet S, Mulligan LM, Pelet A, Dow E, Abel L, Holder S, Nihoul Fekete C, Ponder BAJ, Munnich A. (1994) Mutations of the ret proto-oncogene in

Hirschsprung's disease. *Nature* **367**: 378–380.

Eng C, Smith DP, Mulligan LM, Nagal MA, Healey CS, Ponder MA, Gardner E, Scheumann GFW, Jackson CE, Tunnacliff A, Ponder BAJ. (1994) Point mutations within the tyrosine kinase domain of the RET proto-oncogene in multiple endocrine neoplasia type 2B and related sporadic tumors. *Hum. Mol. Genet.* **3**: 237–241.

Erdheim J. (1903) The normal and pathological histology of glandular thyroidism, parathyroidism and hypophysis. *Beitr. Pathol. Anat.* **33**: 158–236.

Farndon JR, Leight GS, Dilley WG, Baylin SB, Smallridge RC, Harrison TS, Wells SA. (1986) Familial medullary thyroid carcinoma without associated endocrinopathies: a distinct clinical entity. *Br. J. Surg.* **73**: 278–281.

Farrer LA, Goodfellow PJ, White BN, Holden JJA, Kidd JR, Simpson NE, Kidd KK. (1987) Linkage analysis of multiple endocrine neoplasia type 2A (MEN-2A) and three DNA markers on chromosome 20: evidence against synteny. *Cancer Genet. Cytogenet.* **27**: 327–334.

Friedman E, Sakaguchi K, Bale AE, Falchetti A, Streeten E, Zimering MB, Weinstein LS, McBride WO, Nakamura Y, Brandi M-L, Norton JA, Aurbach GD, Spiegel AM, Marx SJ. (1989) Clonality of parathyroid tumors in familial multiple endocrine neoplasia type I. *N. Engl. J. Med.* **321**: 213–218.

Fujimori M, Wells SA, Nakamura Y. (1992) Fine-scale mapping of the gene responsible for multiple endocrine neoplasia type 1 (MEN1). *Am. J. Hum. Genet.* **50**: 399–403.

Gagel RF, Tashijan AH Jr, Cummings T, Papathanasopoulos N, Kaplan MM, De Lillis RA, Wolfe HJ, Reichlin S. (1988). The clinical outcome of prospective screening for multiple endocrine neoplasia type 2a. *N. Engl. J. Med.* **318**: 478–484.

Gagel RF, Cote GJ, Martins Bugalho MJG, Boyd AE, Cummings T, Goepfert H, Evans DB, Cangir A, Khorana S, Schultz PN. (1995) Clinical use of molecular information in the management of multiple endocrine neoplasia type 2A. *J. Intern. Med.*, in press.

Gardner E, Papi L, Easton DF, Cummings T, Jackson CE, Kaplan M, Love DR, Mole SE, Moore JK, Mulligan LM, Norum RA, Ponder MA, Reichlin S, Stall G, Telenius H, Telenius-Berg M, Tunnacliffe A, Ponder BAJ. (1993) Genetic linkage studies map the multiple endocrine neoplasia type 2 loci to a small interval on chromosome 10q11.2. *Hum. Mol. Genet.* **2**: 241–246.

Goldsmith RE, Sizemore GW, Chen IW, Zalme E, Altemeier WA. (1976) Familial hyperparathyroidism: description of a large kindred and a review of the literature. *Ann. Intern. Med.* **84**: 36–43.

Gonsky R, Herman V, Melmed S, Fagin J. (1991) Transforming DNA sequences present in human prolactin-secreting pituitary tumors. *Mol. Endocrinol.* **5**: 1687–1695.

Goodfellow PJ, White BN, Holden JJA, Duncan AMV, Sears EVP, Wang H-S, Berlin L, Kidd KK, Simpson NE. (1985) Linkage analysis of a DNA-marker localised to 20p12 and multiple endocrine neoplasia type 2A. *Am. J. Hum. Genet.* **37**: 890–897.

Graham SM, Genel M, Touloukian RJ, Barwick KW, Gertner JM, Torony C. (1987) Provocative testing for occult medullary carcinoma of the thyroid: findings in seven children with multiple endocrine neoplasia type IIA. *J. Pediatr. Surg.* **22**: 501–503.

Grieco M, Santoro M, Berlingieri MT, Melillo RM, Donghi R, Bongarzone I, Pierotti MA, Della Porta G, Fusco A, Vecchio G. (1990) PTC is a novel rearranged form of the *Ret* proto-oncogene and is frequently detected in vivo in human thyroid papillary carcinomas. *Cell* **60**: 557–563.

Harris H. (1969) Suppression of malignancy by cell fusion. *Nature* **223**: 363–368.

Hofstra RM, Landsvater RM, Ceccherini I, Stulp RP, Stelwagen T, Luo Y, Pasisni B, Hoppener JWM, Ploos van Amstel HK, Romeo G, Lips CJ, Buys CHC. (1994) A mutation in the RET proto-oncogene associated with multiple endocrine neoplasia type 2B and sporadic medullary thyroid carcinoma. *Nature* **367**: 375–383.

Iwasaki H, Stewart PW, Dilley WG, Holt MS, Steinbreux TD, Wells SA JR, Donis-Keller H. (1992) A minisatellite and microsatellite polymorphism within 1.5 kb at the human muscle glycogen phosphorylase (PYGM) locus can be amplified by PCR and have combined informativeness of PIC 0.95. *Genomics* **13**: 7–15.

Jackson CE, Block MA, Greenawald KA, Tashijan AH Jr. (1979) The two-mutational-event theory in medullary thyroid carcinoma. *Am. J. Hum. Genet.* **31**: 704–710.

Khairi MRA, Dexter RN, Burzynski NJ, Johnston CC. (1975) Mucosal neuroma, pheochromocytoma, and medullary thyroid carcinoma: multiple endocrine neoplasia type 3. *Medicine (Baltimore)* **54**: 85–112.

Kidd KK, Kidd JR, Castiglione CM, Pakstis AJ, Sparkes RS. (1986) Progress towards resolving the possible linkage of multiple endocrine neoplasia type 2A to haptoglobin and group-specific loci: use of restriction fragment length polymorphisms extends exclusion region. *Genet. Epidemiol.* **3**: 195–200.

Knudson AG. (1971) Mutation and cancer: statistical study of retinoblastoma. *Proc. Natl Acad. Sci. USA* **68**: 820–823.

Kruger SD, Gertner JM, Sparkes RS, Haedt LE, Crist M, Sparkes MC, Genel M, Kidd KK. (1986) Linkage analysis of multiple endocrine neoplasia, type 2 (MEN-2) with 23 classical genetic markers. *Hum. Hered.* **36**: 6–11.

Larsson C, Skogseid B, Öberg K, Nakamura Y, Nordenskjöld M. (1988) Multiple endocrine neoplasia type 1 gene maps to chromosome 11 and is lost in insulinoma. *Nature* **332**: 85–87.

Larsson C, Shepherd J, Nakamura Y, Blomberg C, Weber G, Werelius B, Hayward N, Teh BT, Tokino T, Seizinger B, Skogseid B, Öberg K, Nordenskjöld M. (1992) Predictive testing for multiple endocrine neoplasia type 1 using DNA polymorphisms. *J. Clin. Invest.* **89**: 1344–1349.

Larsson C, Calender A, Grimmond S, Giraud S, Hayward NK, Teh B, Farnebo F. (1995) Molecular tools for presymptomatic testing in multiple endocrine neoplasia type 1. *J. Intern. Med.*, in press.

Levine MA. (1991) The McCune–Albright syndrome: the whys and wherefores of abnormal signal transduction. *N. Engl. J. Med.* **325**: 1738–1740.

Lips CJM, van der Sluys Veer J, Struyvenberg A, Alleman A, Leo JR, Wittebol P, Minder WH, Kooiker CJ, Geerdink RA, van Waes PFGM, Hackeng WHL. (1981) Bilateral occurrence of pheochromocytoma in patients with the multiple endocrine neoplasia syndrome 2A (Sipple's syndrome). *Am. J. Med.* **70**: 1051.

Lips CJM, Landsvater RM, Höppener JWM, Geerdink RA, Blijham G, Jansen-Schillhorn van Veen JM, van Gils APG, de Wit MJ, Zewald RA, Berends MJH, Beemer FA, Brouwers-Smalbraak J, Jansen RPM, Ploos van Amstel HK, van Vroonhoven TJMV, Vroom TM. (1994) Clinical screening as compared with DNA analysis in families with multiple endocrine neoplasia type 2A. *N. Engl. J. Med.* **331**: 828–835.

Luo Y, Ceccherini I, Pasini B, Matera I, Bicocchi MP, Barone V, Bocciardi R, Kääriäinen H, Weber D, Devoto M, Romeo G. (1993) Close linkage with the RET proto-oncogene and boundaries of deletion mutations in autosomal dominant

Hirschsprung's disease. *Hum. Mol. Genet.* 2: 1803–1808.

Lyonnet S, Bolini A, Pelet A, Abel L, Nihoul-Fékété C, Briad ML, Mok-Siu V, Kääriäinen H, Martucciello G, Lerone M, Pulti A, Luo Y, Weissenbach J, Devoto M, Munnich A, Romeo G. (1993) A gene for Hirschsprung's disease maps to the proximal long arm of chromosome 10. *Nature Genet.* 4: 346–350.

Lyons J, Landis CA, Harsh G, Vallar L, Gunwald K, Feichtinger H, Duh Q-Y, Kawasaki E, Bourne HR, McCormick F. (1990) Two G protein oncogenes in human endocrine tumors. *Science* 249: 655–659.

Martucciello G, Bicocchi MP, Dodero P, Lerone M, Silengo Cirillo M, Pulti A, Gimelli G, Romeo G, Jasonni. (1992) Total colonic aganglionosis associated with interstitial deletion of the long arm of chromosome 10. *Pediatr. Surg.* 7: 308–310.

Mathew CGP, Chin KS, Easton DF, Thorpe K, Carter C, Liou GI, Fong S-L, Bridges CDB, Haak H, Nieuwenhuijzen Kruseman AC, Schifter S, Hansen HH, Telenius H, Telenius-Berg M, Ponder BAJ. (1987) A linked genetic marker for multiple endocrine neoplasia type 2A on chromosome 10. *Nature* 328: 527–528.

Matsuo K, Tang SH, Fagin JA. (1991) Allelotype of human thyroid tumors: loss of chromosome 11q13 sequences in follicular neoplasms. *Mol. Endocrinol.* 5: 1873–1879.

McDermid HE, Goodfellow PJ, Duncan AMV, Brasch KR, Simpson NE, Souza CD, Holden JJA, White BN. (1987) A polymorphic locus, D10S5, at 10q21. *Nucleic Acids Res.* 15: 5499.

Mole SE, Mulligan LM, Healey CS, Ponder BA, Tunnacliffe A. (1993) Localisation of the gene for multiple endocrine neoplasia type 2A to a 480 kb region in chromosome band 10q11.2. *Hum. Mol. Genet.* 2: 247–252.

Mulligan LM, Kwock JBJ, Healey CS, Elsdon MJ, Eng C, Gardner E, Love DR, Mole SE, Moore JK, Papi L, Ponder MA, Telenius H, Tunnacliffe A, Ponder BAJ. (1993a) Germ-line mutations of the ret proto-oncogene in multiple endocrine neoplasia type 2A. *Nature* 363: 458–460.

Mulligan LM, Gardner E, Smith BA, Mathew CGP, Ponder BAJ. (1993b) Genetic events in tumour initiation and progression in multiple endocrine neoplasia type 2. *Genes Chrom. Cancer* 6: 166–177.

Mulligan LM, Eng C, Healey CS, Clayton D, Kwok JBJ, Gardner E, Ponder MA, Frilling A, Jackson CE, Lehnert H, Neumann HPH, Thibodeau SN, Ponder BAJ. (1994) Specific mutations of the ret proto-oncogene are related to disease phenotype in MEN2A and FMTC. *Nature Genet.* 6: 70–74.

O'Halloran DJ, Shalet SM. (1994) A family pedigree exhibiting features of both multiple endocrine neoplasia type 1 and McCune–Albright syndromes. *J. Clin. Endocr. Metab.* 78: 523–535.

Pachnis V, Mankoo B, Constantini F. (1993) Expression of the c-ret proto-oncogene during mouse embryogenesis. *Development* 119: 1005–1017.

Petty EM, Green JS, Marx SJ, Taggart T, Farid N, Bale AE. (1994) Mapping the gene for hereditary hyperparathyroidism and prolactinoma (MEN1BURIN) to chromosome 11q: evidence for a founder effect in patients from New Foundland. *Am. J. Hum. Genet.* 54: 1060–1066.

Ponder BAJ. (1994) The gene causing multiple endocrine neoplasia type 2 (MEN 2). *Ann. Med.* 26: 199–203.

Puffenberger EG, Kauffman ER, Bolk S, Matise TC, Washington SS, Angrist M, Weissenbach J, Garver KL, Mascari M, Ladda R, Slaugenhaupt SA, Chakravarti A. (1994) Identity-by-descent and association mapping of a recessive gene for Hirschsprung disease on chromosome 13q22. *Hum. Mol. Genet.* 3: 1217–1225.

Romeo G, Ronchetto P, Luo Y, Barone V, Seri M, Ceccherini I, Pasini B, Bocciardi R, Lerone M, Kääriäinen H, Martucciello G. (1994) Point mutations affecting the tyrosine kinase domain of the ret proto-oncogene in Hirschsprung's disease. *Nature* **367**: 377–378.

Sandelin K, Larsson C, Falkmer U, Farnebo L-O, Grimelius L, Nordenskjöld M. (1992) Morphology, DNA ploidy and allele losses on chromosome 11 in sporadic hyperparathyroidism and that associated with multiple endocrine neoplasia, type 1. *Eur. J. Surg.* **158**: 199–206.

Scappaticci S, Maraschio P, del Ciotto N, Fossati GS, Zonta A, Fraccaro M. (1991) Chromosome abnormalities in lymphocytes and fibroblasts of subjects with multiple endocrine neoplasia type 1. *Cancer Genet. Cytogenet.* **52**: 85–92.

Schuchard A, D'Agati V, Larsson-Blomberg L, Constantini F, Pachnis V. (1994) Defects in the kidney and enteric nervous system of mice lacking the tyrosine kinase receptor RET. *Nature* **367**: 380–383.

Shepherd JJ. (1991) The natural history of multiple endocrine neoplasia type 1 highly uncommon or highly unrecognized. *Arch. Surg.* **126**: 935–952.

Simpson NE, Falk J. (1982) Exclusion of linkage between the loci for multiple endocrine neoplasia type-II (MEN2) and HLA. *Hum. Genet.* **60**: 157.

Simpson NE, Kidd KK. (1987) Where is a locus for multiple endocrine neoplasia type 2A? *Henry Ford Hosp. Med. J.* **35**: 168–171.

Simpson NE, Kidd KK, Goodfellow PJ, McDermid H, Myers S, Kidd JR, Jackson CE, Duncan AMV, Farrer LA, Brasch K, Castiglione C, Genel M, Gertner J, Greenberg CR, Gusella JF, Holden JJA, White BN. (1987) Assignment of multiple endocrine neoplasia type 2A to chromosome 10 by linkage. *Nature* **328**: 528–530.

Skogseid B, Larsson C, Lindgren PG, Kvanta E, Rastad J, Theodorsson E, Wide L, Wilander E, Öberg K. (1992) Clinical and genetic features of adrenocortical lesions in multiple endocrine neoplasia type 1. *J. Clin. Endocrinol. Metab.* **75**: 76–81.

Streekantaiah C, Jockin H, Brecher ML, Sandberg AA. (1989) Interstitial deletion of chromosome 11q in a pineoblastoma. *Cancer Genet. Cytogenet.* **39**: 125–131.

Teh BT, Cardinal J, Shepherd J, Hayward NK, Weber G, Cameron D, Larsson C. (1995) Genetic mapping of the multiple endocrine neoplasia type 1 locus at 11q13. *J. Intern. Med.*, in press.

Temperani P, Savin E, Aloesio R, Forabosco A. (1989) Ring chromosome in a patient with MEN IIA. *Cancer Genet. Cytogenet.* **37**: 23–27.

Thakker RV, Bouloux P, Wooding C, Chotai K, Broad PM, Spurr NK, Besser GM, Riordan JHLO. (1989) Association of parathyroid tumors in multiple endocrine neoplasia type I with loss of alleles on chromosome 11. *N. Engl. J. Med.* **321**: 218–224.

Thakker RV, Wooding C, Pang JT, Farren B, Harding B, Andersen DC, Besser GM, Bouloux P, Brenton DP, Buchanan KD, Edwards CR, Heath DA, Jackson CF, Jansen S, Lips K, Norum R, Sampson J, Shalet SM, Taggart RT, Tailor D, Wheeler MH, Wollard PM, Yates Y. (1993a) Linkage analysis of 7 polymorphic markers at chromosome 11q12–11q13 in 27 multiple endocrine neoplasia type 1 families. *Ann. Hum. Genet.* **57**: 17–25.

Thakker RV, Pook MA, Wooding C, Boscaro M, Scanarini M, Clayton RN. (1993b) Association of somatotrophinomas with loss of alleles on chromosome 11 and with gsp mutations. *J. Clin. Invest.* **91**: 2815–2821.

Ulrich A, Schlessinger J. (1990) Signal transduction by receptors with tyrosine kinase activity. *Cell* **61**: 203–212.

91

Wassif W, Moniz CF, Friedman E, Weber G, Wong S, Nordenskjöld M, Peters T, Larsson C. (1993) Familial isolated hyperparathyreoidism: a distinct genetic entity with an increased risk of parathyroid cancer. *J. Clin. Endocrinol. Metab.* **77**: 1485–1489.

Weber G, Friedman E, Grimmond S, Hayward N, Phelan C, Skogseid B, Gobels A, Zedenius J, Carson E, Sandelin K, Teh B, White I, Öberg K, Shepherd J, Nordenskjöld M, Larsson C. (1994) The phospholipase C β3 gene located in the MEN1 region shows loss of expression in endocrine tumors. *Hum. Mol. Genet.* **3**: 1775–1781.

Wolfe HJ, Melvin KEW, Cervi–Skinner SJ, Al Saadi AA, Juliar JP, Jackson CE, Tashijan AH Jr. (1973) C-cell hyperplasia preceding medullary thyroid carcinoma. *N. Engl. J. Med.* **289**: 437–441.

Wurster–Hill DH, Noll WW, Devlin JT, Erbe RW, Gibson SH, Dossu J. (1988) Fragile sites and high-resolution chromosome studies in multiple endocrine neoplasia type 2A. *Cancer Genet. Cytogenet.* **35**: 273–277.

Yano T, Linehan M, Anglard P, Lerman MI, Daniel LN, Stein CA, Robertson CN, LaRocca R, Zbar B. (1989) Genetic changes in human adrenocortical carcinomas. *J. Natl Cancer Inst.* **7**: 518–523.

Yoshimoto K, Iwahana H, Kubo K, Saito S, Itakura M. (1991) Allele loss on chromosome 11 in a pituitary tumor from a patient with multiple endocrine neoplasia type 1. *Jpn J. Cancer Res.* **82**: 886–889.

Yuasa H, Tokito S, Nakagaki H, Kitamura K. (1990) Familial pituitary adenoma: report of four cases from two unrelated families. *Neurol. Med. Chir.* **30**: 1016–1019.

Zedenius J, Wallin G, Hamberger B, Nordenskjöld M, Weber G, Larsson C. (1994) Somatic and MEN2A de novo mutations identified in the RET proto-oncogene by screening of sporadic MTCs. *Hum. Mol. Genet.* **3**: 1259–1262.

Genetics of breast and ovarian cancer

D. Timothy Bishop

Family history is well recognized as a risk factor for breast cancer both by epidemiologists and the general public (Kelsey, 1979; Petrakis *et al.*, 1982; Cannon-Albright *et al.*, 1991). Newspaper articles stress the extraordinary family histories and debate the appropriate action to take to alleviate some of the risk and anxiety. Of course, a positive family history, that is having a relative with breast cancer, is common simply because breast cancer is common; overall, breast cancer accounts for 32% of all new cases of cancer in women (Boring *et al.*, 1992). Detailed analysis of family history shows that in most of these cases there is only one affected relative (and often this relative was elderly when the diagnosis was made); such a family history may of course occur by chance. Occasionally, family histories are, however, much stronger than this and are consistent with autosomal dominant inheritance. The other notable feature of such families is that the onset is often particularly young, in many cases before the menopause.

5.1 Population studies

Population studies usually involve the interviewing of a number of cancer cases and request for precise details of family history, such as age and sex of relatives, their current ages or ages at death and their medical history. For scientific credibility, these cases should represent the complete set of new diagnoses in a defined geographical region in a defined time frame. In any study, similar questions are asked of a series of controls who are usually matched to the cancer cases for age, sex and residence. The family histories of the cases and controls are then compared to quantitate the increased frequency when there is a positive family history compared to the control.

The most illuminating observation from such studies is the identification of characteristics of the cases (or families) that predict a familial predisposition. For breast cancer, one of the most useful descriptors of predisposition is the age of onset of the breast cancer case. Basically, the earlier the age of onset of the case, the higher the risk of subsequent breast cancer in a relative. The largest study which has attempted to look at the relevance of family history is the Cancer and Steroid Hormone Study (CASH) of the Centers for Disease Control and National Cancer Institute in the USA (Sattin et al., 1985; Claus et al., 1990, 1991). This study involved interviewing almost 5000 cases and a similar number of controls to establish their family histories. All cases were diagnosed under the age of 55 years and so represent a highly selected group of cases, but complete coverage of a geographical region was achieved. From this study the trend was clear; the younger the age of onset of the index case the greater the risk of breast cancer in a relative. Thus, the risk of breast cancer before age 60 for mothers and sisters of breast cancer patients was estimated to be 0.16 (when case diagnosis occurred before the age of 30 years), 0.10 (for cases diagnosed between 30 and 39 years), 0.06 (40–49 years) and 0.04 (50–54 years), and 0.03 for relatives of controls (which represents the risk in the general population). The risk was found to be even higher for women with two close relatives with breast cancer as compared to those with one. For instance, women with an affected mother and an affected sister had a sevenfold increased risk of breast cancer over that for a woman with no such family history (and with 2.3-fold increased risk with only an affected sister). These results are consistent with those from other studies (see, for instance, Bishop et al., 1994). Finally, a number of studies have shown that the risk of cancer is higher for relatives of patients with bilateral tumours compared with unilateral cases (Anderson and Badzioch, 1985; Bernstein et al., 1992).

Thus, an increased risk of cancer in relatives, corresponding with early age of onset and the presence of more affected relatives is consistent with breast cancer having a major genetic component and gives an indication of the frequency of this susceptibility. The formal proof that genetic predisposition is important in some families is described below (see Section 5.2) but the proportion of patients with an inherited susceptibility due to the clinically recognized syndromes is unknown.

For ovarian cancer, population analysis also shows an increased risk of tumours in the first-degree relatives of ovarian cancer cases (Schildkraut et al., 1989), who also show an increased risk of breast cancer (Claus et al., 1991).

5.1.1 Segregation analysis

Segregation analysis is a statistical technique which attempts to explain the familial occurrence of disease (in this case, breast cancer) as a function of genetic susceptibility, common exposure to (unmeasured) environmental risk factors and random factors. Almost all analyses, so far, have concluded that susceptibility to breast cancer is inherited as an autosomal dominant disease with high penetrance with a lifetime risk of tumour development close to 100% (for a review see Bishop et al., 1994). The precise estimates associated with each study have shown marginal variation. In the analysis performed by Claus et al. (1991), the statistical solution, assuming that inherited predisposition were due to a single gene, was that approximately 6.5% of the general population carry a susceptibility to breast cancer, and that the risk of developing breast cancer in females carrying the mutation is 67% by age 65 years and 100% by age 80 years. If these figures are correct then approximately 25% of breast cancers diagnosed before the age of 40 years would be due to mutations in this postulated gene.

Extrapolation of the above results back to the overall population, however, showed that the model did not fit all age classes with the same precision. Most noticeable was the fact that, at young ages, the observed risk to sisters of early onset cases, who also had affected mothers, was always higher than that predicted by the best fitting statistical model. Whilst not statistically significant in any one analysis, the uniformity of the result, across ages, is striking. One explanation for this is that there are, in fact, two or more genetic phenomena acting (these families would be the ones most likely to have an affected mother), and the solution obtained by the CASH segregation analysis was in fact an amalgam of these distinct risks (segregation analysis is not capable, in general, of detecting genetic heterogeneity whether it leads to phenotypic variation or not). If there were two causes of tumorigenesis, then the obvious difference would be that one was responsible for tumours in younger patients whilst the other was more important in patients of intermediate age; very little information is available from the CASH study on postmenopausal breast cancer because of it being restricted to cases of breast cancer diagnosed under the age of 55 years.

One of the obstacles to understanding the importance of these high penetrance mutations is the difficulty in identifying the number of families in the general population with such overt disease. An estimate can only be made, and then only superficially, when family history includes reference to relatives of second (aunts, uncles, grandparents etc.) and higher degrees of relationship (especially cousins). However, few epidemiological studies include relatives other than first degree (brothers and sisters, parents) with any verified information, so such estimates are not available. There is some limited information from the CASH study regarding the frequency of predisposition, since 343 out of 4570 cases had an affected mother (7.5%) while only 98 had two or more first- or second-degree relatives affected with breast cancer (2%) (Claus, 1988).

5.2 Hereditary syndromes predisposing to breast cancer and ovarian cancer

Patients with three different syndromes are recognized clinically as having a substantially increased risk of developing breast cancer. All of these syndromes show a dominant predisposition such that offspring of mutation carriers have a 50% risk of inheriting the mutation, and those females inheriting the mutation have a high risk of developing breast cancer. In the first of these syndromes, the *site-*

specific breast cancer syndrome, family members are at increased risk of breast cancer alone. Although precise estimates of the risk of breast cancer are difficult to make, crude estimates suggest that it is of the order of 80% in women (Newman *et al.*, 1988; Iselius *et al.*, 1991). In the second, *breast–ovary syndrome*, female mutation carriers are at increased risk of cancer at both of these sites. Families in this category show a wide variation in the risk of ovarian cancer as compared to breast cancer (Gardner and Stephens, 1950; Fraumeni *et al.*, 1975; Lynch *et al.*, 1978). Finally, the *Li–Fraumeni syndrome* is characterized by an increased risk of soft-tissue sarcomas, osteosarcomas, adrenocortical cancers, leukaemia, brain tumours and lung cancer, as well as early onset breast cancer (Li and Fraumeni, 1969; Garber *et al.*, 1991). The increased risk of breast cancer in these families has been estimated to be 18-fold higher than the incidence in the general population. In the majority of cases, the increased risk is evident at a young age (perhaps by the age of 30 years).

The syndromes which predispose to ovarian cancer include the *breast–ovary syndrome* described above, the possibility of a *site-specific ovarian cancer syndrome* and the *Lynch syndromes*, which predispose predominantly not only to cancer of the bowel but also to tumours at a variety of other sites, most notably the endometrium but including the ovary. There are occasional families in which predisposition may simply be to cancer of the ovary (see, for example, Steichen-Gersdorf *et al.*, 1994) although such families may represent the extremes of the breast–ovary syndrome. The Lynch syndromes, usually termed hereditary non-polyposis colorectal cancer (HNPCC), are rare, dominantly transmitted syndromes with increased risk of cancer at a variety of sites. The frequency of the syndrome has been studied but with quite varied results. For instance, St John *et al.* (1993) found that about 0.5% of colorectal cancer cases had a family history consistent with HNPCC while other studies have found rates closer to 5% (Watson and Lynch, 1993).

5.3 Genetic studies of breast and ovarian cancers

The availability of large families with many cases of either breast cancer alone or breast and ovarian cancer, as well as families with the hereditary bowel cancer syndromes, made these cancers natural candidates for genetic mapping studies.

5.3.1 BRCA1

After a number of false reports of linkage for breast cancer, Dr Mary-Claire King and co-workers (Hall *et al.*, 1990) provided the first convincing evidence when they reported linkage to chromosome 17q in some families. The linkage was reported with a DNA marker D17S74, and initial analysis suggested that the gene (labelled *BRCA1*) lay close to this polymorphic locus, but there was clear evidence of heterogeneity; that is, some families showed linkage to this gene while other families did not. The authors suggested that age was the critical factor in determining whether mutations in *BRCA1* explained the family aggregation, with families having early onset (before the age of 45 years) being due to *BRCA1* whereas families with older onset of tumours did not appear to be linked. In fact, reanalysis with a more appropriate statistical model did not provide overwhelming support for heterogeneity since, in the revised analysis, little weight was given to later onset cases as being due to an inherited susceptibility (Margaritte *et al.*, 1992).

The 17q linkage was immediately confirmed by Narod *et al.* (1991) but using families which had been chosen for different reasons to those of Hall *et al.* (1990); these latter families were chosen simply because of the occurrence of a number of early onset breast cancers in close relatives. The collaborative group from Omaha and the IARC (Lyon, France) chose for their study those families with the clearest evidence of a dominantly inherited, highly penetrant disease which were considered to be those families with both breast and ovarian cancer. These families were collected by Dr Henry Lynch and co-workers (Narod *et al.*, 1991) from the Midwest of the USA.

At the end of the 1980s, a collaborative group had been set up under the auspices of the UK Cancer Family Study Group to pool information from linkage studies on possible locations for hereditary breast cancer genes. Subsequent to the reports by Hall *et al.* (1990) and Narod *et al.* (1991), a joint analysis of 214 families was completed. These families were largely chosen because of a number of early onset breast cancer cases but, in some cases, because of the co-occurrence of breast and ovarian cancer in close relatives. All pedigrees were broadly consistent with dominant inheritance of a susceptibility gene; for simplicity in the rest of this discussion, results of the analysis by this group will be referred to as the Breast Cancer Linkage Consortium (BCLC) (Easton *et al.*, 1993a)

Several conclusions were noted from the BCLC (Easton *et al.*, 1993a). First, there was overwhelming statistical support for a locus on 17q. Second, over all, approximately 62% of all breast cancer families were due to *BRCA1* but there was a clear distinction between those families in which only diagnoses of breast cancer were apparent and those in which there was at least one diagnosis of ovarian cancer. For the latter group, it was estimated that all families were due to *BRCA1*, although the confidence interval stretched from 79% to 100%. This estimate of 100% was surprising since a crude definition such as this could easily include a family with a non-genetic case of ovarian cancer in an unlinked breast cancer family; this would have lowered the estimate to below 100%. In fact, subsequently, various families with multiple ovarian cancer diagnoses have been reported (Narod *et al.*, 1995) showing this estimate to be a little excessive. The true figure is now estimated to be 92% when attention is focused on those families with at least two cases of ovarian cancer. This genetic heterogeneity is important to understand as numerous families are undergoing clinical follow-up in specialist genetics groups and the possibility of DNA testing for carrier status would be welcomed. The proportion of families actually linked to 17q is an important factor in risk assessment since the higher the base-line frequency, the easier it is to provide convincing evidence that the family undergoing DNA-based assessment is actually due to *BRCA1*.

For the breast cancer families, approximately 45% of breast cancer-only families (confidence interval 25–66%) were due to *BRCA1*. Linkage was most likely the greater with the greater number of early onset cases present in the family. Approximately 40% of families, with at most four cases occurring before the age of 45 years, were due to *BRCA1*, while about three-quarters of those with at least five cases were so attributable.

Analysis of the location of *BRCA1* by Easton *et al.* (1993a) provided strong evidence that the gene was between D17S250 and D17S588 (see *Figure 5.2*), a distance of approximately 8.3 cM in males and 18.0 cM in females (this region is known to have much higher female than male recombination rates). Such statistical evidence is important because individual families are often not sufficiently convincing in providing unequivocal evidence of location; potential meiotic recombinants can also usually be explained as phenocopies (that is, cases of breast cancer not due to inherited susceptibility).

The definition of close flanking boundaries requires the analysis of large numbers of families until, by chance, a recombination event is observed that delimits the location of the gene. For breast cancer there are added difficulties, since breast cancer can, of course, arise spontaneously and not in the context of inherited susceptibility. *Figure 5.1* shows an important recombination event in a family with high probability of being due to a *BRCA1* mutation. The maternally derived chromosome is seen to be recombinant between D17S857 and D17S846, with the disease-causing mutation being transmitted with the distal portion of the chromosome. This observation makes D17S857 the proximal boundary for *BRCA1*. The certainty of the implication of this observation depends upon the age of onset of the cancer in this woman. If the onset were at an age typical in the general population (60s or 70s) then it is feasible that this woman developed breast cancer sporadically (i.e. independently of inherited susceptibility) and would not therefore provide any further information on the *BRCA1* location; the earlier the onset, the clearer the information. In fact, she developed bilateral breast cancer at ages 36 and 41, both ages being much younger than the population norm, and the possibility of genetic susceptibility is strengthened by the observation of bilateral disease. This observation therefore provides convincing evidence of location. The complexity of these analyses is exemplified, however, by another woman in this pedigree who developed ovarian cancer at age 57 years. Since this is a clear breast–ovary family with a *BRCA1* mutation, and taking into account the comparative rareness of ovarian cancer (as compared to breast cancer), there would be considerable weight to any observation concerning her genotype. In fact, after extensive work, it was possible to show that this woman's genotype is not consistent with her carrying any portion of the D17S250–D17S293 region from her maternal chromosome containing the

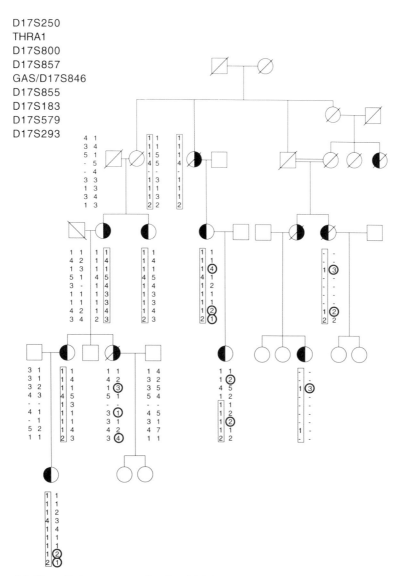

Figure 5.1. Part of a breast–ovary family which by linkage mapping was shown to be due to *BRCA1* (Kelsell *et al.*, 1993). The average age of breast cancer onset in this family is 42 years. Left shading of females in the pedigree indicates breast cancer diagnosis; right shading indicates ovarian cancer. The DNA markers typed in these family members are shown on the top left of the pedigree in order from centromere to telomere. The alleles co-segregating with the *BRCA1* mutation are shown in a box. The woman with ovarian cancer in the fourth generation who does not share the *BRCA1* mutation-bearing chromosome is a sporadic (i.e. non-genetic) case. The woman, also in the fourth generation, with breast cancer who shares only part of the chromosome with her affected relatives had early onset bilateral breast cancer and so is important in delimiting the location of *BRCA1*. A circle round an allele indicates that this allele was deleted in the tumour from this woman; this information supports the notion that *BRCA1* is a tumour suppressor gene.

BRCA1 mutation. This woman has therefore developed ovarian cancer sporadically. The only clue to this observation was that her age of onset was towards the high end for ovarian cancer in *BRCA1* families (Easton *et al.*, 1993a).

Analysis of individual families suggested that the distal boundary of the *BRCA1* region was D17S183 (Bowcock *et al.*, 1993) and, subsequently, D17S78 (Simard *et al.*, 1993), while the proximal boundary was defined by D17S857 (Kelsell *et al.*, 1993) and insulin-like growth factor-binding protein 4 (*IGFBP4*) (Tonin *et al.*, 1993). The strength of the observations contributing to this fine-scale mapping are presented in *Figure 5.2*, under each vertical bar which delimits the region enclosing *BRCA1*. In fact, the systematic reduction in the size of the candidate region is notable especially with the observation that *BRCA1* was eventually found to be essentially in the middle of the region (*Figure 5.2*).

5.3.2 Loss of heterozygosity (LOH) around BRCA1

In keeping with a number of other predisposition genes for cancer syndromes, *BRCA1* appears to be a tumour suppressor gene. This was shown by both Smith *et al.* (1992) and Kelsell *et al.* (1993) who examined tumours from *BRCA1* mutation carriers. For both breast and ovarian tumours, loss of the wild-type chromosome was observed (i.e. the chromosome carrying the *BRCA1* mutation was retained in the tumour). The majority of these tumours had detectable LOH.

Some sporadic tumours, that is tumours which arise outside the context of these families, also show LOH around the *BRCA1* region, although various studies suggest that there are multiple regions of 17q involved which complicates interpretation. For instance, Saito *et al.* (1993) found that two regions of 17q were deleted in breast tumours, one at 17q12–21, the other at 17q25. Similar results were obtained by Cornelius *et al.* (1994). Deletions identified in these studies were often found to be large (often either a whole chromosome arm or other significant sections) and so did not assist in delimiting the region containing *BRCA1* (Brown and Solomon, 1994). In a detailed analysis of sporadic and familial ovarian tumours, Godwin *et al.* (1994) found evidence for a focus of LOH distal in *BRCA1* and in the region between *GH* and D17S74. Allele loss for *BRCA1* was found in 58% of tumours but in 73% for the more distal region.

The difficulty of interpreting *BRCA1* LOH studies is exemplified by the study of Borg *et al.* (1994) who found that their LOH rates for D17S855 (now known to lie adjacent to *BRCA1*) were 47% in sporadic tumours and 67% in familial tumours (these families were selected to include two or more first-degree relatives with breast or ovarian cancer). The proximity of other potential tumour suppressor genes in the region complicates interpretation with respect to *BRCA1*; in this series, 27% of tumours showed LOH at both D17S855 and *NME1* (which codes for a protein expressed in non-metastatic cells), while only 5% had lost D17S855 and retained *NME1*, and only 2% had lost *NME1* but not D17S855.

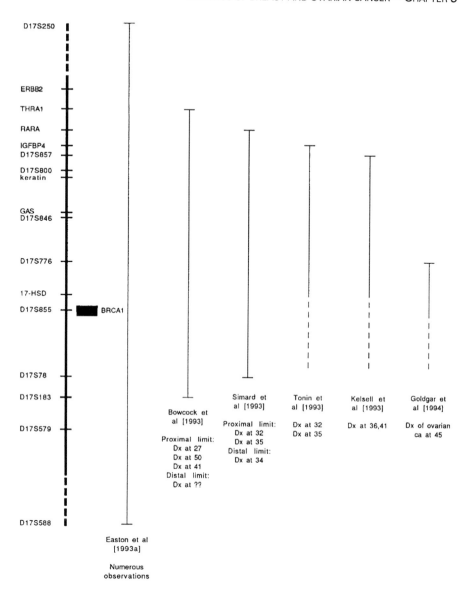

Figure 5.2. The critical recombination events which successfully delimited the *BRCA1* region. The importance of each event together with the publication describing the result and critical observations are indicated. Since age of onset is the crucial factor in determining the importance of an observation of a recombination, the figure lists such information (Dx at *n* indicates a diagnosis of breast cancer at *n* years). The solid line indicates the region remaining after the given observation. The dashed line indicates that the publication did not add to information on distal boundaries. The map on the left which places relevant DNA markers and some of the genes of interest on the chromosome 17 map is based on Albertsen *et al.* (1994a,b). *BRCA1* is now known to lie adjacent to D17S855.

101

5.3.3 Risk of cancer associated with BRCA1 mutations

A detailed study of families reported to the BCLC and which were unequivocally linked to *BRCA1* showed that mutation carriers were at increased risk of a variety of cancers. Most notably, of course, was the observation of the high risks of breast cancer and ovarian cancer. Analysis was conducted in a number of different ways, since the selection of the families was biased by the investigators collecting those families with the highest apparent risks of breast or ovarian cancer. These analyses take advantage of the carrier status which is inferred from the linkage analysis. In one method, information was obtained on recurrence of breast cancer in a female mutation carrier who had received surgery to remove a tumour in the contralateral breast. Such information can be used to re-estimate the age-specific rates in mutation carriers by translating the risk for the remaining breast with that which would be expected if there were two 'at-risk' breasts.

Although the BCLC analysis concluded an increased risk of ovarian cancer, further analysis suggested the presence of two classes of mutations: the first group would have a risk of breast cancer of 71% and of ovarian cancer of 87% (both risks to age 70 years), while in the second group the corresponding risks would be 86% and 18% (Ford *et al.*, 1994). Essentially, these estimates suggest that the mutations may differ in the implied risk of ovarian cancer while lifetime risks of breast cancer are independent of the specific mutations. Goldgar *et al.* (1994) analysed a large Utah pedigree and estimated the risk of breast or ovarian cancer to be 40% by age 40 years and 90% by age 70 years, similar to that estimated by BCLC.

Increased risk of other cancers has also been shown for *BRCA1* mutation carriers. Ford *et al.* (1994) found that the risk of prostate cancer was 3.3-fold increased in male mutation carriers while the risk of bowel cancer was 4.1-fold increased in males and females combined.

On the basis of the BCLC figures, the plausible frequency of mutation carriers was estimated by Easton *et al.* (1994) to be between one in 500 and one in 2000. Also, on the basis of the BCLC-estimated penetrance for *BRCA1* mutations, and assuming the higher population frequency (i.e. at birth, four in 1000 women (and men) would be heterozygotes for a *BRCA1* mutation), Easton *et al.* (1994) estimated that *BRCA1* explained approximately a quarter of all breast cancers diagnosed before the age of 30 years but perhaps only 1% of those in women over the age of 70 years.

5.3.4 Genetic and physical mapping around BRCA1

Linkage mapping successfully narrowed the region containing *BRCA1* to approximately 1.0–1.5 Mb between D17S857 and D17S78 (in fact, as shown in *Figure 5.2,* a later recombination event dramatically narrowed the size of the region; however, the majority of the physical mapping work was based on this larger region). Approaches to producing a physical map included: (i) constructing chromosome 17 hybrid mapping panels using either chromosome-specific radiation-reduced somatic cell hybrids (Abel *et al.*, 1993; Black *et al.*, 1993; O'Connell *et al.*, 1994) or chromosome-mediated gene transfer (Xu *et al.*, 1988); (ii) yeast artificial chromosomes; and (iii) pulsed-field gel electrophoresis (PFGE) using probes from the *BRCA1* region. Solomon and co-workers (unpublished data, reported in Brown and Solomon, 1994) performed PFGE of the *BRCA1* region and created a long-range restriction map which connected many of the known markers in that region. Further details of these approaches are given by Brown and Solomon (1994). Albertsen *et al.* (1994b) constructed a physical map comprising 137 overlapping yeast artificial chromosomes and P1 clones. During the construction of this map, as with other physical mapping approaches, the map order for polymorphic markers was determined, a process which requires genetic analysis of exceptionally high numbers of families. Albertsen *et al.* (1994b) managed to locate 112 polymorphic DNA markers on their map as well as 20 genes. Finally, Albertsen *et al.* (1994a) constructed a high resolution genetic map of the *BRCA1* region.

5.3.5 Candidate genes for BRCA1

One of the notable features of the mapping of *BRCA1* was that the region was already known to contain a number of genes which were biologically plausible candidates, and a number of others were also identified during the search for *BRCA1*. Initially, the candidate genes included the retinoic acid receptor α (*RARA*), *IGFBP4*, prohibitin (*PHB*) and the metalloprotease-like, disentigrin-like, cysteine-rich protein (*MDC*). These candidates were all excluded by informative recombinant events (*Figure 5.2*). Perhaps the strongest initial candidate within the region was the oestradiol 17β hydroxysteroid dehydrogenase II (*EDH*) gene. Since this gene was known to be expressed in both the ovary and breast it was plausibly consistent with the observations from the family studies (Easton *et al.*, 1993a). However, nucleotide sequencing failed to identify germline mutations in any clearly linked families (Kelsell *et al.*, 1993; Simard *et al.*, 1993). Similar results were obtained for *1A1*, the gene for a novel B-box protein (Campbell *et al.*, 1994). Finally, the observation of a family with segregated susceptibility to breast and ovarian cancer as well as a skin disorder, epidermolytic palmoplantar keratosis, suggested a common aetiology. The skin manifestations were, however, shown to be due to a mutation in the keratin 9 gene which presumably co-segregated with a mutation in *BRCA1*, since the keratin cluster was excluded from consideration (*Figure 5.2*).

5.3.6 BRCA1 cloning

BRCA1 was cloned by a collaborative group from the University of Utah, Myriad Genetics and the National Institutes of Health (Miki *et al.*, 1994). These investigators found a further recombinant that limited the size of the region containing *BRCA1* to 600 kb (*Figure 5.2*). Transcripts within this region were mapped and 65 candidate sequences were identified. These sequences were characterized by DNA sequence determination in both 'normal' individuals and obligate mutation carriers as determined by linkage analysis and by database comparisons and genomic structure. In fact, three of the expressed sequences were found to be components of a larger transcribed region which, in turn, was shown to be the *BRCA1* gene. The gene was found to consist of 24 exons distributed over 100 kb of genomic DNA and in close proximity to D17S855. The encoded protein has 1863 amino acids with a 7.8-kb transcript on Northern blots. The gene is expressed in various tissues including the breast and the ovary.

In the first analysis of *BRCA1* mutations, one family was found to have a nonsense coding mutation, another had a single nucleotide insertion, while a third had a missense mutation (Futreal *et al.*, 1994). A fourth family was found to have a regulatory mutation (although the precise mutation had not been identified) since only one cDNA product could be identified even though the germline was heterozygous for two distinct polymorphisms. The lack of this same product could be shown in a number of family members who, by linkage analysis, should be mutation carriers.

In a review of 80 germline *BRCA1* mutations identified by a number of groups, 36 distinct mutations have been recognized (Shattuck-Eidens *et al.*, 1995). The majority of the mutations are either frameshift or nonsense mutations, while a few are missense. Mutations were found throughout the coding sequence with little evidence of clustering. Finally, the authors suggested that the locations of the mutations in families with higher ovarian cancer risk were distinct from the distribution in lower risk families, although the evidence for this was not significant; such evidence would support the statistical analysis (see above) which suggested the presence of multiple classes of mutations with respect to ovarian cancer risk.

5.3.7 BRCA2 *mapping*

The strong evidence of other highly penetrant dominant genes indicated that a search for another gene, or genes, was worthwhile in those families shown by linkage analysis not to be due to *BRCA1*. Wooster *et al.* (1994) typed 15 such families with a large number of polymorphic markers, eventually finding linkage to markers on 13q12–13. The locus, now called *BRCA2*, is known to lie between D13S267 and D13S289, on the basis of several informative recombinants. At present, there is little information regarding the relative frequencies of *BRCA1* and *BRCA2*, or their relative penetrances, although *BRCA2* does appear to have a lower ovarian cancer risk than *BRCA1*. To date, there is little evidence for a third high penetrance locus.

5.3.8 Male breast cancer

Male breast cancer is rare (the incidence being about 1% of that of female breast cancer). An implication of the classical models of inherited susceptibility is that families in which members of the gender with the lower population incidence are actually affected are more likely to be segregating the inherited susceptibility. With this in mind, various studies have shown that the risk of female breast cancer in close relatives of male breast cancer cases is increased approximately twofold but that, in about 20% of families of male breast cancer cases, there is a noticeable history of female breast cancer, which is suggestive of inherited susceptibility. *Figure 5.3* shows one such family identified during a recent study in the north of England. Among the known genetic factors predisposing to male breast cancer, males with Kleinfelter syndrome have an increased risk while two families have been reported with mutations in the DNA binding domain of the androgen receptor (Wooster *et al.*, 1992; Lobaccaro *et al.*, 1993). Since both the males with Kleinfelter syndrome and those with mutations in the androgen receptor share phenotypic features of androgen insufficiency, these two susceptibilities appear to be related.

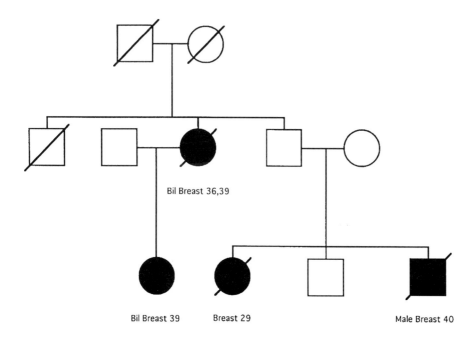

Figure 5.3. A male breast cancer family now thought to be due to *BRCA2* on the basis of linkage mapping. Ages of onset of breast cancer are included under the shaded symbols indicating affected family members. 'Bil', bilateral breast cancer.

The BCLC consortium, and a larger subsequent study, found that none of the families containing male breast cancer cases appeared to be linked to 17q (although there were only five such families) and, indeed, the best estimate of linked families was 0%. *BRCA1* is therefore excluded as having a major role in male breast cancer susceptibility (Stratton *et al.*, 1994). However, several of the families due to *BRCA2* did contain examples of male breast cancer cases; preliminary evidence suggests that the majority of families containing male breast cancer may well be due to *BRCA2*.

5.3.9 Genes for HNPCC

Families with hereditary predisposition to bowel cancer through the HNPCC syndromes also have an increased risk of ovarian cancer. This risk is much less than with *BRCA1* since families rarely have more than one female affected with ovarian cancer. The genes responsible for the majority of HNPCC syndromes were mapped by Peltomäki *et al.* (1993) to chromosome 2p and by Lindblom *et al.* (1993) to chromosome 3p. Preliminary analysis suggests that the two genes explain approximately equal proportions of HNPCC. The specific genes involved were found to be components of a mismatch repair pathway studied extensively in yeast (Strand *et al.*, 1993) and including two genes labelled *MSH2* and *MLH1*. The 2p gene for HNPCC was shown to be the human homologue of *MSH2* (*hMSH2*) while the 3p gene was related to *MLH1* of yeast (*hMLH1*) (Fishel *et al.*, 1993; Leach *et al.*, 1993; Bronner *et al.*, 1994; Papadopoulos *et al.*, 1994). The rapid identification of the genes involved, which followed within months of mapping data, was based on an observation of genetic changes in colorectal tumours from within the HNPCC families. Aaltonen *et al.* (1993) and, subsequently, Lindblom *et al.* (1993) showed that, in the majority of these tumours, the allele sizes produced by microsatellite repeat markers [consisting of repetitions of cytosine (C) and adenine (A) nucleotides] were distinct from those identified in the germline. In some cases, the number of CA units had increased over that seen in the germline while, in others, the number had decreased. The original germline bands could also be seen. These changes were observed in microsatellite markers from all over the genome and suggested that the HNPCC genes were also involved in faithful replication of DNA, and were similar to observations made by Strand *et al.* (1993) for yeast which was deficient in one or other of these genes. The observation of the varying allele sizes is commonly termed 'replication error'.

5.4 Involvement of other genes

5.4.1 p53

In 1990, Malkin *et al.* found germline mutations in the *p53* gene in five families with the Li–Fraumeni syndrome, three of which contained early onset breast cancer cases. Such germline mutations are rare; for instance, only three mutations were identified in over 400 samples in two studies (Borresen *et al.*, 1992; Sidransky *et al.*, 1992) and these cases were all diagnosed before the age of 41 years. Two of the women had a family history consistent with the Li–Fraumeni syndrome. *p53* mutations are, of course, commonly found in sporadic breast tumours as well as many other human cancers (Hollstein *et al.*, 1991).

5.4.2 HRAS

While *BRCA1* and *BRCA2* are important in the few high-risk families, there remains a distinct possibility of genetic susceptibility to breast cancer through lower penetrance but higher frequency mutations in other genes. In 1985, Krontiris *et al.* found surprising evidence for an association between risk of cancer and variant alleles at the locus for the proto-oncogene *HRAS1*. The DNA marker examined involves a 28-bp repeat sequence, and typical samples contain between 30 and 100 repeats of this unit. The DNA marker is located approximately 1000 bp downstream from the coding sequence of *HRAS1*. While four variant alleles dominate the system, there are other alleles which, together, are rare but as a unit make up approximately 6% of all alleles at this locus. These variants are usually single repeat units different in size from a 'common' allele. Krontiris *et al.* (1985) found that these 'rare' alleles were more common in cancer patients than controls. This study has been repeated by many other groups and the resulting data submitted to meta-analysis by Krontiris *et al.* (1993). Overall, the strength of the result is more convincing, with women having a single rare allele having an increased risk of breast cancer of 1.7 times the population rate, while homozygotes have an estimated increased risk of 4.6-fold for a common cancer (including cancer of the breast, bowel, bladder and melanoma). The mechanism of the increased risk is unknown but there has been some speculation that the rare alleles disrupt the expression of *HRAS1* or other local genes. The interest in this result is based not so much on the magnitude of the increased risk but, rather, the observation that this genetic make-up is found in approximately 10% of the general population and so alone probably explains more breast cancer cases than *BRCA1* and *BRCA2* mutations.

5.4.3 Ataxia telangiectasia

While ataxia telangiectasia is an extremely rare medical condition due to homozygosity of the *AT* gene [approximately one case per 100 000 births (Pippard *et al.*, 1988)], heterozygotes for one or more of the variants are considerably more common (at least 1% of the general population). In the absence of a test for carrier status, the only known heterozygotes are parents of cases. Swift *et al.* (1991) found a fivefold increased risk of breast cancer for mothers of AT homozygotes. Easton *et al.* (1993b) estimated that AT heterozygosity could explain 7% of all breast cancer cases. However, this increased risk of breast cancer, and the frequency of AT heterozygotes, may not be sufficient to explain any significant proportion of familial breast cancer (Wooster *et al.*, 1992).

5.4.4 Cytogenetic and LOH studies

Cytogenetic studies of sporadic breast cancers have been reviewed by Devilee and Cornelisse (1995). No characteristic cytogenetic abnormalities have been detected although translocations often involve chromosome 1q. LOH studies show that at least 11 chromosomal arms are involved (Devilee and Cornelisse, 1995). Of particular note are changes observed for chromosomes 1, 3p, 11p, 16q, 17p, 17q, 18q, 22q and Xp.

References

Aaltonen LA, Peltomäki P, Leach FS, *et al.* (1993) Clues to the pathogenesis of familial colorectal cancer. *Science* 260: 812–816.

Abel KJ, Boehnke M, Prahalad M, *et al.* (1993) A radiation hybrid map of the BRCA1 region of chromosome 17q12–q21. *Genomics* 17: 632–641.

Albertsen H, Plaetke R, Ballard L, *et al.* (1994a) Genetic mapping of the BRCA1 region on chromosome 17q21. *Am. J. Hum. Genet.* 54: 516–525.

Albertsen H, Smith SA, Mazoyer S, *et al.* (1994b) A physical map and candidate genes in the BRCA1 region on chromosome 17q12–21. *Nature Genet.* 7: 472–479.

Anderson DE, Badzioch MD. (1985) Bilaterality in familial breast cancer patients. *Cancer* 56: 2092–2098.

Bernstein JL, Thompson WD, Risch N, Holford TR. (1992) The genetic epidemiology of second primary breast cancer. *Am. J. Epidemiol.* 136: 937–948.

Bishop DT, Eby N, Chang-Claude J. (1994) Recent progress in breast cancer genetic epidemiology. In: *Endocrine Therapy of Breast Cancer*, Vol. VI (ed. A. Howell). Springer-Verlag, Berlin, pp. 3–15.

Black DM, Nicolai H, Borrow J, Solomon E. (1993) A somatic cell hybrid map of the long arm of human chromosome 17, containing the familial breast cancer locus (BRCA1). *Am. J. Hum. Genet.* 52: 702–710.

Borg A, Zhang QX, Johannsson O, Olsson H. (1994) High-frequency of allelic imbalance at the BRCA1 region on chromosome 17q in both familial and sporadic ductal breast carcinomas. *J. Natl Cancer Inst.* 86: 792–794.

Boring CC, Squires TS, Tong T. (1992) Cancer statistics. *CA Cancer J. Clin.* 42: 19–38.

Borresen AL, Anderson TI, Garber J, *et al.* (1992) Screening for germline TP53 mutations in breast cancer patients. *Cancer Res.* 52: 3234–3236.

Bowcock AM, Anderson LA, Friedman LS, *et al.* (1993) THRA1 and D17S183 flank an interval of <4 cM for the breast–ovarian cancer gene (BRCA1) on chromosome 17q21. *Am. J. Hum. Genet.* 52: 718–722.

Bronner CE, Baker SM, Morrison PT, *et al.* (1994) Mutation in the DNA mismatch repair gene homologue hMLH1 is associated with hereditary non-polyposis. *Nature* 368: 258–261.

Brown MA, Solomon E. (1994) Towards cloning the familial breast-ovarian cancer gene on chromosome 17. *Curr. Opin. Genet. Devel.* 4: 439–445.

Campbell IG, Nicolai HM, Foulkes WD, Senger G, Stamp GW, Allan G, Boyer C, Jones K, Bast RC, Solomon E. (1994) A novel gene encoding a b-box protein within the BRCA1 region at 17q21.1. *Hum Mol. Genet.* 3: 589–594.

Cannon-Albright LA, Bishop DT, Goldgar DE, Skolnick MH. (1991) Genetic predisposition to cancer. In: *Important Advances in Oncology* (eds VT DeVita, S Hellman, SA Rosenberg). JB Lippincott, Philadelphia, pp. 39–55.

Claus EB. (1988) Age of onset and the inheritance of breast cancer. Ph.D. thesis, Yale University, New Haven, CT.

Claus EB, Risch NJ, Thompson WD. (1990) Age at onset as an indicator of familial risk of breast cancer. *Am. J. Epidemiol.* **131**: 961–972.

Claus EB, Risch NJ, Thompson WD. (1991) Genetic analysis of breast cancer in the cancer and steroid hormone study. *Am. J. Hum. Genet.* **48**: 232–242.

Cornelius RS, van Vliet M, Vos CBJ, et al. (1994) Evidence for a gene on 17p13.3, distal to TP53, as a target for allele loss in breast tumors without p53 mutations. *Cancer Res.* **54**: 4200–4206.

Devilee P, Cornelisse CJ. (1995) Somatic genetic changes in human breast cancer. *Biochim. Biophys. Acta* **1198**: 113–130.

Easton DF, Bishop DT, Ford D, Crockford GP, and the Breast Cancer Linkage Consortium. (1993a) Genetic linkage analysis in familial breast and ovarian cancer: results from 214 families. *Am. J. Hum. Genet.* **52**: 678–701.

Easton D, Ford D, Peto J. (1993b) Inherited susceptibility to breast cancer. *Cancer Surv.* **18**: 1–17.

Easton DF, Narod SA, Ford D, Steel M, on behalf of the Breast Cancer Linkage Consortium. (1994) The genetic epidemiology of BRCA1. *Lancet* **344**: 761.

Fishel R, Lescoe MK, Rao MRS, et al. (1993) The human mutator gene homolog *MSH2* and its association with hereditary nonpolyposis colon cancer. *Cell* **75**: 1027–1038.

Ford D, Easton DF, Bishop DT, Narod SA, Goldgar DE, Breast Cancer Linkage Consortium. (1994) Risks of cancer in BRCA1-mutation carriers. *Lancet* **343**: 692–695.

Fraumeni Jr JF, Grundy GW, Creagan ET, Everson RB. (1975) Six families prone to ovarian cancer. *Cancer* **36**: 364–369.

Futreal PA, Liu Q, Shattuck-Eidens D, et al. (1994) BRCA1 mutations in primary breast and ovarian carcinomas. *Science* **266**: 120–122.

Garber JE, Goldstein AM, Kantor AF, Dreyfus MG, Fraumeni JF, Li FP. (1991) Follow-up study of twenty-four families with the Li–Fraumeni syndrome. *Cancer Res.* **51**: 6094–6097.

Gardner EJ, Stephens FE. (1950) Breast cancer in one family group. *Am. J. Hum. Genet.* **2**: 30–40.

Godwin AK, Vanderveer L, Schultz DC, et al. (1994) A common region of deletion on chromosome 17q in both sporadic and familial epithelial ovarian tumors distal to BRCA1. *Am. J. Hum. Genet.* **55**: 666–677.

Goldgar DH, Fields P, Lewis CM, et al. (1994) A large kindred with 17q linked breast and ovarian cancer: genetic, phenotypic, and genealogical analysis. *J. Natl Cancer Inst.* **86**: 200–209.

Hall JM, Lee MK, Newman B, et al. (1990) Linkage of early-onset familial breast cancer to chromosome 17q21. *Science* **250**: 1684–1689.

Hollstein M, Sidransky D, Vogelstein B, Harris CC. (1991) p53 mutations in human cancers. *Science* **253**: 49–53.

Iselius L, Slack J, Littler M, Morton NE. (1991) Genetic epidemiology of breast cancer in Britain. *Ann. Hum. Genet.* **55**: 151–159.

Kelsell DP, Black DM, Bishop DT, Spurr NK. (1993) Genetic analysis of the BRCA1 region in a large breast/ovarian family: refinement of the minimal region containing

BRCA1. *Hum. Mol. Genet.* 2: 1823–1828.

Kelsey JL. (1979) A review of the epidemiology of human breast cancer. *Epidemiol. Rev.* 1: 74–109.

Krontiris TG, DiMartino NA, Colb M, Parkinson DR. (1985) Unique allelic restriction fragments of the human Ha-ras locus in leukocyte and tumour DNAs of cancer patients. *Nature A* 313: 369–372.

Krontiris TG, Devlin B, Karp DD, Robert NJ, Risch N. (1993) An association between the risk of cancer and mutations in the HRAS1 minisatellite locus. *N. Engl. J. Med.* 329: 517–523.

Leach FS, Nicolaides NC, Papadopoulos N, *et al.* (1993) Mutations of a *mutS* homolog in hereditary nonpolyposis colorectal cancer. *Cell.* 75: 1215–1225.

Li FP, Fraumeni Jr JF. (1969) Soft-tissue sarcomas, breast cancer, and other neoplasms. A familial syndrome? *Ann. Intern. Med.* 71: 747–752.

Lindblom A, Tannergård P, Werelius B, *et al.* (1993) Genetic mapping of a second locus predisposing to hereditary non-polyposis colon cancer. *Nature Genet.* 5: 279–282.

Lobaccaro J-M, Lumbroso S, Belon C, *et al.* (1993) Male breast cancer and the androgen receptor gene. *Nature Genet.* 5: 109–110.

Lynch HT, Harris RE, Guigis HA, Maloney KM, Carmody LL, Lynch JF. (1978) Familial association of breast/ovarian carcinoma. *Cancer* 41: 1543–1549.

Malkin D, Li FP, Strong LC, *et al.* (1990) Germ line p53 mutations in a familial syndrome of breast cancer, sarcomas, and other neoplasms. *Science* 250: 1233–1238.

Margaritte P, Bonaiti-Pellie C, King M-C, Clerget-Darpoux F. (1992) Linkage of a familial breast cancer to chromosome 17q21 may not be restricted to early-onset disease. *Am. J. Hum. Genet.* 50: 1231–1234.

Miki Y, Swensen J, Shattuck-Eidens D, *et al.* (1994) A strong candidate for the breast and ovarian cancer susceptibility gene BRCA1. *Science* 266: 66–71.

Narod SA, Feuteun J, Lynch HT, *et al.* (1991) Familial breast-ovarian cancer locus on chromosome 17q12-q23. *Lancet* 338: 82–83.

Narod SA, Ford D, Devilee P, *et al.* (1995) An evaluation of genetic heterogeneity in 145 breast–ovarian cancer families. *Am. J. Hum. Genet.* 56: 254–264.

Newman B, Austin MA, Lee M, King M-C. (1988) Inheritance of human breast cancer: evidence for autosomal dominant transmission in high-risk families. *Proc. Natl Acad. Sci. USA* 85: 3044–3048.

O'Connell P, Albertsen H, Matsunami N, *et al.* (1994) A radiation hybrid map of the BRCA1 region. *Am. J. Hum. Genet.* 54: 526–534.

Papadopoulos N, Nicolaides NC, Wei Y-F, *et al.* (1994) Mutation of a mutL homolog in hereditary colon cancer. *Science* 263: 1625–1629.

Peltomäki P, Aaltonen LA, Sistonen P, *et al.* (1993) Genetic mapping of a locus predisposing to human colorectal cancer. *Science* 260: 810–812.

Petrakis NL, Ernster VL, King M-C. (1982) Breast. In: *Cancer Epidemiology and Prevention* (eds D Schottenfeld, JF Fraumen Jr). WB Saunders, Philadelphia, pp. 855–870.

Pippard EC, Hall AJ, Barker DJP, Bridges BA. (1988) Cancer in homozygotes and heterozygotes of ataxia-telangiectasia and xeroderma pigmentosum in Britain. *Cancer Res.* 48: 2929–2932.

Saito H, Inazawa J, Saito S, *et al.* (1993) Detailed deletion mapping of chromosome 17q in ovarian and breast cancers: 2-cM region on 17q21.3 often and commonly deleted in tumors. *Cancer Res.* 53: 3382–3385.

Sattin RW, Rubin GL, Webster LA, *et al.* (1985) Family history and the risk of breast cancer. *J. Am. Med. Assoc.* **253**: 1908–1913.

Schildkraut JM, Risch N, Thompson WD. (1989) Evaluating genetic association among ovarian, breast, and endometrial cancer: evidence for a breast/ovarian cancer relationship. *Am. J. Hum. Genet.* **45**: 521–529.

Shattuck-Eidens D, McClure M, Simard J, *et al.* (1995) A collaborative survey of 80 mutations in the BRCA1 breast and ovarian cancer susceptibility gene: implications for presymptomatic testing and screening. *J. Am. Med. Assoc.* **273**: 535–541.

Sidransky D, Tokino T, Helzsouer K, *et al.* (1992) Inherited p53 gene mutations in breast cancer. *Cancer Res.* **52**: 2984–2986.

Simard J, Feunteun J, Lenoir G, *et al.* (1993) Genetic mapping of the breast-ovarian cancer syndrome to a small interval on chromosome 17q12–21: exclusion of candidate genes EDH17B2 and RARA. *Hum. Mol. Genet.* **2**: 1193–1199.

Smith SA, Easton DF, Evans DGR, Ponder BAJ. (1992) Allele losses in the region 17q12–21 in familial breast and ovarian cancer involve the wild-type chromosome. *Nature Genet.* **2**: 128–131.

Steichen-Gersdorf E, Gallion HH, Ford D, *et al.* (1994) Familial site-specific ovarian cancer is linked to BRCA1 on 17q12–21. *Am. J. Hum. Genet.* **55**: 870–875.

St John DJB, McDermott FT, Hopper JL, Debney EA, Johnson WR, Hughes ESR. (1993) Cancer risk in relatives of patients with common colorectal cancer. *Ann. Intern. Med.* **118**: 785–790.

Strand M, Prolla TA, Liskay RM, *et al.* (1993) Destabilization of tracts of simple repetitive DNA in yeast by mutations affecting DNA mismatch repair. *Nature* **365**: 274–276.

Stratton MR, Ford D, Neuhausen S, *et al.* (1994) Familial male breast cancer is not linked to the BRCA1 locus on chromosome 17q. *Nature Genet.* **7**: 103–107.

Swift M, Morrell D, Massey R, Chase CL. (1991) Incidence of cancer in 161 families affected by ataxia-telangiectasia. *N. Engl. J. Med.* **325**: 1831–1836.

Tonin P, Ehrenborg E, Lenior G, *et al.* (1993) The human insulin-like growth factor-binding protein 4 gene maps to chromosome region 17q12–q21.1 and is close to the gene for hereditary breast-ovarian cancer. *Genomics* **18**: 414–417.

Watson P, Lynch HT. (1993) Extracolonic cancer in hereditary nonpolyposis colorectal cancer. *Cancer* **71**: 677–685.

Wooster R, Mangion J, Eeles R, *et al.* (1992) A germline mutation in the androgen receptor gene in two brothers with breast cancer and Reifenstein syndrome. *Nature Genet.* **2**: 132–134.

Wooster R, Neuhausen SL, Mangion J *et al.* (1994) Localization of a breast cancer susceptibility gene, BRCA1, to chromosone 13q 12-13. *Science* **265**: 2088–2090.

Xu W, Gorman PA, Rider SH, *et al.* (1988) Construction of a genetic map of human chromosome 17 by use of chromosome-mediated gene transfer. *Proc. Natl Acad. Sci. USA* **85**: 8563–8567.

Molecular genetics of colon cancer

M.G. Dunlop

Understanding of the genetic basis of colorectal cancer has advanced at a truly remarkable pace in recent years. Although a number of important discoveries in the field were made prior to 1987, when the gene for the autosomal dominant syndrome of familial adenomatous polyposis (FAP) was regionally localized, it was the discovery of this first colorectal cancer gene which heralded a new and exciting period of research endeavour. Since then, identification of a number of genes involved in both the development of, and susceptibility to, colorectal cancer has resulted in an unprecedented expansion in knowledge of the fundamental genetic aspects of colorectal cancer biology. A number of key discoveries have been made which have already had an impact on clinical management of patients at risk of colorectal cancer. Greater understanding of the fundamental processes involved in colorectal carcinogenesis will undoubtedly lead to the development of novel treatment modalities for both FAP and for patients affected by colorectal cancer. This review does not set out to discuss all aspects of the molecular genetics of colorectal cancer in depth but concentrates on the role of the adenomatous polyposis coli (*APC*) gene and DNA mismatch repair genes in colorectal carcinogenesis. The identification of these genes and their functional significance in the neoplastic process is discussed and the relevance of such discoveries to future research and to clinical management is explored.

6.1 Introduction

Although breast cancer is more common in females, colorectal cancer is by far the most common cause of early death due to malignancy in the non-smoking male and female population in Westernized countries, and consumes considerable health care resources. There has been scarcely any change in the death rate from the disease in the last 50 years despite improvements in surgical technique and anaesthesia. Although these grim statistics mask an underlying improvement in survival from the disease, there has been a concomitant increase in the incidence of colorectal cancer. There can be little doubt that current treatment modalities of surgery, with or without adjuvant chemotherapy or radiotherapy, are essentially damage-limitation exercises since patients usually present with late stage disease. The relative 5-year survival in population studies is only around 37%. There is a clear need to improve understanding of the fundamental mechanisms of colorectal carcinogenesis and to develop new genetic and bio-chemical ways to interrupt the malignant process. A further, attractive proposition is to identify those individuals who are at high risk of developing the disease to allow prevention of the onset of frank malignancy. Such prophylaxis may involve chemoprevention with rational use of drugs acting centrally in the neoplastic process or by removal of premalignant polyps during endoscopic examination of the colon and rectum. Identification of the genes responsible for the syndromes of FAP and a total of four genes responsible for hereditary non-polyposis colorectal cancer (HNPCC) has already had a direct impact on the prevention of colorectal cancer. Patients who carry mutations in the respective genes can be identified and offered appropriate screening and/or prophylactic surgery.

There are a number of conditions which result in predisposition to colorectal cancer. This review discusses the two genetic syndromes which predominate, namely FAP and HNPCC. Although patients with FAP now make up only a small proportion of all cases of colorectal cancer due to effective prophylaxis, HNPCC constitutes a major input into the overall incidence of the disease. Only around 10% of all cases of colorectal cancer achieve the strictest criteria for defi-nition of the syndrome of HNPCC (Vasen et al., 1991) but such inclusion criteria are undoubtedly too restrictive, and there is evidence to suggest that a substan-tial proportion of all cases of large bowel benign and malignant tumours arise as a result of mutations in heritable susceptibility genes (Cannon-Albright et al., 1988). Indeed, it has been calculated that one in every 200 of the entire popula-tion may carry an HNPCC allele (Lynch et al., 1993). Identifying the gene(s) responsible would be a major step forward in identifying those individuals at high risk of developing colorectal cancer. If the population frequency of HNPCC is sufficiently great, the question of screening the entire population for HNPCC mutations may also have to be considered.

6.2 The role of the *APC* gene in colorectal cancer

6.2.1 Familial adenomatous polyposis

FAP is an autosomal dominant heritable disorder with a population frequency of around 1/7000 and penetrance of almost 100%. The syndrome is characterized by the development of hundreds or thousands of neoplastic polyps in the colon and rectum during the second decade of life with more than 100 adenomatous polyps being diagnostic of the disease. Progression of one or more of the colorectal polyps to carcinoma is almost inevitable unless prophylactic surgical excision of the large bowel is performed. Presymptomatic colonic screening of at-risk relatives and appropriate prophylactic surgery considerably reduce morbidity and mortality from FAP (Jarvinen, 1992).

Extracolonic features of FAP include epidermoid cysts of the skin, retinal pigmentation, benign craniofacial and long bone tumours, polyposis of the upper gastrointestinal (GI) tract and fibrous tumours (desmoids) of the retroperitoneum and abdominal wall. Upper GI malignancy, especially of the region around the entry of the bile duct to the duodenum (the ampulla) is a particular problem as it is the major cause of premature death in FAP patients who have undergone prophylactic removal of the colon.

6.2.2 The APC gene and FAP

The gene for FAP (*APC*) was regionally localized by genetic linkage analysis to the long arm of chromosome 5 following a clue from cytogenetic studies (Bodmer *et al.*, 1987; Leppert *et al.*, 1987). A positional cloning strategy was then adopted by a number of groups and eventually the gene was cloned, sequenced and mutations identified in FAP kindreds (Groden *et al.*, 1991; Joslyn *et al.*, 1991; Kinzler *et al.*, 1991; Nishishio *et al.*, 1991). Germline mutations have since been reported in around 250 FAP families (Nagase *et al.*, 1992; Nagase and Nakamura, 1993; Mandl *et al.*, 1994) (see *Figure 6.1*). All confirmed *APC* mutations produce a premature stop codon by base substitution or by deletion/insertion resulting in frameshift of the coding sequence 3' of the mutation. Although truncation of the APC protein product is the universal mechanism of inactivation, there are only two mutations which occur in more than 5% of cases. Somatic mutational inactivation of *APC* has also been demonstrated in almost all sporadic (non-FAP) carcinomas and in even the very smallest premalignant adenomas (Powell *et al.*, 1992; Ichii *et al.*, 1993). The combination of the causative role of *APC* mutations in the FAP syndrome and the detection of *APC* mutations at the very earliest stages of colorectal neoplasia indicates the crucial role of *APC* mutations in the genesis of colorectal cancer.

6.2.3 APC function and genotype–phenotype correlation

APC is a fairly large gene consisting of 15 exons encoding a 2843-residue protein product. Exons 1–14 are small while exon 15 is large, accounting for 77% of the coding region (see *Figure 6.1*). There are alternatively spliced forms of *APC* and some reports have suggested that there are more than 15 exons. Short repeat sequences, predicted to form coiled-coil structures, indicate that normal APC product may be functional in dimeric configuration. APC function remains unknown but recently several avenues of research have shed some light on the interactions and cellular localization of the protein. APC has been shown to associate with α- and β-catenins. These proteins bind to the cell surface molecule,

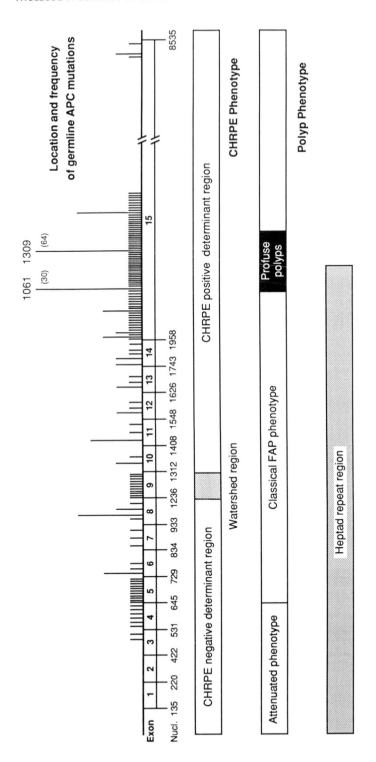

Figure 6.1. Location and frequency of germline *APC* mutations in unrelated families collated from the literature (Nagase and Nakamura, 1993; Olschwang *et al.*, 1993; Spirio *et al.*, 1993; Mandl *et al.*, 1994). Functional domains illustrated in terms of phenotypic markers of FAP in mutation carriers. One vertical line indicates one reported mutation site, and the number of mutations at that site are indicated proportionately by the height of the vertical line (one mutation being the smallest number reported at any one locus). The number of reported mutations at the two common mutation sites (codon 1061 and 1309) are given by number. CHRPE, congenital hypertrophy of the retinal pigment epithelium.

116

E-cadherin, and are essential for its role in cellular adhesion. It is proposed that the APC protein may affect the interaction between catenins and E-cadherin thus influencing cellular adhesion and possibly intercellular communication (Rubinfeld et al., 1993; Su et al., 1993a). Antibodies to the APC protein have identified a 300-kDa cytoplasmic protein expressed in epithelial cells in the upper portions of the colonic crypts, suggesting it is functional in the mature colonocyte (Smith et al., 1993). Further subcellular localization has demonstrated that wild-type APC is closely associated with microtubule formation (Munemitsu et al., 1994; Smith et al., 1994). However, mutant APC product (missing the carboxy terminal part of the molecule due to premature truncation) does not bind to microtubules. Indeed, in vitro studies indicate that wild-type APC not only binds to microtubules but promotes their formation (Munemitsu et al., 1994). The organization and structure of microtubules are vital to cell division and so it is possible that the association of APC with the catenins, and hence cadherin, may afford a direct intracellular line of communication from the cell surface to the site of microtubule formation. It will be of great interest to determine the underlying function of APC since it may be possible to restore or augment the effects of deleterious mutations by pharmacological means.

Much can be learned about the in vivo functional significance of portions of the APC by observation of the phenotypic effect of mutations at various locations within the gene. Mutant APC protein has been shown to interact with normal APC, forming a heterodimer which may result in defective function in a dominant-negative manner (Su et al., 1993b). The extent to which normal APC function is impaired may be related to the structure and residual ability of the mutant protein to form dimers. Truncated APC protein due to mutations occurring in the most 5' part of the gene may not be capable of dimerization with wild-type APC whereas more 3' mutations may allow translation of repeat sequences which can allow dimer formation. Thus, it might be predicted that a more aggressive phenotype may result from the more 3' mutations. Genotype–phenotype correlations do indeed bear this out, suggesting a functional boundary between codons 157 and 168. FAP families with an attenuated, usually late onset, disease have truncating mutations 5' of this boundary (Spirio et al., 1993). Similarly, there is evidence that mutations in the region between codons 1250 and 1464 result in a more extreme phenotype with profuse polyps in the colon and rectum (Nagase et al., 1992). Such findings support the notion of a dominant-negative effect of truncating APC mutations. However, in contradistinction to this, when one copy of the APC gene is lost in a constitutional chromosomal deletion, the classical FAP phenotype with profuse, early onset polyposis is the norm, suggesting that the dominant-negative effect is not the only phenotypic determinant.

One interesting aspect of the phenotypic effects of mutations at different locations in APC is the effect on the retinal pigmentation seen in FAP, congenital hypertrophy of the retinal pigment epithelium (CHRPE). The critical boundary lies in exon 9 with a striking polarization of mutations 3′ to this region being associated with CHRPE lesions and mutations 5′ being associated with a normal retina (Olschwang *et al.*, 1993) (see *Figure 6.1*). As the number of reported mutations increases, and the quality of clinical data improves, complete elucidation of the clinical effects of different *APC* mutations will be of great interest and may well yield clues to the functional significance of various domains of the APC protein.

There is little doubt that phenotypic expression is modified by other factors and identical *APC* mutations have been shown to be associated with diverse FAP phenotypes (Groden *et al.*, 1993; Paul *et al.*, 1993). Environmental interaction has an undoubted role in modulating expression of the FAP phenotype. Evidence for such environmental modulation comes from studies which show reduction in the number of colorectal polyps by treatment with the non-steroidal anti-inflammatory agent, sulindac (Giardiello *et al.*, 1993) and regression of rectal polyps after colectomy and ileorectal anastomosis (Nicholls *et al.*, 1988). However, there is also evidence of strong genetic modifiers. Phenotypic expression in the *Min* (multiple intestinal neoplasia) mouse model of FAP (Moser *et al.*, 1990) has been shown to be modified by an unlinked genetic locus in terms of numbers of tumours arising in several different genetic crosses in mice (Dietrich *et al.*, 1993). The modifier locus has been mapped to mouse chromosome 4, corresponding to the human chromosomal region 1p35–36. Interestingly, this region shows frequent allele loss in sporadic colorectal tumours. Cloning and sequencing of the murine and homologous human modifier genes will be of great importance. Pharmacological modulation of the FAP phenotype and indeed of the processes involved in colorectal carcinogenesis may be possible via manipulation of such a modifier gene.

6.2.4 APC *mutational spectrum in FAP*

In FAP, and in sporadic colorectal neoplasms, 97% of mutations occur in the 5′ half of the gene, 60% in a region of 600 codons within exon 15 (see *Figure 6.1*). Of the mutations responsible for FAP, 20% are short deletions at codons 1061 and 1309 (Nagase and Nakamura, 1993) and the mutation at codon 1309 is the most frequently detected in new (or sporadic) cases of FAP. Mutation analysis of over 800 FAP kindreds has been reported in the world literature and the overall frequency of identifying the causative mutation is only 30% (Nagase and Nakamura, 1993; Mandl *et al.*, 1994), ranging from 21% (Mandl *et al.*, 1994) to 67% (Nagase *et al.*, 1992). This reflects a number of factors including the selection criteria for the diagnosis of FAP, the sensitivity and robustness of the mutation detection technique employed, the assiduousness of the search for mutations, the presence of causative gene alterations influencing *APC* expression outwith the coding sequence in promoter or intronic sequences, and the possibility of genetic heterogeneity in FAP. Hence, it is unwise to be overly didactic about the type of mutation responsible for FAP and indeed for sporadic colorectal neoplasia since bias in the reported literature is likely to be due to any one or a combination of the above confounding factors.

Mutations so far detected are mainly deletions or insertions of short sequences, suggestive of errors of replication rather than the action of mutagens (Nagase and Nakamura, 1993). Single nucleotide substitutions (most commonly C→T transitions) account for around 40% of germline mutations already identified. Although two deletion mutations at codons 1061 and 1309 have been detected in a total of around 10% of cases, the majority show no evidence of clustering. Hence, identification of the *APC* mutation responsible for the FAP syndrome in any particular family is time consuming and laborious. This has led to the development of functional assays for *APC* mutation detection which capitalize on the core defect of causative *APC* mutations, namely premature truncation of the protein product. One such technique involves *in vitro* synthesis and radiolabelling of surrogate APC protein from polymerase chain reaction (PCR)-amplified patient DNA (or cDNA from the first 14 exons) (Powell *et al.*, 1993). The presence of an *APC* mutation is indicated by truncated protein when electrophoresed in standard SDS–PAGE. While the technique does not delineate the underlying defect at the DNA level, it is a rapid and efficient means of obtaining *de facto* evidence of the presence of a truncating and therefore causative mutation. The mutation can be fairly accurately localized using well-characterized and localized mutations as controls, and then sequencing of the respective segment if so desired. Another functional assay for *APC* mutations, which allows screening of large stretches of the gene in one analysis, involves cloning of a PCR-amplified *APC* fragment from patient material in-frame with the *LacZ* gene of a suitable vector to allow colour selection of bacterial colonies after transformation (Varesco *et al.*, 1993). Bacteria expressing a wild-type APC–LacZ fusion protein produce deep blue colonies plated on X-gal agar while colonies with mutant APC do not produce LacZ efficiently due to the presence of a premature stop codon, and so produce light blue or white colonies. These assays are elegant examples of the manner in which knowledge of the mutational spectrum in a specific disease can be used to practical effect. These strategies will be applicable to other genes where mutations alter the protein structure through premature truncation.

6.2.5 Predictive genetic testing for FAP

Predictive genetic testing in FAP, using the relatively crude tool of genetic linkage analysis, has been shown to be of considerable practical value by its identification of a number of affected individuals who had previously been inappropriately discharged from clinical follow-up (Dunlop *et al.*, 1991). Highly polymorphic microsatellite markers increase the predictive power of linkage analysis substantially but the availability of DNA from (preferably) two affected individuals in the family severely restricts the applicability of the test. Problems such as early death of gene carriers, uncertain paternity within families and the fact that around 25% of FAP patients carry a new mutation with no family history substantially reduce the value of linkage studies in many clinical situations. In addition, linkage analysis-derived testing provides an estimation of the risk of being a gene carrier rather than the definitive diagnosis afforded by mutation analysis. Hence, detection of the underlying *APC* mutation must be considered the optimal predictive testing in FAP.

As discussed above, there is no marked clustering of mutations in the *APC* gene and so the causative lesion is initially localized within the gene by screening techniques which can detect single base pair changes in a sequence of a few hundred nucleotides or by the protein truncation test described above. Mutation screening techniques, including single-stranded conformation polymorphism analysis, gel electrophoresis in a gradient of temperature or denaturant, heteroduplex analysis and chemical cleavage of mismatch analysis, allow the identification of the region containing the mutation (Grompe, 1993). DNA sequencing is then required to characterize the mutation and determine the exact location of the defect. Once the *APC* mutation has been detected in an affected family member, user-friendly techniques can be designed to screen other at-risk relatives, such as mutation-specific PCR or exploitation of alteration in restriction endonuclease sites induced by the mutation. Relatives who do not carry the mutant gene need not be subjected to any further clinical screening. Already, in some centres, documented *APC* mutation carriers are now the only family members under surveillance; this is to allow assessment of the optimum timing for prophylactic surgery. Mutation analysis is already the most frequently used molecular technique for presymptomatic testing in FAP. Linkage analysis is reserved for confirmation of carrier status predictions for those in whom no *APC* mutation can be detected. Although laborious, identification of every mutation responsible for FAP is desirable in elucidating possible genotype–phenotype correlations and providing ultimate diagnostic reliability.

6.2.6 The role of the APC *gene in sporadic colorectal tumorigenesis*

FAP has long been considered a model of sporadic colorectal carcinogenesis, and this notion has been borne out by demonstration of somatic *APC* mutations in sporadic colorectal adenomas and carcinomas. Indeed, it appears that such mutations are ubiquitous in colorectal tumours and may well be an absolute requirement for the initiation of the neoplastic process.

Loss of heterozygosity at markers around the *APC* gene occurs in around 50% of tumours (Vogelstein *et al.*, 1988; Ashton-Rickardt *et al.*, 1989) while truncating *APC* mutations have been identified in 60% of sporadic colorectal cancers and adenomas examined (Powell *et al.*, 1992; Ichii *et al.*, 1993; Nagase and Nakamura, 1993). Both copies of *APC* are inactivated either by mutation on each allele, or by mutation on one allele and structural deletion resulting in hemizygosity or mitotic recombination and endo-reduplication of the mutant *APC*. In patients with FAP, one *APC* allele is constitutionally inactivated. The adenomas in the colon and rectum acquire somatic mutations in the remaining wild-type (normal) copy of the *APC* gene with increasing frequency as adenomas increase in size and dysplasia. Approximately half of these somatic events are extensive genetic rearrangements or deletions manifest as allele loss or cytogenetic abnormalities (Miyaki *et al.*, 1990). The remainder are mutations with a similar profile to those occurring in the germline, and all result in truncation of the protein. In addition to the virtually ubiquitous presence of *APC* mutations in colorectal tumours, inactivation of *APC* appears to be one of the very earliest events in sporadic colorectal tumorigenesis. Mutations have been detected in adenomas as small as 3 mm in diameter (Powell *et al.*, 1992; Ichii *et al.*, 1993). The frequency of detection of mutations in adenomas is the same as the frequency of mutation detection in late stage carcinomas, indicating that *APC* mutations are likely to be involved in the initiation of the neoplastic process.

The mutational spectrum of somatic *APC* mutations in sporadic colorectal adenomas and carcinomas is remarkably similar to the germline changes in patients with FAP. Approximately 60% of the mutations are short deletions or insertions resulting in frameshift and a premature stop codon 3′ of the mutation. The remainder are point mutations, frequently C→T transitions and especially at CpG dinucleotides (Miyoshi *et al.*, 1992). The mutational mechanism is of considerable interest since generation of somatic *APC* mutations in the colon could occur due to mutagens in the faecal stream or could be a result of defects in mismatch repair (see Section 6.3). There is no convincing evidence for the presence of any mutagens in the stool and so it is tempting to suggest that impairment of DNA repair mechanisms, such as occurs in *hMSH2* mutations, may be the underlying cause in at least a proportion of cases. The net result of *APC* mutation is selective growth advantage of epithelial cells which carry such a defect but, until the details of APC function have been elucidated, the exact mechanism of such clonal expansion is unlikely to be fully understood.

6.3 The role of DNA mismatch repair genes in colorectal carcinogenesis

6.3.1 Phenotype and penetrance of HNPCC

HNPCC is an autosomal dominant disorder characterized by early onset of colorectal cancer and familial aggregation of cases. The myriads of adenomas (benign tumours) seen in FAP are not a feature of HNPCC. There is a propensity for tumours to develop in the proximal part of the colon. In some families, inheritance of the HNPCC gene results only in cancer of the colon or rectum while, in other families, in addition to large bowel malignancy there is also segregation of uterine, ovarian, gastric, upper urinary tract, pancreatic, small bowel and skin cancers. *Figure 6.2* shows an HNPCC kindred which highlights some areas of

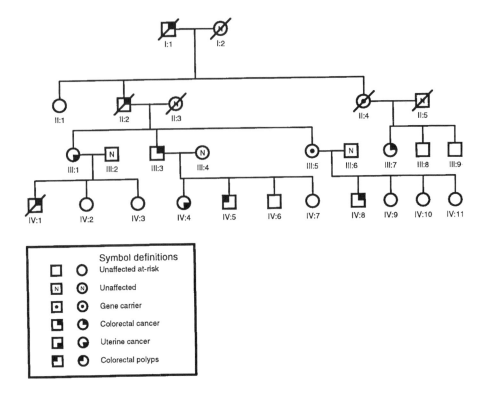

Figure 6.2. A four generation family with HNPCC with several cases of colorectal cancer, two cases of uterine cancer and evidence for non-penetrance in two females. Symbol definitions are given in the key. Although adenomatous polyps are not a predominant feature of HNPCC, one individual does have a small number of colorectal polyps.

interest. All cases of colorectal cancer were of early onset, before the age of 55 years. There are at least two unaffected presumptive obligate gene carriers (II:4 and III:5), one of whom was elderly when she died of an unrelated condition and another who is as yet unaffected. This pedigree provides strong evidence that HNPCC gene mutations are not so highly penetrant as is widely believed. In addition, female gene carriers seem more protected from colorectal cancer than male gene carriers, suggesting the intriguing possibility of a sex-linked modifier gene. There is also evidence of co-segregation of early onset colorectal polyps and uterine and colorectal cancers.

HNPCC accounts for around 5–15% of cases of colorectal cancer (Dunlop *et al.*, 1992) but this proportion is determined largely by empirically derived criteria to allow a diagnosis of the syndrome. Thus, when diagnostic criteria are loosened, more cases of colorectal cancer will be attributed to HNPCC and vice versa. Not only does the issue of diagnostic criteria have a major effect on the apparent prevalence of HNPCC, it also considerably influences the apparent penetrance of the gene defect(s). Since there are no biomarkers for HNPCC, currently accepted criteria for diagnosis effectively predetermine the penetrance of HNPCC genes. Thus, HNPCC is said to be a disorder with high penetrance but it is the diagnostic criteria which demand a highly penetrant disease! It can be seen that diagnostic criteria create considerable circularity in the assignment of penetrance and indeed the prevalence of HNPCC. The criteria proposed by the International Collaborative Group on HNPCC are: (i) three or more relatives with histologically proven colorectal cancer, one being a first-degree relative of the other two; (ii) two or more generations affected; and (iii) at least one family member affected before the age of 50 years (Vasen *et al.*, 1991). Such criteria will only identify families where the gene defect is highly penetrant and which are of sufficient size to allow the appropriate number of cases to arise, such that many small families will be excluded inappropriately. With the recent identification of several causative genes, it will be of great interest to calculate the true penetrance of HNPCC. Our own unpublished data from probands who were selected only on the basis of being affected by colorectal cancer and not because of a known or suspected family history strongly support the notion that HNPCC is a relatively low penetrance condition, and that the gene prevalence is much higher than previously suspected.

6.3.2 Genes responsible for HNPCC

Several areas of apparently unrelated research have recently come together to identify a number of genes responsible for HNPCC. These studies have also gone a considerable way to showing a common functional deficit at the molecular level caused by mutation of such genes in humans, bacteria and yeast. A total of four human genes have been identified which, when mutated, result in constitutional predisposition to colorectal cancer transmissible through the germline. These genes are human homologues of yeast and bacterial DNA repair genes and are known as *hMSH2* on chromosome 2p (Fishel *et al.*, 1993; Leach *et al.*, 1993), *hMLH1* on chromosome 3p (Bronner *et al.*, 1994; Papadopoulos *et al.*, 1994), *hPMS1* on chromosome 2q (Nicolaides, 1994) and *hPMS2* on chromosome 7q (Nicolaides *et al.*, 1994). Identification of the genes has been as a result of a number of different approaches to gene mapping.

hMSH2 was originally localized by linkage studies in two large, unrelated HNPCC pedigrees (Peltomaki *et al.*, 1993a). A systematic linkage mapping strategy was embarked upon and a total of 345 restriction fragment length polymorphisms and dinucleotide repeat markers spread evenly throughout the human genome were analysed for evidence of linkage to the disease phenotype. An anonymous marker, *D2S123*, mapping to the short arm of chromosome 2, was found to co-segregate with the disease in both families. Fourteen smaller families were also analysed and one third showed no evidence of linkage to

D2S123. In addition, a second locus was identified shortly afterwards by a report of linkage to a marker on chromosome 3p (Lindblom *et al.*, 1993), clearly establishing genetic heterogeneity in HNPCC. Apparently diverse areas of research then came together to give the answer to the underlying defect in HNPCC and gave clues to the genes responsible.

Studies on repeat sequences in genomic DNA from colorectal tumours identified widespread instability in short repetitive tracts which was very suggestive of a defect of DNA mismatch repair (Aaltonen *et al.*, 1993; Ionov *et al.*, 1993; Thibodeau *et al.*, 1993). The most well-defined mismatch repair system is in *Escherichia coli* (reviewed in Grilley *et al.*, 1990) and a number of gene products are required, namely MutL, MutH, MutS, MutU. Yeast also has a similar mismatch repair pathway which requires a homologue of *MSH* and two *MutL* homologues, *MLH1* and *PMS1* (Prolla *et al.*, 1994a,b). With the finding that there was a dramatic increase in repetitive tract instability in yeast when mutations were induced in *MLH* or *PMS* (Strand *et al.*, 1993), such a group of genes became a strong candidate for the human condition of HNPCC with the characteristic change in repeat DNA.

A combination of positional cloning and candidate gene approaches to gene isolation was employed by two groups to identify the human homologues of the yeast and bacterial DNA mismatch repair genes. The first to be isolated was *hMSH2* on chromosome 2p (Fishel *et al.*, 1993; Leach *et al.*, 1993). Using degenerate PCR primers for the yeast *MSH* genes, Fishel *et al.* (1993) identified the human homologue *hMSH2* and localized it to the same region on chromosome 2p as the markers linked in HNPCC families. However, the mutation data were not strong and it now seems certain that a polymorphism was identified and not a causative mutation. Vogelstein's group (Leach *et al.*, 1993) employed a combination of positional cloning and candidate gene strategies, generating multiple markers within a 25-cM region defined by linkage studies. When analysed in HNPCC families which were linked to the 2p gene, recombination events were identified which designated a 0.8-Mb region containing the locus of interest. Candidate genes mapping to this 0.8-Mb region were screened for the presence of germline mutations in HNPCC kindreds and eventually causative mutations demonstrated in *hMSH2*. A 2802-bp cDNA from *hMSH2* was found to contain a highly conserved region between codons 615 and 788 with considerable cross-species homology between human, yeast and bacteria. Several mutations were identified within the highly conserved region, including stop codons resulting in premature truncation of the protein product and a splice site alteration. Mutations were also shown to co-segregate with the disease in HNPCC families.

Once it became clear that human homologues of DNA mismatch repair genes were involved, the human homologues of other genes involved in the mismatch repair pathway were obvious candidates and these were quickly identified and mutations demonstrated to be responsible for the HNPCC syndrome (Bronner *et al.*, 1994; Papadopoulos *et al.*, 1994). *hMLH1* was shown to be the gene segregating with chromosome 3p markers (Lindblom *et al.*, 1993) and *hPMS1* and *hPMS2* lie on chromosomes 2q and 7q, respectively (Nicolaides *et al.*, 1994). One interesting aspect of the identification of these genes has been the use of commercial genome databases which are built up on random cDNA sequencing (Nicolaides *et al.*, 1994; Papadopoulos *et al.*, 1994). Searching for sequence homologies to expressed genes can identify cross-species homology and also other related genes. It seems likely that such an approach may become more widespread since this first demonstration of the power of the approach.

At this early stage, the mutational spectrum in *hMLH*, *hMSH* and *hPMS* genes in HNPCC is very much affected by the selection of cases. However, reported mutations include base substitutions, short deletions and deletions of a few hundred base pairs. The tendency for generation of stop codons to be the underlying mechanism of inactivation has been exploited by Vogelstein's group to screen for mutations using the protein truncation test (Nicolaides *et al.*, 1994; Papadopoulos *et al.*, 1994), used to such effect in *APC* mutation screening.

The fact that there are at least four human genes involved in DNA mismatch repair, and that inactivation of either *MLH1* or *PMS1* in yeast results in a mutator phenotype (Strand *et al.*, 1993), has important implications for human population genetics. As discussed in Section 6.3.1, the true penetrance of HNPCC gene mutations is unknown but our data suggest that it is relatively low and that HNPCC gene frequency is substantially higher than previously suspected. Indeed, Lynch *et al.* (1993) has calculated that the population gene frequency may be around 0.005. Hence, it is perfectly reasonable to expect progeny from (say) an *hMLH1*– and *hMSH2*– pairing. This possibility has very substantial implications indeed for predictive testing in such offspring, particularly since very early onset, highly penetrant disease has been selected for by most centres interested in the field. Since only 25% of offspring will be normal at both loci, it would seem wise to avoid any question of predictive testing until the true prevalence and penetrance of HNPCC gene mutations is clearly defined, unless mutations in each of the other mismatch repair genes have been excluded by sequencing in an affected individual. While it is possible to be confident of the gene of major effect in large HNPCC families, because linkage can be established with a high degree of certainty, the smaller families are of greater concern. It is possible that a causative mutation in (say) *hMSH2* could be identified and non-carriers then reassured inappropriately while they *actually* carry a mutation in *hMLH*. Such a possibility demands extreme caution before widespread introduction of genetic screening for HNPCC. It would seem wise to counsel at-risk individuals who are shown to be non-gene carriers in smaller families, where the causative mutation has been identified in affected individuals, but to recommend that colonoscopic screening continues.

The possibility of an *hMLH1*– and *hMSH2*– pairing also has intriguing implications for the DNA repair pathway of the progeny of such a pairing. It is possible that the hMLH1–/hMSH2– genotype may be lethal, but all the evidence suggests that there is no effect of heterozygous inactivation of either gene and that homozygous inactivation is required for tumour formation (Parsons *et al.*, 1993). In addition, MLH1/PMS1 double mutants in yeast have very similar phenotypes to either an MLH1 or a PMS1 single mutant (Strand *et al.*, 1993; Prolla, 1994a). Clearly, such individuals carrying such a genotype would be at risk of HNPCC-related cancers. It is also exciting to speculate that such a genotype may have a dramatic effect on repeat sequences in the germline of such double mutant gene carriers. Expansion of repeated sequences is known to cause diseases such as Huntington's disease and Fragile X, and perhaps it is passage through the germline of such a gene carrier, or even of the germline of a single mutant, that causes the expansion of triplet repeats characteristic of such genetic disorders.

6.3.3 DNA instability in tumours from HNPCC gene carriers

As discussed in Section 6.3.2, the characteristic change in repetitive tract instability is detectable in tumours from HNPCC gene carriers (Aaltonen *et al.*, 1993, 1994; Peltomaki *et al.*, 1993b). However, such changes are also detectable in around one in six of apparently sporadic colorectal cancers (Aaltonen *et al.*, 1993; Ionov *et al.*, 1993; Lothe *et al.*, 1993; Thibodeau *et al.*, 1993). Such 'sporadic' tumours with microsatellite instability tend to be right sided, diploid, with an inverse relationship with *p53* mutation, and to be associated with a strong family history of colorectal cancer. Surprisingly, there is no age difference between patients with tumours which exhibit microsatellite instability and those with tumours which do not (Ionov *et al.*, 1993; Lothe *et al.*, 1993). The inverse relationship between microsatellite instability and aneuploidy and *p53* mutation is very suggestive of one of two mechanisms of genomic instability being required in colorectal cancer. One involves defects in mismatch repair in diploid cells while the other involves *p53* mutation and the development of aneuploidy (Carder *et al.*, 1993).

Although it is possible that some tumours with microsatellite instability arise because of double somatic mutations of DNA repair genes, it seems likely that a substantial proportion of patients with such tumours will carry constitutional HNPCC gene mutations. Identification of microsatellite instability in tumours on a prospective basis from the general population may be a practical means of identification of HNPCC gene carriers. Of tumours from patients with proven germline mismatch repair gene defects, 100% have microsatellite instability (Aaltonen *et al.*, 1994). Hence, by employing an initial screen of all colorectal cancers coming into pathology departments for microsatellite instability, screening of the entire population of colorectal cancer cases for mismatch repair gene defects may well become a practical possibility. *Figure 6.3* shows typical findings of microsatellite instability in a sporadic colorectal cancer. The autoradiograph shows the banding patterns at the *D2S119* locus from normal (N) and tumour

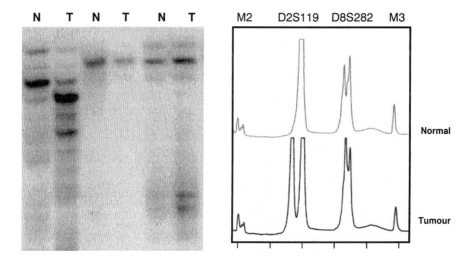

Figure 6.3. Two different techniques for detection of microsatellite instability. In each case, matched normal and tumour DNA from the same individual are compared after amplification by the PCR. The autoradiograph on the left shows one tumour in which the banding pattern shows a clear reduction in the electrophoretic mobility of the CA repeat in tumour (T) compared to normal (N) DNA. On the right is a representative analysis using fluorescence-labelled PCR primers of the tumour and normal DNA from one individual at two different CA repeat loci. There has been a change at one of the loci and not at the other. This emphasizes the fact that analysis of replication error (RER) status in tumour cannot rely on the results from only one CA repeat locus as some tumours only exhibit microsatellite instability at a proportion of loci.

(T) tissue in three separate cases, one which shows a clear alteration from the normal to tumour tissue. The data from the automated sequencer on the right show analysis at two loci from a single case. While *D2S119* does show an alteration, there is no change between normal and tumour at *D8S282*. This has important implications for screening tumours for microsatellite instability since it demonstrates the importance of analysis of multiple loci.

The detection of microsatellite instability in endometrial cancer (Risinger *et al.*, 1993) and in a number of cancers of different tissue origin in HNPCC gene carriers (Aaltonen *et al.*, 1994) is of great interest as it demonstrates that the underlying HNPCC gene mutation is indeed responsible for the mutator phenotype. In addition, it has been shown that microsatellite instability occurs in tumours from patients with multiple different primary cancers (Horii *et al.*, 1994), suggesting that such individuals may well carry DNA repair gene mutations even without a family history of cancer.

The global importance of instability in short tandem repeats in cancer in general is indicated by the finding of instability in tri- and tetranucleotide repeats in breast cancer. Interestingly, dinucleotide repeat instability was far less common

(Wooster *et al.*, 1994). These findings suggest the intriguing possibility that some aspect of DNA mismatch repair may be involved in the genesis and predisposition to breast and other cancers.

6.3.4 Mismatch repair defects in humans and the adenoma carcinoma sequence

The human mismatch repair system appears to be similar to that in yeast with *MSH*, *MLH* and *PMS* homologues. There does not appear to be a homologue of the *E. coli MutH* gene in either humans or in yeast. Mismatch repair involves recognition and binding of mismatches by *MSH2*. The correct strand for excision is recognized by the fact that nascent DNA is transiently under-methylated. *MLH* and *PMS* then appear to work in concert to excise a segment of DNA some 1–2 kb in length which is then repaired by a DNA polymerase. There may well be further mismatch repair systems. One such candidate is a nucleotide-specific mismatch repair system which recognizes G:T mismatches (Wiebauer and Jiricny, 1990). Such a system is of great interest since G:T mismatches will result in a C→T transition if allowed to progress through mitosis. As previously discussed, such transitions are frequently seen in *APC* in colorectal tumours in addition to short deletions or insertions at repetitive sequences. Hence, this opens the intriguing possibility that many of the somatic changes that are involved in the very genesis of colorectal cancer, and indeed of FAP, are caused by defects in mismatch repair.

The possibility that mismatch repair defects are causal in colorectal carcinogenesis is also supported by the finding that microsatellite instability is detectable in early adenomas in patients carrying HNPCC gene mutations (Aaltonen *et al.*, 1994). In addition, colorectal cancer cell lines which exhibit microsatellite instability (and hence are assumed to have a defect in mismatch repair) have been shown to have a continuing defect in mismatch repair with progressive alterations in repetitive tracts (Parsons *et al.*, 1993; Bhattacharyya *et al.*, 1994; Shibata *et al.*, 1994). The data also strongly support the notion that the change occurs early in the neoplastic process (Shibata *et al.*, 1994). Although loss of heterozygosity at the *hMSH2* or chromosome 3p locus is not the rule (Aaltonen *et al.*, 1993), second somatic inactivating mutations have been identified in tumours from patients with germline mutations in *hMLH1* and *hMSH2* (Leach *et al.*, 1993; Papadopoulos *et al.*, 1994).

The finding that a lymphoblastoid cell line from a patient with HNPCC was repair proficient (Parsons, 1993) is suggestive that homozygous inactivation of a DNA repair gene is required for tumour formation. However, it is clear that there is a phenotype for heterozygous carriers since abnormal colonic crypt cell production rate and DNA repair has been reported in the apparently normal colorectal mucosa of affected and at-risk HNPCC family members (Pero *et al.*, 1983; Lipkin *et al.*, 1984). Only one HNPCC lymphoblastoid cell line has been examined for DNA repair proficiency and so it remains an interesting possibility that HNPCC gene carriers do have a low level of instability in the normal colorectal mucosa and in other tissues, which is manifest in subtle ways such as alteration in cell turnover.

6.4 Perspective

This review has not touched on the role of a number of other important genes involved in colorectal cancer, including *p53*, *KRAS* and *DCC*, and the events described above must be taken in the context of alterations in these genes (Vogelstein *et al.*, 1988). However, the pivotal role of *APC* and of the mutator genes in colorectal cancer is clear. Indeed, it is possible that changes in *p53* and *KRAS* could be induced by defects in mismatch repair in a similar fashion to that discussed above for *APC*. The practical clinical importance of *APC*, *hMSH2*, *hMLH1* and *hPMS1/2* cannot be overemphasized. Two applications of current knowledge are already under investigation, namely screening for FAP and HNPCC, and gene therapy for FAP.

6.4.1 Screening for gene carriers in FAP and HNPCC

It is already possible to identify gene carriers in HNPCC and FAP and recommend appropriate screening but, as discussed in Section 6.3.2, it would be prudent not to deny any at-risk individual screening on the basis of a negative test for an HNPCC gene. However, the benefits of genetic testing in FAP are clear and such screening has established a central place in the management of at-risk individuals.

In numerical terms, HNPCC is of far greater importance than FAP and the possibility of population screening for the condition has been alluded to (see Section 6.1). However, the number of genes involved and the variety of mutations responsible mitigate against non-selective screening of the general population. The most practical approach would be to screen all colorectal tumours for microsatellite instability and then to target patients with tumours showing evidence of defects in mismatch repair for mutation screening of *hMSH2*, *hMLH1* and *hPMS1/2*. While this approach would be of no value to those affected patients from whom the tumours were obtained, those shown to have germline mutations could be counselled with a view to offering screening to their relatives. This strategy seems to offer the most practical approach to combatting what could be a major problem in proving that a mutation in one of the mismatch repair genes would actually result in cancer.

6.4.2 Gene therapy for FAP

FAP is a potentially attractive candidate for gene therapy where the delivery and expression of the normal *APC* gene may allow correction of the phenotype and reduction of cancer risk. However, prophylactic colectomy in FAP is associated with such low operative mortality and provides such a drastic reduction in the risk of cancer, that there would have to be considerable evidence of potential benefit before surgical prophylaxis could be abandoned. Nonetheless, gene therapy for FAP has many attractive therapeutic targets including local therapy to the retained rectum after colectomy and ileorectal anastomosis, the control of upper GI polyposis and malignancy (which is now the major cause of mortality in patients who have undergone prophylactic colectomy), and the control of desmoid disease which is resistant to all forms of treatment. Since gene therapy to the retained rectum to control residual polyps is likely to be the first target, it is important that surgeons are aware that avoidance of rectal excision for FAP patients is desirable unless rectal polyposis is uncontrollable.

6.5 Conclusion

This review has summarized some exciting aspects of current research in the genetics of colorectal cancer. Already the fruits of this work have entered into the clinical setting and are having an appreciable effect. It seems likely that this new understanding of the problem may at last lead to an improvement in survival from colorectal cancer and in particular to a reduction in the incidence of the disease.

References

Aaltonen LA, Peltomaki P, Leach FS, *et al.* (1993) Clues to the pathogenesis of familial colon cancer. *Science* **260**: 812–816.

Aaltonen LA, Peltomaki P, Mecklin J-P. (1994) Replication errors in benign and malignant tumours from hereditary non-polyposis colorectal cancer patients. *Cancer Res.* **54**: 1645–1648.

Ashton–Rickardt PG, Dunlop MG, Nakamura Y, *et al.* (1989) High frequency of *APC* loss in sporadic colorectal carcinoma due to breaks clustered in 5q21–22. *Oncogene* **4**: 1169–1174.

Bhattacharyya NP, Skandalis A, Groden J, Meuth M. (1994) Mutator phenotypes in human colorectal carcinoma cell lines. *Proc. Natl Acad. Sci. USA* **91**: 6319–6323.

Bodmer WF, Bailey CJ, Bodmer J, *et al.* (1987) Localisation of the gene for familial polyposis coli on chromosome 5. *Nature* **328**: 614–616.

Bronner EC, Baker SM, Morrison PT, *et al.* (1994) Mutation in the DNA mismatch repair gene homologue hMLH1 is associated with hereditary non-polyposis colon cancer. Nature 368: 258–261.

Cannon–Albright LA, Skolnick MH, Bishop DT, *et al.* (1988) Common inheritance of susceptibility to colonic adenomatous polyps and associated colorectal cancers. *N. Engl. J. Med.* **319**: 533–537.

Carder P, Wyllie AH, Purdie CA, *et al.* (1993) Stabilised p53 facilitates aneuploid clonal divergence in colorectal cancer. *Oncogene* **8**: 1397–1401.

Dietrich WF, Lander ES, Smith JS, *et al.* (1993) Genetic identification of Mom-1 a major modifier locus affecting Min-induced intestinal neoplasia in the mouse. *Cell*

75: 631–639.

Dunlop MG, Wyllie AH, Steel CM, Piris J, Evans HJ. (1992) Linked DNA markers for presymptomatic diagnosis of familial adenomatous polyposis. *Lancet* **337**: 313–316.

Fishel R, Lescoe MK, Rao MRS, *et al.* (1993) The human mutator gene homolog *MSH2* and its association with hereditary nonpolyposis colon cancer. *Cell* **75**: 1027–1038.

Giardiello FM, Hamilton SR, Krush AJ, *et al.* (1993) Treatment of colonic and rectal adenomas with sulindac in familial adenomatous polyposis. *N. Engl. J. Med.* **328**: 1313–1316.

Grilley M, Holmes J, Yashar B, Modrich P. (1990) Mechanisms of DNA-mismatch correction. *Mutat. Res.* **26**: 253–267.

Groden J, Thliveris A, Samowitz W, *et al.* (1991) Identification and characterization of the Familial Adenomatous Polyposis Coli gene. *Cell* **66**: 589–600.

Groden J, Gelbert L, Thliveris A, *et al.* (1993) Mutational analysis of patients with adenomatous polyposis: identical inactivating mutations in unrelated individuals. *Am. J. Hum. Genet.* **52**: 263–272.

Grompe M. (1993) The rapid detection of unknown mutations in nucleic acids. *Nature Genet.* **5**: 111–117.

Horii A, Han H-J, Shimada M, *et al.* (1994) Frequent replication errors at microsatellite loci in tumours of patients with multiple primary cancers. *Cancer Res.* **54**: 3373–3382.

Ichii S, Takeda S, Horii A, *et al.* (1993) Detailed analysis of genetic alterations in colorectal tumours from patients with and without familial adenomatous polyposis (FAP). *Oncogene* **8**: 2399–2405.

Ionov Y, Peinado MA, Malkhosyan S, Shibata D, Perucho M. (1993) Ubiquitous somatic mutations in simple repeated sequences reveal a new mechanism for colonic carcinogenesis. *Nature* **363**: 558–561.

Jarvinen HJ. (1992) Epidemiology of familial adenomatous polyposis in Finland: impact of family screening on colorectal cancer rate and survival. *Gut* **33**: 357–360.

Joslyn L, Carlson M, Thliveris A, *et al.* (1991) Identification of deletion mutations and three new genes at the Familial Polyposis locus. *Cell* **66**: 601–613.

Kinzler KW, Nilbert MC, Su L-K, *et al.* (1991) Identification of FAP locus genes from chromosome 5q21. *Science* **253**: 661–665.

Leach FS, Nicolaides NC, Papadopolous N, *et al.* (1993) Mutations of a MutS homolog in hereditary non-polyposis colorectal cancer. *Cell* **75**: 1215–1225.

Leppert M, Dobbs M, Scambler P, *et al.* (1987) The gene for familial polyposis maps to the long arm of chromosome 5. *Science* **238**: 1411–1413.

Lindblom A, Tannergard P, Werelius B, Nordenskjold M. (1993) Genetic mapping of a second locus predisposing to hereditary non-polyposis colon cancer. *Nature Genet.* **5**: 279–282.

Lipkin M, Blattner WA, Gardner EJ, *et al.* (1984) Classification and risk assessment of individuals with familial polyposis, Gardner's syndrome and familial non-polyposis colon cancer from [³H]thymidine labelling patterns in colonic epithelial cells. *Cancer Res.* **44**: 4201–4204.

Lothe RA, Peltomaki P, Meling GI, *et al.* (1993) Genomic instability in colorectal cancer: relationship to clinicopathological variables and family history. *Cancer Res.*

53: 5849–5852.

Lynch HT, Watson P, Smryk TC, *et al*. (1992) Colon cancer genetics. *Cancer* 70: 1300–1312.

Lynch HT, Smryk TC, Watson P, *et al*. (1993) Genetics, natural history, tumor spectrum, and pathology of hereditary nonpolyposis colorectal cancer: an updated review. *Gastroenterology* 104: 1535–1549.

Mandl M, Paffenholz R, Friedl W, Caspari R, Sengteller M, Propping P. (1994) Frequency of common and novel inactivating *APC* mutations in 202 families with familial adenomatous polyposis. *Hum. Mol. Genet.* 3: 181–184.

Miyaki M, Seki M, Okamoto M, *et al*. (1990) Genetic changes and histopathological types in colorectal tumours from patients with familial adenomatous polyposis. *Cancer Res.* 50: 7166–7173.

Miyoshi Y, Nagase H, Ando H, *et al*. (1992) Somatic mutations in the APC gene in colorectal tumours: mutation cluster region in the APC gene. *Hum. Mol. Genet.* 1: 229–233.

Moser AR, Pitot HC, Dove WF. (1990) A dominant mutation that predisposes to multiple intestinal neoplasia in the mouse. *Science* 247: 322–324.

Munemitsu S, Souza B, Muller O, Albert I, Rubinfeldt B, Polakis P. (1994) The APC gene product associates with microtubules *in vivo* and promotes their assembly *in vitro*. *Cancer Res.* 54: 3676–3681.

Nagase H, Nakamura Y. (1993) Mutations of the *APC* (adenomatous polyposis coli) gene. *Hum. Mutat.* 2: 425–434.

Nagase H, Miyoshi Y, Horii A, *et al*. (1992) Screening for germ-line mutations in familial adenomatous polyposis patients: 61 new patients and a summary of 150 unrelated patients. *Hum. Mutat.* 1: 467–473.

Nicholls RJ, Springdall RG, Gallagher P. (1988) Regression of rectal adenomas after colectomy and ileorectal anastomosis for familial adenomatous polyposis. *Br. Med. J.* 296: 1707–1708.

Nicolaides NC, Papadopoulos BL, Wei Y-F, *et al*. (1994) Mutations of two PMS homologues in hereditary nonpolyposis colon cancer. *Nature* 371: 75–80.

Nishishio I, Nakamura Y, Miyoshi Y, *et al*. (1991) Mutations of chromosome 5q21 genes in FAP and colorectal cancer patients. *Science* 253: 665–669.

Olschwang S, Tiret A, Laurent-Puig P, Muleris M, Parc R, Thomas G. (1993) Restriction of ocular fundus lesions to a specific subgroup of *APC* mutations in adenomatous polyposis coli patients. *Cell* 75: 959–968.

Papadopoulos N, Nicolaides NC, Wei Y-F, *et al*. (1994) Mutation of a MutL homolog in hereditary colon cancer. *Science* 263: 1625–1629.

Parsons R, Li G-M, Longley MJ, *et al*. (1993) Hypermutability and mismatch repair deficiency in RER+ tumour cells. *Cell* 75: 1227–1236.

Paul P, Letteboer T, Gelbert L, Groden J, White R, Coppes MJ. (1993) Identical *APC* exon 15 mutations result in a variable phenotype in familial adenomatous polyposis. *Hum. Mol. Genet.* 2: 925–931.

Peltomaki P, Aaltonen LA, Sistonen P, *et al*. (1993a) Genetic mapping of a locus predisposing to human colorectal cancer. *Science* 260: 810–812.

Peltomaki P, Lothe RA, Aaltonen LA, *et al*. (1993b) Microsatellite instability is associated with tumours that characterize the hereditary non-polyposis colorectal carcinoma syndrome. *Cancer Res.* 53: 5853–5855.

Pero RW, Miller DG, Lipkin M, *et al*. (1983) Reduced capacity for DNA repair synthesis in patients with or genetically predisposed to colorectal cancer. *J. Natl Cancer Inst.* **70**: 867–875.

Powell SM, Zilz N, Beazer–Barclay Y, *et al*. (1992) *APC* mutations occur early during colorectal tumorigenesis. *Nature* **359**: 235–237.

Powell SM, Petersen GM, Krush AJ, Booker S, Jen J, Giardello FM, Hamilton SR, Vogelstein B, Kinzler KW. (1993) Molecular diagnosis of Familial Adenomatous Polyposis. *N. Engl. J. Med.* **329**: 1982–1987.

Prolla TA, Christie D-M, Liskay RM. (1994a) Dual requirement in yeast DNA mismatch repair for MLH1 and PMS1, two homologs of bacterial MutL Gene. *Mol. Cell. Biol.* **14**: 407–415.

Prolla TA, Pang Q, Alani E, Kolodner RD, Liskay RM. (1994b) MLH1, PMS1 and MSH2 interactions during the initiation of DNA mismatch repair in yeast. *Nature* **265**: 1091–1093.

Risinger JI, Berchuck A, Kohler MF, *et al*. (1993) Genetic instability of microsatellites in endometrial carcinoma. *Cancer Res.* **53**: 5100–5103.

Rubinfeld B, Souza B, Albert I, *et al*. (1993) Association of the *APC* gene product with β-catenin. *Science* **262**: 1731–1734.

Shibata D, Peinado MA, Ionov Y, Malkhosyan S, Perucho M. (1994) Genomic instability in repeated sequences is an early somatic event in colorectal tumorigenesis that persists after transformation. *Nature Genet.* **6**: 273–281.

Smith KJ, Johnson KA, Bryan T, *et al*. (1993) The *APC* gene product in normal and tumour cells. *Proc. Natl Acad. Sci. USA* **90**: 2846–2850.

Smith KJ, Levy DB, Maupin P, Pollard TD, Vogelstein B, Kinzler KW. (1994) Wild-type but not mutant APC associates with microtubule cytoskeleton. *Cancer Res.* **54**: 3672–3675.

Spirio L, Olshwang S, Groden J, *et al*. (1993) Alleles of the *APC* gene: an attenuated form of familial polyposis. *Cell* **75**: 951–957.

Strand M, Prolla TA, Liskay RM, Petes TD. (1993) Destabilization of tracts of simple repetitive DNA in yeasts by mutations affecting DNA mismatch repair. *Nature* **365**: 274–276.

Su L-K, Vogelstein B, Kinzler K. (1993a) Association of the APC tumour suppressor protein with catenins. *Science* **262**: 1734–1737.

Su L-K, Johnson KA, Smith KJ, Hill DE, Vogelstein B, Kinzler KW. (1993b) Association between wild type and mutant *APC* gene products. *Cancer Res.* **53**: 2728–2731.

Thibodeau SN, Bren G, Schaid D. (1993) Microsatellite instability in cancer of the proximal colon. *Science* **260**: 816–819.

Varesco L, Groden J, Spirio L, *et al*. (1993) A rapid screening method to detect nonsense and frameshift mutations: identification of disease-causing alleles. *Cancer Res.* **53**: 5581–5584.

Vasen HFA, Mecklin J-P, Meera–Khan P, *et al*. (1991) The International Collaborative Group on Hereditary Non-Polyposis Colorectal Cancer (ICG-HNPCC). *Dis. Colon Rectum* **34**: 424–425.

Vogelstein B, Fearon ER, Hamilton SR, *et al*. (1988) Genetic alterations during colorectal tumour development. *N. Engl. J. Med.* **319**: 525–532.

Wiebauer KM, Jiricny J. (1990) Mismatch-specific thymine DNA glycosylase and DNA polymerase B mediate the correction of G.T mispairs in nuclear extracts from human cells. *Proc. Natl Acad. Sci. USA* **87**: 5842–5845.

Wooster R, Cleton–Jansen A-M, Collins N, *et al.* (1994) Instability of short tandem repeats (microsatellites) in human cancers. *Nature Genet.* **6**: 152–156.

The *p53* tumour suppressor gene: from molecular biology to clinical investigation

Thierry Soussi

7.1 Introduction

The discovery in 1979 of the p53 protein was the culmination of two types of study involving a virological and a serological approach. It had been shown in simian virus 40 (SV40)-transformed cells that the SV40 large T antigen (AgT) is specifically complexed with a 53-kDa cellular protein. This protein was subsequently found in association with the E1b protein of adenovirus type 5. Polyclonal and monoclonal antibodies raised against this protein showed that it accumulates in the nucleus of various transformed cell lines: chemically induced murine sarcoma, virus-induced leukaemia, and spontaneously or retrovirus-transformed fibroblasts. However, this 53-kDa protein was not found in different types of normal embryonic or adult cells.

Antibodies directed against a 53–55-kDa protein demonstrated its presence in the serum of animals with chemically or retrovirus-induced tumours, and subsequently in the serum of patients with breast cancer. Analysis of the protein recognized by these antibodies showed that it was the same protein found in association with SV40 AgT or adenovirus E1b. This protein became known as p53 (Crawford, 1983).

The observation that the p53 protein accumulates specifically in the nucleus of many transformed or tumour cell lines suggested that it might play an important role in cell transformation. This led many groups to initiate studies on the gene coding for p53. In 1983, the first published cDNA sequence of murine *p53* appeared (Oren and Levine, 1983). Shortly afterwards, in 1985, the human cDNA and *p53* gene were identified, yet it was not until 1989 that interest in the *p53* gene really began to flourish, after reports that this gene was altered in human cancers (Baker *et al.*, 1989; Takahashi *et al.*, 1989).

This review can be divided into two main parts: the first (Sections 7.2 and 7.3) deals with the analysis of the properties of p53 in the cell and its potential role in the maintenance of the genetic integrity of the genome. The second part (Sections 7.4 and 7.5) describes the alterations found in *p53* in human cancers, and the relationship between these alterations and various factors such as environmental carcinogens and clinical prognosis.

7.2 p53 protein: activities and functions

7.2.1 p53 and the cell cycle

p53 is undetectable in normal cells. It has a half-life of about 20 min and is generally located in the cell nucleus (Oren *et al.*, 1981). In serum-stimulated, non-transformed Balb/c 3T3 cells, the cellular localization of the p53 protein varies during the cell cycle (Shaulsky *et al.*, 1990a). The protein is found in the cytoplasm during G1, then enters the nucleus during the G1/S transition where it remains until the end of the G2/M phase. After DNA synthesis, it is again found in the cytoplasm. As a rule, in transformed or tumour cells, the p53 protein is strictly nuclear but some exceptions have been observed.

Several observations suggest that the p53 protein is required for the progression of the cell cycle. In murine fibroblasts whose growth is arrested by serum deprivation, *p53* RNA and protein levels are low (Reich and Levine, 1984). When these cells are stimulated to grow by the addition of serum, RNA and protein levels increase to a maximum that occurs at the G1/S transition, just before the start of DNA replication (Reich and Levine, 1984). Likewise, blockage of the p53 protein by micro-injection of anti-p53 antibodies at the time of stimulation of quiescent 3T3 cells prevents these cells from entering the S phase (Mercer *et al.*, 1982; Deppert *et al.*, 1990). Inhibition of *p53* transcription by antisense RNAs in transformed or non-transformed cells also leads to complete growth arrest (Shohat *et al.*, 1987).

Paradoxically, the introduction of wild-type *p53* into different tumour cell lines may have different consequences, depending on the cell type under study. In a large number of tumour cells expressing a mutant *p53*, the overexpression of wild-type *p53* has a negative effect on cell growth (Baker *et al.*, 1990; Chen *et al.*, 1990; Diller *et al.*, 1990; Mercer *et al.*, 1990). The wild-type p53 protein is not tolerated, and stable clones either express a mutant *p53* through *de novo* selection of mutations (Diller *et al.*, 1990) or have acquired amplification of the murine double minute (*MDM2*) gene (Otto and Deppert, 1993) whose product inactivates the function of wild-type p53 (see Sections 7.2.3 and 7.2.4). Blockage of the cell cycle takes place between the G1 and S phase, more precisely at the end of G1 near the restriction point (R-point) (Diller *et al.*, 1990; Lin *et al.*, 1992).

Conditional expression of wild-type *p53* in a glioblastoma cell line shows that the G0 to G1 passage is not inhibited by p53 but cells are arrested at the end of the G1 phase with inhibition of the expression of certain cellular genes such as those coding for proliferating cell nuclear antigen, polymerase-α and histone H3 (Mercer *et al.*, 1991). This phenotype is the most common and forms the basis of the tumour suppressor function of p53.

In other cases, the reintroduction of *p53* into cell lines that no longer synthesize p53, such as mouse erythroleukaemia cells, chronic myeloid leukaemia cells or osteosarcoma cell lines, induces a phenomenon of cell death characteristic of apoptosis (Yonish-Rouach *et al.*, 1991; Shaw *et al.*, 1992). The role of p53 in apoptosis is strengthened by the studies of Lowe *et al.* (1993). It is known that irradiation of murine thymocytes induces apoptosis, which is a reflection of the maturation of these cells. This phenomenon is totally abolished when mouse thymocytes deleted for the two *p53* alleles are used. Finally, overexpression of wild-type *p53* in L12 cells (murine preB line) or in K562 cells (chronic myeloid leukaemia line) can induce differentiation (Shaulsky *et al.*, 1991; Feinstein *et al.*, 1992), as visualized by the expression of markers such as immunoglobulin and haemoglobin production. Taken together, these data suggest that p53 is an important element in the regulation of cell growth and that its role may vary according to cell type. In view of the contradictory observations indicating a role for p53 in both progression of the cell cycle and negative regulation of growth, it is important to keep in mind that all experiments involving growth inhibition were conducted by overexpression of wild-type *p53* expressed under the control of strong transcriptional promoters under conditions far from the normal physiological conditions of the cell.

The production of transgenic mice harbouring deletions of the two *p53* alleles generated some excitement, since these *p53–/p53–* mice are fully viable and fertile, and no embryonic malformations have been observed (Donehower *et al.*, 1992). The only major phenotype is the very high frequency of cancers that appear from the 6th month (in 75% of the animals) until the 10th month (100% of animals). Heterozygous *p53+/p53–* mice also develop cancer, but at a much lower rate (30% of animals) and at a later age (after 15 months). Many different tumour types are observed, with a large majority of lymphomas in *p53–/p53–* animals compared with osteosarcomas or soft tissue sarcomas in *p53+/p53–* heterozygotes. This latter observation may be related to the phenotype of patients with Li–Fraumeni syndrome, who carry germline mutations in the *p53* gene, who also develop a large number of soft tissue sarcomas. Studies in mice with different haplotypes show that the spectrum of tumour types can vary depending on the mouse strain used (Harvey *et al.*, 1993). p53 is therefore not essential for cell function. However, its inactivation is an essential factor in the development of some types of neoplasia.

7.2.2 p53, DNA damage and genetic stability

In 1984, Maltzman and Czyzyk reported that UV irradiation of mouse cells induces stabilization of the p53 protein *in vivo*. These results were repeated by Kastan *et al.* (1991) who observed that gamma radiation produces a similar phenomenon. These authors also showed that p53 protein accumulation induces a transient blockage of the cell cycle at the G1 phase, just before DNA replication (Kastan *et al.*, 1991; Kuerbitz *et al.*, 1992). It is generally agreed that this blockage of cell division following DNA damage gives the cell time to induce an SOS response to repair the lesions. The most interesting finding in this work is that this phenomenon is absent in cells expressing a mutant *p53*, i.e. these cells do not undergo growth arrest after DNA damage (Kastan *et al.*, 1991; Kuerbitz *et al.*, 1992). Transfection of wild-type *p53* in cells lacking functional p53 restores the G1 arrest following irradiation. Fritsche *et al.* (1993) showed that this phenomenon is not specific to UV or gamma irradiation, but applies to any type of DNA lesion. In fact, agents such as ethidium bromide and vincristine, which do not cause DNA strand breaks, are incapable of inducing a G1 arrest whereas agents such as mitomycin D and cis-platin, which cause DNA lesions directly, can induce the overexpression of *p53*. The addition of a DNA restriction enzyme to eukaryotic cells can also induce p53 accumulation (Lu and Lane, 1993). This role of p53 in encouraging genetic stability is strengthened by the finding that gene amplification in the presence of PALA [*N*-(phosphonacetyl)-L-aspartate, a uridine synthesis inhibitor] is much more frequent in cells lacking p53 (Livingstone *et al.*, 1992; Yin *et al.*, 1992).

The studies described above led to the proposal that p53 helps to preserve the genetic integrity of the cell (Lane, 1992). p53 would therefore act as a kind of 'stop light' that would induce an arrest of cell division to give the cell time to repair DNA damage. If, for reasons not yet known, the cell is incapable of DNA repair, it might be imagined that p53 would induce cell death by inducing apoptosis. Tumour cells containing mutant *p53* are no longer able to maintain their genetic integrity because they no longer receive a growth arrest signal, resulting in a cell whose genome is less stable and which will accumulate various mutations giving rise to clones with greater malignancy. This might also better explain the phenotype of *p53*– mice which, although viable, have a higher incidence of cancer.

The mechanism responsible for the stabilization of p53 following DNA damage is not yet known. The most seductive and simplest hypothesis would be a post-translational modification of p53 by phosphorylation. Indeed, kinases specific for the free ends of DNA exist and it is possible to envisage a phosphorylation cascade that culminates in the activation of p53. As will be seen below (see Sections 7.3.2 and 7.3.3), p53 is a substrate for several kinases.

7.2.3 The protein partners of p53

In 1993, Vogelstein suggested that there was 'no more room at the p53 inn', meaning that the number of proteins that can complex with p53 was ever increasing (Pietenpol and Vogelstein, 1993). In view of the literature since that time, the inn is going to have to become a hotel. No fewer than 17 viral or cellular proteins have been described which are apparently capable of interacting with p53 (*Figure 7.1* and *Table 7.1*). In the case of viral proteins, the primary role of this interaction is to inactivate p53. Indeed, the principal concern of a virus is to ensure its survival in the cell it infects. DNA tumour viruses require that their host cells be dividing in order to optimize virus multiplication. To this end, these different viruses have developed a similar strategy; namely, the inactivation of the two proteins involved in the negative regulation of cell growth, the products of the retinoblastoma (*RB*) and *p53* genes (Levine and Momand, 1990). In the case of SV40, it is the AgT that binds both p53 (at the carboxy terminal) and the RB protein (at the amino terminal). In the case of adenovirus type 5, the viral E1b protein specifically binds p53 and inhibits its transactivational activity, while the E1A protein sequesters the RB protein. Finally, in the case of the human papillomaviruses (HPVs), the E6 protein of HPV16 and HPV18 can induce ubiquitin-dependent degradation of p53 while the E7 protein specifically binds the RB protein (Scheffner *et al.*, 1990). Binding of E6 to p53 takes place by means of a cellular protein called E6-AP (Huibregtse *et al.*, 1991). It should be noted that mutations affecting the immortalizing or transforming properties of these viral proteins co-localize exactly with the p53 or RB binding sites. This observation suggests that the oncogenicity of these viruses occurs via

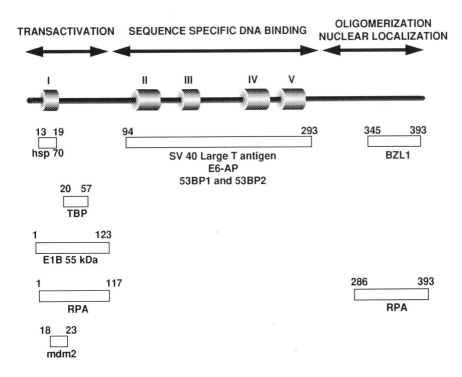

Figure 7.1. Functional domains of p53. The coordinates correspond to human p53.

139

Table 7.1. Characteristics of proteins which have reportedly been shown to associate with p53

Protein	Wild-type p53	Mutant p53	p53 domain involved	Effect	References
SV40 AgT	+	Varies	Central	p53 stabilization	Lane and Crawford (1979); Linzer and Levine (1979)
Ad5 E1b	+	Varies	Amino	Induce a cytoplasmic localization of p53	Sarnow et al. (1982)
HBx	+	Varies	?	Inhibits p53-dependent transactivation / Inhibits p53/ERCC3 interaction	Feitelson et al. (1993) / Wang et al. (1994)
E6	+	Varies	?	Induce p53 degradation	Werness et al. (1990)
EBNA5	+	+	?	?	Szekely et al. (1993)
BZL1	+	+	Carboxy	Inhibits p53-dependent transactivation	Zhang et al. (1994)
E6-AP	+	?	Central	Mediate p53–E6 interaction?	Huibregtse et al. (1991)
hsp70	–	Varies	Amino and/or carboxy	Stabilize mutant p53?	Pinhasi-Kimhi et al. (1986)
MDM2	+	+	Amino	Inhibits p53-dependent transactivation	Momand et al. (1992)
TBP	+	–	Amino	Inhibit TFIID transcriptional activity	Seto et al. (1992)
WT1	+	+	?	Repress WT1 activity	Maheswaran et al. (1993)
CBF	+	?	?	Repress transcription	Agoff et al. (1993)
RPA	+	+	Amino and/or carboxy	Inhibit DNA binding activity of RPA	Dutta et al. (1993)
S100	+	?	?	?	Baudier et al. (1992)
ERCC3	+	+	?	?	Wang et al. (1994)
BPi1	+	–	Central	Inhibit DNA binding activity of p53	Iwabuchi et al. (1994)
BPi2	+	–	Central	Inhibit DNA binding activity of p53	Iwabuchi et al. (1994)

inactivation of RB and p53 function. The interaction of p53 with the hepatitis B virus (HBV) HBx protein or the Epstein–Barr virus (EBV) Epstein–Barr nuclear antigen 5 (EBNA5) or immediate-early BZLF1 proteins has not yet been sufficiently studied to propose an analogous mechanism.

The interaction between p53 and the heat-shock protein, hsp70, was first detected in cells transformed by a mutant murine *p53* (Pinhasi-Kimhi *et al.*, 1986). It was subsequently shown that only mutant p53 is able to interact with hsp70 (Hinds *et al.*, 1987; Stürzbecher *et al.*, 1987), the wild-type protein being totally devoid of this property. The role of this interaction is not clear. These hsp70 chaperone molecules not only aid in the transportation of various proteins in certain cell compartments but also specifically bind partially denatured proteins when the cell is subjected to stress. Considering that some p53 mutants have undergone a conformational change, it is possible that their central, hydrophobic regions are more exposed and that they are recognized by hsp70 as being partially denatured. In fact, the most hydrophobic regions of p53 are located in its central region and correspond to the domains most commonly mutated in human cancers. As will be seen in Section 7.3.6, only some p53 mutants (those having undergone a conformational change) can bind hsp70. Hainaut and Milner (1992) showed, on the one hand, that the carboxy terminal of p53 is required for this interaction. On the other hand, it has been shown that a synthetic peptide, corresponding to the highly conserved block I of p53 (amino terminal region), can specifically bind hsp70 (Lam and Calderwood, 1992).

The MDM2 protein has been identified as a 90-kDa protein that coprecipitates with wild-type or mutant p53 (Hinds *et al.*, 1990; Barak and Oren, 1992). This 90-kDa protein was purified and sequenced and corresponds to the product of the MDM2, or murine double minute, gene (Momand *et al.*, 1992). This gene was first described as being amplified in a spontaneously transformed mouse cell line (Fakharzadeh *et al.*, 1991). Sequence analysis of the MDM2 protein indicates that it is a transcription factor, but no targets have yet been characterized. The p53–MDM2 interaction takes place at the amino terminal of p53 (Chen *et al.*, 1993; Oliner *et al.*, 1993) and induces a decrease in p53 transactivational activity (Barak *et al.*, 1993; Wu *et al.*, 1993).

The interactions between p53 and the other viral or cellular proteins have not yet been analysed in detail. Since many of these interactions have been identified either *in vitro* or in systems involving the overexpression of *p53*, it remains to be proved that they truly play a role in the various signalling pathways involving p53 *in vivo*.

7.2.4 Transactivational activity of p53

Wild-type human p53 binds to DNA specifically. The first binding sequence to be identified contains two or three repetitions of a TGCCT motif and is called the ribosomal gene cluster (RGC) (Kern *et al.*, 1991). Subsequently, a specific consensus binding sequence was identified by enrichment of DNA–protein complexes through immunoprecipitation followed by polymerase chain reaction (PCR) amplification of DNA fragments specifically immunoprecipitated with wild-type p53 (El-Deiry *et al.*, 1992; Funk *et al.*, 1992). This 10-base pair (bp) motif has an internal symmetry with two repeats of 5'-PuPuPuC(A/T)(T/A)GPyPyPy-3' separated by 0–13 bp (CONS sequence). The double symmetry of this sequence is compatible with the hypothesis that p53 binds to DNA as a tetramer. These sequences are artificial in the sense that their selection was based on their affinity for p53. Nevertheless, addition of these sequences to a vector carrying a transcriptional promoter controlling the expression of a reporter gene shows that wild-type p53 can specifically activate transcription. Mutant p53 lacks this activity. As will be seen below, the central region of p53 is responsible for this specific DNA binding activity, while the amino terminal contains the transactivator domain.

Several cellular genes possess a p53 binding site, which in the *MDM2* gene is located in intron I (CONS-type sequence) (Wu *et al.*, 1993). *In vivo* studies show that there is cross-regulation between p53 and *MDM2*. The p53 protein activates transcription of the *MDM2* gene (by binding to the regulatory element in intron I) and the *MDM2* protein regulates p53 activity by masking its amino terminal transactivator domain (Barak *et al.*, 1993; Wu *et al.*, 1993). The biological function of this phenomenon is not known at present.

The p53 protein can also bind to the murine creatine phosphokinase (*MCK*) gene promoter (Weintraub *et al.*, 1991) or to a GC-rich region of the SV40 replication origin (SP1 sites) (Bargonetti *et al.*, 1991). Although these promoters are specifically activated by wild-type p53, and not by mutant p53, it is difficult to place this observation within the scope of our current understanding of its cellular function.

Using a differential screening approach between cells expressing wild-type and mutant *p53*, El Deiry *et al.* (1993) were able to characterize a gene that is specifically induced by wild-type p53: the *WAF1* gene. The transcriptional promoter of this gene contains a DNA sequence homologous to the consensus sequence defined above. Overexpression of the p21[WAF1] protein in various cell types induces growth arrest at the G1 phase that is dependent on wild-type p53 (El-Deiry *et al.*, 1994). This p21 protein has also been identified by different groups as being able to associate specifically with cyclin-dependent kinase (cdk2) and inhibit the kinase activity of cdk2–cyclin complexes (Gu *et al.*, 1993; Harper *et al.*, 1993; Xiong *et al.*, 1993) (*Figure 7.2*). The induction of DNA lesions in cells expressing wild-type p53 causes overexpression of *WAF1* RNA and protein, but this is not observed in the presence of mutant p53 (El-Deiry *et al.*, 1994). Irradiation of human fibroblasts induces an arrest of cell division concomitant with a loss in activity of the cdk2–cyclin E complex (Dulic *et al.*, 1994). The expression levels of these two proteins and their complex is not affected by the treatment. Using antibodies specific for p21[WAF1/CIP1], it could be shown that the expression of this protein is responsible for this loss in activity of the cdk2–cyclin E complex. This result is a good demonstration that p53 intervenes at a later stage of the G1 phase, at the time of the G1/S transition (*Figure 7.2*).

The studies described in this section demonstrate a direct role of p53 in the

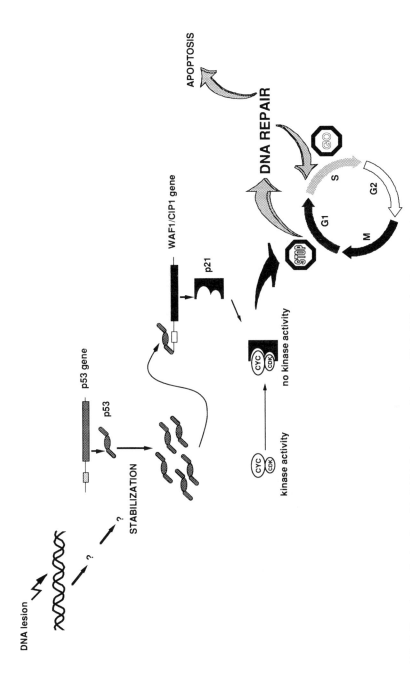

Figure 7.2. Cell cycle regulation by p53. DNA lesions can induce p53 stabilization by mechanisms that are still unclear. This stabilization allows the protein to transactivate the *WAF1/CIP1* gene. p21$^{WAF1/CIP1}$ binds to the cdk–cyclin complex and inhibits the kinase activity required to activate the genes needed in the G1/S transition. Cells blocked in late G1 can then repair the DNA lesions, after which the cell cycle resumes. In some cases (too many lesions?), the cycle does not resume and the cell dies by apoptosis.

transactivation of cellular genes. p53 is also a negative transcriptional regulator of certain cellular and viral transcriptional promoters such as *c-jun, c-fos, β-actin*, growth arrest DNA damage inducible 45 (*GADD45*), multi-drug resistance (*MDR*), *RB*, interleukin-6 (*IL-6*), Rous sarcoma virus (*RSV*), human immuno-deficiency virus (*HIV*), herpes simplex virus (*HSV-1*), *SV40* and cytomegalovirus (*CMV*) (Ginsberg *et al.*, 1991; Santhanam *et al.*, 1991; Chin *et al.*, 1992; Deb *et al.*, 1992, 1994; Kley *et al.*, 1992; Shiio *et al.*, 1992; Subler *et al.*, 1992, 1994; Agoff *et al.*, 1993). This regulation does not occur through binding of p53 to these pro-moters, but rather through its interaction with certain components of the transcription complex. Seto *et al.* (1992) showed that wild-type p53, but not mutant p53, can interact specifically with the TATA binding protein (TBP). These findings were confirmed by the demonstration that this repression depends on the presence of a TBP binding site in the target promoter of p53 (Mack *et al.*, 1993). The biological consequence of p53 binding to these different promoters is not yet clear.

7.2.5 p53 and differentiation

Several lines of evidence indicate that p53 is involved in early embryonic devel-opment and during differentiation. The p53 protein was detected in primary cells from 10–14-day-old mouse embryos but not 16-day-old mouse embryos (Chandrasekaran *et al.*, 1981, 1982). A similar observation was made in primary cultures of rat and hamster embryos at mid-gestation, showing that *p53* expres-sion decreases with increasing developmental stage. A marked reduction in the amount of mouse *p53*-specific mRNA was observed from day 11 onward, which is correlated with progress in differentiation (Rogel *et al.*, 1985). Using *in situ* hybridization, Schmid *et al.* (1991) showed the expression over time of *p53* mRNA in all cells of a mouse embryo from the age of 8.5 to 10.5 days old. Upon differentiation, the amount of *p53* mRNA declined sharply. A similar observa-tion was made during embryonic development in chickens (Louis *et al.*, 1988). In *Xenopus laevis*, large amounts of *p53* mRNA are stored in oocytes and behave like maternal RNA (Tchang *et al.*, 1993).

More recently, *p53* expression was shown to be very high during spermatoge-nesis (Schwartz *et al.*, 1993) and, using *in situ* hybridization, this expression was shown to be specific for the tetraploid primary spermatocytes. Using specific p53 monoclonal antibodies, an accumulation of the p53 protein was found at this spe-cific phase. This expression at a phase in the cell cycle that involves pairing of chromosomes, recombination and repair of DNA suggests that p53 could play a direct role in these activities. This view is reinforced by the notion of p53 as a 'guardian' of the integrity of the genome.

Undifferentiated embryonic carcinoma cells, thought to be analogous to nor-mal embryonic stem cells, contain relatively high levels of *p53* mRNA and protein. Upon differentiation *in vitro*, there is a marked decrease in mRNA levels, and it was shown that a post-transcriptional mechanism is involved in this regulation, since the rate of *p53* expression does not change upon differentiation (Dony *et al.*, 1985). A similar observation was made in another model of cellular differentiation, the virus-transformed murine erythroleukaemia cell line (Ben-Dori *et al.*, 1987; Khochbin *et al.*, 1988).

Taken together, these data suggest that *p53* expression may be involved during either embryonic development or cell differentiation. On the other hand, null mice (without *p53* expression) apparently have normal embryonic development and are able to produce normal progeny. There is no clear explanation for these two observations. It is easy to invoke 'alternative pathways' or 'redundancy of function' so as to hide our ignorance, but it is clear that an improved knowledge of p53 function is needed in order to build a model capable of integrating all available data.

7.3 The p53 protein and *p53* gene

7.3.1 The p53 *gene: structure and conservation*

In contrast to many proto-oncogenes, whose normal or modified sequences can be transduced by retroviruses (and thus be easily characterized), *p53* was never transduced by a retrovirus. This is not surprising in light of our current knowledge of the mechanism of action of the tumour suppressor genes, but placed back into a context where 'oncogenes' were in fashion, this observation seemed to be paradoxical. Its greatest disadvantage was that it deprived researchers of a p53 probe. However, many monoclonal antibodies directed against murine p53 were available. Immunoprecipitation of polysomes during translation enabled Oren and Levine (1983) to isolate a 300-bp fragment corresponding to the cDNA of mouse p53: the cDNA and gene were subsequently isolated and sequenced (Bienz *et al.*, 1984). The use of murine probes made it possible to characterize p53 in many different species, from man to trout (Matlashewski *et al.*, 1984; Soussi *et al.*, 1987, 1988a, 1988b; Rigaudy and Eckhart, 1989; Caron de Fromentel *et al.*, 1992; Legros *et al.*, 1992; Kay *et al.*, 1994) (*Table 7.2, Figure 7.3*). Screening for p53 in lower organisms such as invertebrates has only recently become possible by PCR amplification, using degenerate primers and very low stringency during amplification (Winge and Friend, personal communication).

As early as 1987, the characterization of p53 from *X. laevis* revealed a certain number of features common to all p53 proteins (Soussi *et al.*, 1987). The subsequent identification of new p53 proteins only served to confirm these observations (Soussi *et al.*, 1990).

The p53 protein (see *Figure 7.4*) can be divided into three regions: (i) the amino terminal which contains a large number of acidic residues, no basic residues and a large number of prolines (including many Pro–Pro pairs); (ii) the carboxy terminal which is very hydrophilic and contains many charged residues; and (iii) the central region of the protein which contains several very hydrophobic regions and

Table 7.2. Characteristics of p53 from various species

Species	Probe	p53 clone available	mRNA size (kb)	Gene structure (exons)	p53 protein (amino acids)	Accession number	Reference
Human	Mouse p53 whole cDNA	Gene and whole cDNA	2.8	11	393	X02469(E) X54156(E)	Matlashewski et al. (1984)
African green monkey	Human p53 whole cDNA	Whole cDNA	n.a.	n.a	393	X16384(E)	Rigaudy and Eckhart (1989)
Rhesus monkey	PCR amplification	Whole cDNA	n.a.	n.a.	393	L20442	Kay et al. (1994)
Rabbit	Mouse p53 coding sequence	Whole cDNA	n.a.	n.a.	391	n.a.	Caron de Fromentel et al. (unpublished data)
Hamster	Mouse p53 coding sequence	Gene and whole cDNA	n.a.	11	396	M75144 U08134	Legros et al. (1992)
Rat	Mouse p53 whole cDNA	Gene and whole cDNA	2.0	10 (fusion of exons 6 and 7)	391	X13058 L07903–910	Soussi et al. (1988b)
Mouse[a]	Monoclonal antibody	Gene and whole cDNA	2.0	11	387	X00741 X008875–885	Finlay et al. (1988)
Dog	PCR amplification	Partial gene exons 6 and 7	n.a.	n.a.	n.a.	L27630	Devilee et al. (unpublished data)
Cat	PCR amplification	Whole cDNA	n.a.	n.a.	386	457439(N)	Okuda et al. (1994)
Pig	PCR amplification	Partial cDNA	n.a.	n.a.	n.a.	n.a.	Soussi et al. (1988a)
Chicken	Mouse p53 whole cDNA	Whole cDNA	1.8	n.a.	367	X13057	Soussi et al. (1988a)
X. laevis	Mouse p53 coding sequence	Gene and whole cDNA	2.2 and 3	11	363	M36962	Soussi et al. (1987)
Rainbow trout	Xenopus p53 domains IV and V	Whole cDNA	2.4	n.a.	396	M75145	Caron de Fromentel et al. (1992)

n.a., not available.

[a] The first murine p53 cDNAs (Bienz et al., 1984; Jenkins et al., 1984; Pennica et al., 1984; Arai et al., 1986) were shown to contain activating mutations. Finlay et al. (1988) compared all these cDNAs to wild-type murine cDNA cloned from non-transformed cells.

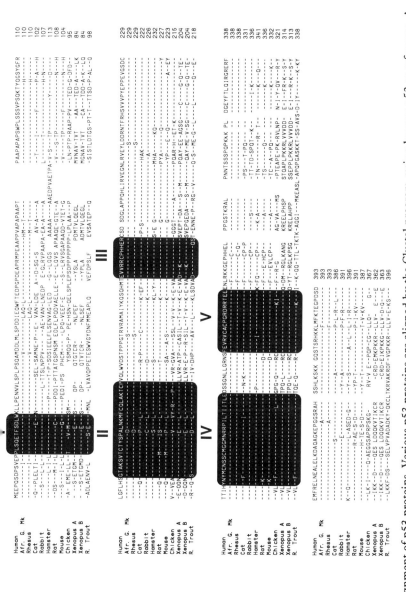

Figure 7.3. Alignment of p53 proteins. Various p53 proteins are aligned by the Clustal program, using human p53 as reference. Amino acids homologous to human p53 are indicated by dashes. Highly conserved blocks I–V are identical to those defined by Soussi *et al.* (1987), apart from block I which was extended by four amino acids towards the carboxy terminal.

very few charged amino acids. Sequence analysis of various p53 proteins shows no motifs characteristic of a particular function. On the other hand, comparison of the various proteins shows that their homology is not uniformly distributed, leading to the identification of five blocks highly conserved during evolution. These conserved blocks, defined in 1987 by the comparison of p53 from *X. laevis* with human p53, have been found in all the p53 proteins characterized to date (Soussi *et al.*, 1990). Four of these five blocks (II–V) are found in the central region of the protein, while block I is located at the amino terminal (*Figure 7.3*). Taken together, these data suggest that the three regions of the p53 protein may have clearly defined roles in its function, with special importance for the central region containing blocks II–V. This model has been widely confirmed by recent data on p53 function (see Section 7.3.6).

7.3.2 Carboxy terminal of p53

This region contains the three nuclear localization signals (NLSs) of the p53 protein (Dang and Lee, 1989; Shaulsky *et al.*, 1990b). Mutagenesis of NLS1 induces the synthesis of a totally cytoplasmic p53 protein, while alteration of NLS2 or NLS3 leads to mixed cytoplasmic and nuclear localization (Shaulsky *et al.*, 1990). Besides heterologous protein–protein interactions between p53 and other cellular or viral proteins, p53 forms homologous oligomers. It was first shown *in vivo* that murine p53 from F9 cells sediments mainly at 8S, which represents tetramers of p53 (McCormick and Harlow, 1980). High molecular weight oligomers of p53 have subsequently been found in most cells studied so far (Kraiss *et al.*, 1988). Mouse and human p53 expressed in insect cells (O'Reilly and Miller, 1988) or expressed in an *in vitro* translation–transcription system (Schmieg and Simmons, 1988) also form high molecular weight oligomers. Using gradient gel electrophoresis and chemical cross-linking, Stenger *et al.* (1992) showed that, under non-denaturing conditions, murine p53 forms mainly tetramers or multiples of tetramers. These oligomers are very stable in the presence of high salt (1 M NaCl) or reducing agents. Pulse-chase analysis shows that oligomerization is a very rapid process (2 min). Truncation of the carboxy terminal of p53 prevents the oligomerization process (Milner *et al.*, 1991). Detailed analysis of this region shows that the tetramerization domain in fact includes two domains, an α-helix whose hydrophobic amino acids are directly involved in the protein–protein interaction, and a basic region that stabilizes the tetramers (Stürzbecher *et al.*, 1992).

The carboxy terminal is phosphorylated by casein kinase II at the serine in position 392 of the human p53 (386 for mouse p53), which is highly conserved in all p53 proteins (Meek *et al.*, 1990; Herrmann *et al.*, 1991; Hupp *et al.*, 1992). In addition, these residues represent the binding site of a small RNA (5.8S rRNA) whose role is not known at present (Samad and Carroll, 1991). One of the cyclin-dependent kinases (cdk) has been shown to phosphorylate human p53 at residue

315 (312 for murine p53) (Bischoff *et al.*, 1990; Stürzbecher *et al.*, 1990). Given the proximity of this site to NLS1, there may be a link between this phosphorylation and transport of p53 into the nucleus, although so far no direct proof of this exists.

The carboxy terminal of p53 is able to bind non-specifically to DNA but it cannot bind to the RGC or CONS sequences, for which the central region is required. Yet, the carboxy terminal has been shown to play an important role in specific DNA binding. Indeed, specific DNA binding is very strongly activated when the C-terminal is phosphorylated by casein kinase II, activated by binding of a monoclonal antibody such as PAb421 or, when this region is deleted, by proteolysis (Hupp *et al.*, 1992, 1993). The reason for this is not known.

The carboxy terminal of p53 is the target of several viral or cellular proteins (BZLF1, RPA or HBx) but, unlike proteins that bind to the amino terminal of p53, no biological function has been assigned to these interactions.

7.3.3 Amino terminal of p53

This acidic, proline-rich region contains the transcriptional transactivator domain of the protein (Fields and Jang, 1990; Raycroft *et al.*, 1990). Deletion of amino acids 20–42 completely abolishes this activity, which can be restored by substituting the transactivator domain of the VP16 protein at the same location (Pietenpol *et al.*, 1994). On the one hand, functional analysis of this region using a library of point mutations at various residues has shown, surprisingly, that no single point mutation is able to abolish transactivational activity completely (Lin *et al.*, 1994). On the other hand, double or triple mutants are totally inactive. Residues Leu-22 and Trp-23 are crucial for this transactivational activity (Lin *et al.*, 1994). This result suggests that the transactivator domain is particularly stable and that its overall structure, rather than its sequence, is important for transactivation. This result also explains why mutations in the transactivator domain are never found in human cancers. Unlike the DNA binding region, which is highly sensitive to any point mutations, the transactivator domain is sufficiently stable to resist inactivation by a single mutational event.

The amino terminal of p53 contains the target for phosphorylation by various kinases (Meek, 1994), one of which is particularly interesting in light of the properties of p53. This is a DNA-dependent kinase which phosphorylates serines 15 and 37 in human p53 (4 and 15 in murine p53) (Lees-Miller *et al.*, 1992; Wang and Eckhart, 1992). Mutation of serine 15 abolishes the antiproliferative power of p53, suggesting that modification of this residue is important in the activity of the protein (Fiscella *et al.*, 1993). The involvement of a DNA-dependent kinase in the activation of p53 is intellectually stimulating because, at present, there are no known links between the time when the lesions alter the DNA and the stabilization of p53. Participation of one or several kinases that would be activated by the presence of the free end of the DNA, or by the presence of small DNA fragments, could explain this activation.

Other kinases also phosphorylate the amino terminal of p53. Casein kinase I phosphorylates serines 4, 6 and 9 of mouse p53 (Milne *et al.*, 1992) while the mitogen-activated protein (MAP) kinase phosphorylates threonines 73 and 83 of murine p53 (Milne *et al.*, 1994). The role of these post-translational modifications remains to be elucidated.

The amino terminal of p53 is the binding site of a large number of viral and cellular proteins, including the Elb protein from adenovirus type 5, the MDM2 protein, the RPA protein (which also binds at the carboxy terminal), the hsp70 protein and the TBP. The interaction of MDM2 and p53 has been studied more closely, either by using point mutations (Lin *et al.*, 1994) or a series of synthetic peptides (Picksley *et al.*, 1994). These studies enabled the definition of a very limited region (amino acids 18–23 in human p53) that includes the residues Leu-22 and Trp-23 shown to be important for transactivation (Lin *et al.*, 1994). This result, therefore, explains how the MDM2 protein inhibits the transactivational activity of p53 by masking the transactivator domain.

7.3.4 Central region of p53

The importance of this region in p53 function has been suggested by several observations: (i) the presence of four of the five evolutionary conserved blocks (Soussi *et al.*, 1990); (ii) the high concentration of mutations in this region (Caron de Fromentel and Soussi, 1992); and (iii) the fact that it is the binding site of AgT (Tan *et al.*, 1986; Jenkins *et al.*, 1988). The finding that p53 is never mutated in SV40-transformed cells suggests that its alteration occurs through interaction with AgT. It was long suggested that this central region could be the target of certain cellular proteins, based on our knowledge of the Rb protein, but more recent work shows that it contains the domain that binds to DNA and, more particularly, to the target DNA sequences of p53. The first indication of the role of the central region in DNA binding came from Halazonetis *et al.* (1993), who demonstrated that the highly conserved evolutionary blocks IV and V were directly involved in DNA contact. Proteolytic digestion of wild-type p53 by enzymes such as thermolysin and subtilysin generates a 27-kDa fragment containing the entire central portion of the protein (amino acids 92/102–306/292, according to the protease used) (Bargonetti *et al.*, 1993; Pavletich *et al.*, 1993). This fragment of p53 is able to bind specifically to the RGC or CONS sequence if it comes from wild-type p53, whereas digestion fragments from mutant p53 can no longer do so (Bargonetti *et al.*, 1993; Pavletich *et al.*, 1993). Using truncated p53 produced in insect cells, Wang *et al.* (1993) defined a similar region (amino acids 80–290) necessary and sufficient for specific DNA binding. It should be noted that this central region does not contain the oligomerization domain, which brings into question its role in p53 activity.

7.3.5 p53 conformation

During mitogenic stimulation of untransformed lymphocytes, p53 can be detected in two forms that react in a mutually exclusive manner with two monoclonal antibodies (Milner, 1984). This observation laid the groundwork for a model (Milner, 1991) proposing the existence of two different conformations of p53 based on their immunoreactivity with certain monoclonal antibodies: a 'suppressor' conformation, which can inhibit cell division (PAb240– and PAb1620+ epitopes), and a 'promoter' conformation (PAb240+ and PAb1620– epitopes). The equilibrium between these two forms would regulate p53 function during the cell cycle. This hypothesis is supported by the existence of a thermosensitive mutant p53 protein from mouse (Ala → Val135) (Michalovitz *et al.*, 1990). At 37°C, this protein is altered in terms of both function and conformation, while at 32°C it exhibits wild-type behaviour (Milner and Medcalf, 1990; Martinez *et al.*, 1991). The change in temperature induces a conformational change detectable by a change in immunoreactivity. The 'promoter' conformation is generally characteristic of the mutant p53 proteins found in human cancers (Gannon *et al.*, 1990). Some mutants are able to induce a mutant conformation to a p53 co-translated from a wild-type gene (Milner and Medcalf, 1991), thereby explaining the transdominant phenomenon observed in certain cancers (see Section 7.4.4). At present, the mechanisms responsible for this conformational change are totally unknown. Some authors have suggested that phosphorylation of the carboxy terminal region could be involved. More recently, this model was extended to include a third 'sensor' conformation corresponding to PAb421+/PAb1620+ (Milner, 1994). This conformation is intermediate between the promoter and suppressor forms described above. In addition, Hainaut and Milner (1993a,b) showed that incubation of wild-type p53 with a metal chelating agent can induce a reversible conformational change to give a protein that is not only 'mutant' in terms of conformation, but also in terms of DNA binding activity. These authors postulated the presence of a zinc molecule bound to p53 via its cysteine residues. This hypothesis has been confirmed by the crystallographic data (Cho *et al.*, 1994). Regardless of the validity of these models, they strongly suggest that p53 is an extremely flexible molecule whose function is governed by very subtle conformational variations due to modifications of either the protein or its environment.

7.3.6 p53 structure

X-ray crystallography of p53 has been an important step in increasing the understanding of the structure of this protein (Cho *et al.*, 1994). The central region (amino acids 102–292) has been crystallized in the form of a protein–DNA complex. This core region has been shown to include the following motifs: (i) two antiparallel β-sheets composed of four and five β-strands. These two sheets form a kind of compact sandwich that holds the other elements; (ii) a loop–sheet–helix

(LSH) motif containing three β-strands, an α-helix and the L1 loop; (iii) an L2 loop containing a small helix; and (iv) an L3 loop mainly composed of turns. It is quite remarkable to note the very good agreement between these various structural elements and the four evolutionary conserved blocks (II–V). The LSH motif and the L3 helix are involved in direct DNA interaction (LSH with the major groove and L3 with the minor groove). The L2 loop is presumed to provide stabilization by associating with the L3 loop. These two loops are held together by a zinc atom tetracoordinated to the following amino acids: Cys176 and His179 on the L2 loop, and Cys278 and Cys242 on the L3 loop (Cho *et al.*, 1994).

Analysis of the distribution of mutations in p53 shows a high concentration in the central region of the protein, especially in the four blocks, II–V, which have been identified as the DNA binding region (*Figure 7.4*). In view of the three-dimensional structure of the protein, it has been proposed that two classes of mutations can be predicted: class I mutations, which affect the amino acids directly involved in the protein–DNA interaction (residues in the LSH and L3), and class II mutations, which affect the amino acids involved in stabilization of the three-dimensional structure of the protein (residues in L2).

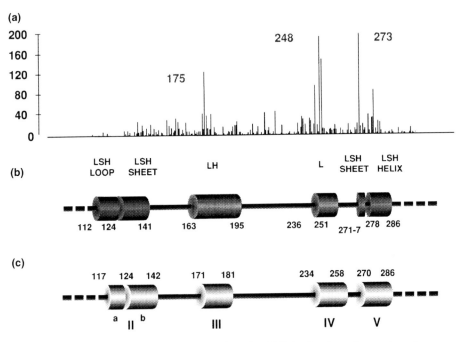

Figure 7.4. Structure–function relationship for p53. (a) Distribution of mutations in the central region of the protein. (b) Structural regions of p53 defined by X-ray crystallography. (c) Blocks highly conserved during evolution. Block II DNA is divided into two regions by an intron.

In fact, as described above, it has been established that mutant p53 can undergo conformational changes leading to its interaction with hsp70 but also to altered accessibility to certain monoclonal antibodies such as PAb1620, which recognizes a specific conformational epitope of wild-type p53, or PAb240, which recognizes a cryptic epitope revealed in mutant p53 (Milner, 1991). In reality the situation is not so simple, and there have been various reports of a certain degree of heterogeneity in the behaviour of different p53 mutants. In particular, some p53 mutants have been shown to be PAb1620+/PAb240–/hsp70–. With the aim of analysing a possible correlation between conformational changes and loss in activity, Ory *et al.* (1994) studied a series of 23 different p53 proteins mutated in the three hot-spot codons, Arg175, 248 and 273. The results show that these mutants may be classified into two different phenotypes, corresponding to the two classes discussed above. The phenotype PAb1620–/PAb240+/hsp70+ corresponds to all mutations found in codon 175 and to a single mutant in codon 273 (Arg → Pro) whereas the PAb1620+/PAb240–/hsp70– phenotype corresponds to mutations in codons 248 and 273. No intermediate cases were found and each of these mutant p53 proteins had lost its transactivation and growth inhibition activities (Ory *et al.*, 1994). In fact, the conformational changes in p53 have been dissected using a new battery of monoclonal antibodies directed against the central region of the protein (Legros *et al.*, 1994b). All these antibodies recognize different epitopes in the central region of p53. Like PAb240, none of these antibodies was able to recognize native, wild-type p53. On the other hand, they were all able to recognize the class I mutants described above, suggesting that these mutants all undergo an overall conformational change that loosens up the compact structure of the protein (Legros *et al.*, 1994b).

7.4 Oncogene or anti-oncogene?

Is *p53* an oncogene or an anti-oncogene? Experiments showing that p53 and the activated *HRAS* oncogene could cooperate to produce transformed foci led to the classification of *p53* as a nuclear oncogene for some time, yet studies in other systems contradicted this notion and suggested that p53 belongs among the tumour suppressor genes. These genes are generally considered as negative regulators of cell growth. They exhibit the following characteristics: (i) loss of function occurring through inactivation of both alleles of the gene; (ii) reintroduction of a tumour suppressor gene in a tumour cell abolishes the tumorigenic power of that cell; (iii) because they are recessive, the tumour suppressor genes are generally implicated in hereditary cancers.

7.4.1 Inactivation of p53 *in Friend murine erythroleukaemia*

In tumours induced by the Friend virus, the *p53* gene found in the tumour cells is very often rearranged, leading to an absence of expression or the synthesis of a truncated or mutant protein (Mowat *et al.*, 1985). The mutation often affects one of the conserved blocks of the protein (Munroe *et al.*, 1988). In all cases studied, the second allele is either lost as a result of total chromosome loss, or partial deletion. In this tumour model, functional inactivation of the *p53* gene seems to confer a selective growth advantage to erythroid cells during the development of Friend leukaemia *in vivo*.

7.4.2 Wild-type p53 does not cooperate with HRAS

Comparison of the different murine p53 sequences used for transfection experiments shows that they carry one or several mutations in the central region of the protein. The use of these mutant p53s derives from the fact that the p53 cDNAs were all isolated from transformed cell banks in which the *p53* gene is rearranged. It has been demonstrated that these mutations are required for the protein to cooperate with the activated *HRAS* oncogene (Finlay *et al.*, 1988). Wild-type p53 is not only unable to cooperate with activated *HRAS*; in addition, it inhibits the cooperation between mutant p53 and this oncogene (Eliyahu *et al.*, 1989; Finlay *et al.*, 1989). An analogous effect is observed when mutant p53 is replaced by the adenovirus E1A antigen.

7.4.3 p53 *is inactivated in many human cancers*

The expression of *p53* in different human cancers or in tumour cell lines has been under study by several different investigators for some time. Expression levels are often high, but no precise explanations exist for this phenomenon because, apart from the case of several osteosarcomas, no gene rearrangements, detectable by Southern blotting, have been detected. Genetic analysis of colorectal cancer reveals a very high rate of heterozygous loss of the short arm of chromosome 17, which carries the *p53* gene (Vogelstein *et al.*, 1988). PCR analysis and sequencing of the remaining *p53* allele shows that it often contains a point mutation (Baker *et al.*, 1989). Similar observations have been made in the case of lung cancer (Takahashi *et al.*, 1989). On the heels of these initial observations have come several hundred reports of alterations of the *p53* gene in all types of human cancer (see Section 7.5). In many cases these mutations are accompanied by a heterozygous loss of the short arm of chromosome 17 (Hollstein *et al.*, 1991; Caron de Fromentel and Soussi, 1992).

7.4.4 p53 *and hereditary cancers*

Transgenic mice carrying a mutant *p53* gene develop many types of cancer, with a high proportion of sarcomas (Lavigueur *et al.*, 1989). This observation led various authors to study patients with Li–Fraumeni syndrome. This syndrome presents as a familial association of a broad spectrum of cancers including osteosarcomas, breast cancer, soft tissue sarcoma and leukaemias, appearing at a very early age. Statistical analysis predicts that 50% of these individuals will have a tumour before the age of 30, and 90% before the age of 70. Germline mutations in the *p53* gene have been found in several families with this syndrome (Malkin *et al.*, 1990; Srivastava *et al.*, 1990). In all cases there is a strict correlation between transmission of the mutant allele and development of a cancer.

Taken together, these data made it possible to define the *p53* gene as a tumour suppressor gene. However, unlike the *RB* gene, which is the archetype of the tumour suppressor genes, the *p53* gene has some original features. In particular, more than 95% of alterations in the *p53* gene are point mutations that produce a mutant protein, which in all cases has lost its transactivational activity (see above). Nevertheless, the synthesis of these mutant p53 proteins is not without consequence for the cell. In particular, it has been shown that some p53 mutants (depending on the site of mutation) exhibit a transdominant phenotype and are able to associate with wild-type p53 (expressed by the remaining wild-type allele) to induce the formation of an inactive heteroligomer (Milner and Medcalf, 1991). Moreover, co-transfection of mutant *p53* with an activated *RAS* gene shows that some p53 mutants have pronounced dominant oncogenic activity (Halevy *et al.*, 1990). These observations led to the proposal that several classes of mutant p53 exist, according to the site of mutation and its phenotype (Michalovitz *et al.*, 1991): (i) null mutations which totally inactivate p53 and which do not directly intervene in transformation; (ii) dominant negative mutations with a totally inactive p53 which is still able to interfere with p53 expressed by the wild-type allele; and (iii) positive dominant mutations where the normal function of p53 is altered but, in this case, the mutant p53 acquires an oncogenic activity that is directly involved in transformation.

7.5 *p53* and human cancers

Mutations in the *p53* gene have been found in most human cancers. Their frequency varies from one cancer to another but, taken together for all cancers, it is of the order of 45–50% (Hollstein *et al.*, 1991; Caron de Fromentel and Soussi, 1992). In general, these mutations are associated with a loss of the second allele of the gene. Analysis of mutations in these different types of cancer reveals some very interesting differences with respect to their site within the gene and the type of mutational event that caused them. Moreover, the relationship between the presence of a mutation in the *p53* gene and different clinical parameters shows that these mutations may have a bearing on clinical prognosis.

7.5.1 Analysis of mutations in the p53 *gene*

As of September 1994, there are over 400 literature reports of mutations in the *p53* gene in various types of human cancer. Two databases of *p53* gene mutations

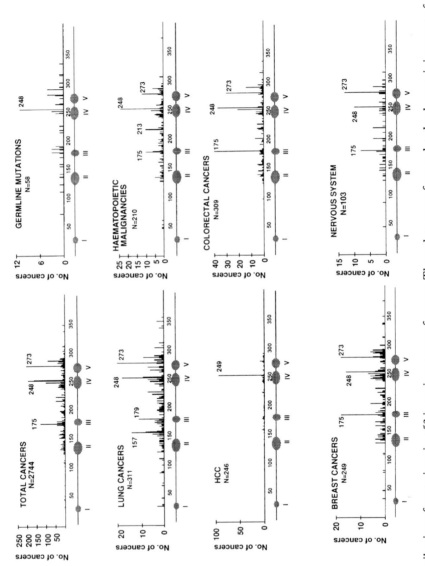

Figure 7.5. Distribution of mutations in *p53* in various types of cancer. These data come from a data bank containing most of the *p53* gene mutations reported in the literature up to March 1994. HCC, hepatocarcinomas.

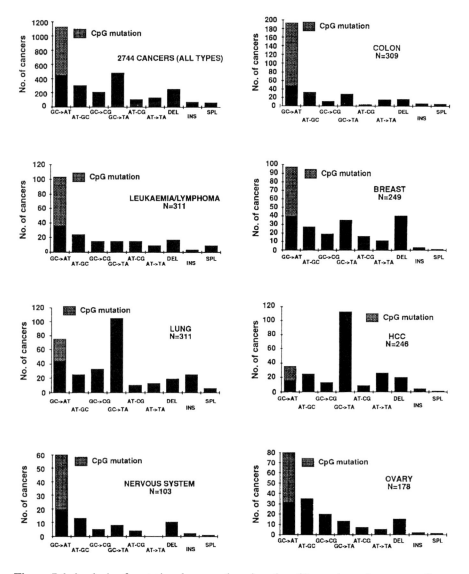

Figure 7.6. Analysis of mutational events that alter the *p53* gene in various types of cancer. These data come from a data bank containing most of the *p53* gene mutations reported in the literature up to March 1994.

have been set up and these data will be transmitted to the EMBL for general availability to the scientific community. This database contains records of 3400 mutations in the *p53* gene. The analysis of these mutations may be carried out at various levels: position of the amino acid mutated in the p53 protein (*Figure 7.5*), and at the level of the mutational events that altered the gene itself (*Figure 7.6*).

Mutations or variations? It is, of course, important to show that these mutations are truly deleterious and inactivate p53 function. Different observations prove that they are truly acquired mutations: (i) they are present only in tumour tissue and absent in healthy tissue from the same patient; (ii) sequence analysis of the *p53* gene from many healthy tissues shows that there is very little polymorphism in the human *p53* gene; (iii) these alterations are found only in the highly conserved blocks of p53, i.e. in the functionally important regions of the protein; (iv) analysis of the properties of various p53 mutants frequently found in human cancers shows that they have lost their transactivational activity. However, a more recent study of more than 20 mutants indicates that some of the very low frequency mutations do not alter any of the properties of the p53 protein (transactivation or growth inhibition) (Ory *et al.*, 1994).

Distribution of mutations in the p53 *gene.* *Figure 7.5* shows that 95% of mutations are located in the central region of p53, more particularly in four hot spots that co-localize with four of the five evolutionary conserved blocks (70% of the mutations are found in the 76 amino acids which make up these blocks). Four amino acids (codons 175, 248, 249 and 273) contain 28% of all mutations. Analysis of the distribution of mutations in several types of cancer shows certain biases in some types of cancer: (i) in lung cancer there are no mutations in codon 175 but many in the region 151–159; (ii) in hepatocarcinomas and radon-induced lung cancer, more than 80% of the mutations are located in codon 249. We will see below that these trends are a reflection of the mode of action of various carcinogens.

p53 *mutations and exposure to chemical carcinogens.* Analysis of the nature of mutations in different types of tumours defines the mutational spectra according to the cancer under study. Generally speaking, there are two types of genetic alterations, those derived from endogenous processes resulting from errors occurring during the various biological processes linked to DNA metabolism, and those of exogenous origin involving environmental factors. The location and type of substitution occurring as a result of these two types of alteration are different. It is therefore possible to use the spectra of these mutations to study the aetiology of a cancer.

Endogenous mutations. Analysis of all mutational events affecting the *p53* gene shows that 42% are G:C $\rightarrow$ A:T transitions, 60% of which affect a CpG dinucleotide (*Figure 7.6*). It is well known that spontaneous deamination of 5-methylcytosine at these nucleotides may be an important cause of this type of transition. In fact, the three hot-spot codons, 175, 248 and 273, contain such a dinucleotide. More than 90% of the mutational events in these codons are compatible with a deamination phenomenon. This observation is confirmed by various studies showing that codons 248 and 273 are methylated *in vivo* (Rideout *et al.*, 1990; Magewu and Jones, 1994). Analysis of the mutational events in cancers such as colon cancer, malignant haemopathies and brain cancer (cancers known to be unrelated to exogenous carcinogens) shows that the mutation rate at the CpG dinucleotide is very high, thus suggesting that most of the mutations that alter the *p53* gene in these cancers are due to endogenous processes related to the deamination of 5-methylcytosine.

Hepatocarcinoma and aflatoxin B1.   In 1990, there were two reports of mutations in the *p53* gene in hepatocarcinomas (HCC), with a predominance of the GC → TA transversion at the third base of codon 249 (Arg → Ser) (Bressac *et al.*, 1991; Hsu *et al.*, 1991). In one case, the patient series was from Mozambique while, in the second, it was from the Qidong province in China. These two regions are known for their consumption of food contaminated by the fungus *Aspergillus flavis*, producer of aflatoxin B1, which is a very potent hepatic carcinogen implicated in the development of HCC, and known to interact synergistically with the HBV. A world-wide epidemiological study showed that the mutation in codon 249 is strictly specific to countries in which the food is contaminated by aflatoxin B1 (Ozturk *et al.*, 1991). In Mozambique, for example, more than 50% of the mutations were found in codon 249 while in Transkei, which borders on Mozambique (and which has a similar rate of chronic HBV infection but no contamination by aflatoxin B1), the mutation rate at codon 249 is less than 10%. In fact, in countries which do not consume contaminated food (including European countries and the USA), the rate of *p53* mutations in HCC is low. It has been demonstrated both *in vitro* and *in vivo* that this phenomenon is due to a very high sensitivity of codon 249 to the action of aflatoxin B1 (Puisieux *et al.*, 1991; Aguilar *et al.*, 1993, 1994). This observation, together with the fact that this mutation is deleterious for p53 function, explains the existence of this mutational hot spot (Ponchel *et al.*, 1994).

Skin cancer and UV radiation.   In squamous cell carcinoma of the skin, C → T mutations predominate in pyrimidine dimers (Brash *et al.*, 1991; Ziegler *et al.*, 1993). It is well known that UV radiation, an aetiological agent of most skin cancers, acts directly on these dimers. A particular characteristic of the action of UV radiation is the change in the bases CC → TT observed in the series (reported by Brash) but also in other skin cancer series such as basocellular cancers (Rady *et al.*, 1992). In patients with genetic DNA repair deficiencies, such as xeroderma pigmentosum (XP), the phenotype is much more marked (Dumaz *et al.*, 1993; Sato *et al.*, 1993). All mutations found in skin cancers are located on the pyrimidine dimers and 55% are tandem mutations CC → TT (Dumaz *et al.*, 1994). This type of mutation is only very rarely found in internal cancers (less than 1%). In skin cancers from XP patients, more than 95% of the mutations are located on the noncoding strand of the *p53* gene while, in other skin tumours or in internal cancers, no special trends are observed. This result therefore suggests that there is preferential repair of the coding strand, which has been confirmed by Tornaletti and Pfeifer (1994) who showed that the repair rate of pyrimidine dimers in the *p53* gene is highly variable, with an especially low rate in the codons that are often mutated in skin cancer. These results taken together (predominance of CC → TT lesions on the non-coding strand) were experimentally confirmed in animals carrying UV-induced tumours (Kress *et al.*, 1992).

Bronchopulmonary cancers and smoking.   Mutational events that alter the *p53* gene in lung cancer (Chiba *et al.*, 1990; Iggo *et al.*, 1990; Takahashi *et al.*, 1991; D'Amico *et al.*, 1992; Mitsudomi *et al.*, 1992) are frequently GC → TA transversions with a minority of transition mutations (less than in other cancers). Moreover, most of the mutated guanines are on the non-coding strand of the gene. This observation is altogether compatible with the role of exogenous carcinogens such as benzo(*a*)pyrene, present in cigarette smoke, and the observation that the non-coding strand is less efficiently repaired. A recent study on bronchial cancers in radon miners revealed a mutational hot spot in codon 249 (16 of 29 mutations) (Taylor *et al.*, 1994). The mutation differs from that seen in HCC because it affects the second base of codon 249 (AGG → ATG). This suggests that radon is responsible for this particular signature, since this mutation is found in less than 1% of other lung cancers. Nonetheless, this result should be interpreted with caution since some authors have suggested that this mutation is due to a mycotoxin synthesized by a fungus often found in the bronchi of radon miners (Venitt and Biggs, 1994).

p53 *gene: a model for molecular epidemiology?*   In order for a particular gene to be used to study the origin of mutagenesis in the human population, it must exhibit the following properties: (i) it must be mutated in a large number of cancers; (ii) the mutation rate must be high; (iii) it must be altered mainly by point mutations; and (iv) molecular analysis of the gene must be relatively easy to carry out (small size gene). At present these characteristics are found in two genes, the *HRAS* oncogene and the *p53* gene. One of the disadvantages of HRAS is the small number of codons (three) that are the target of mutations. In contrast, more than 100 of the 393 codons in the p53 gene can be modified. Moreover, the *p53* gene is mutated in more than 50% of cancers. It is therefore possible to undertake molecular epidemiological studies with the aim of seeking specific signatures of certain carcinogens and demonstrating their role in the development of neoplasia (Dumaz *et al.*, 1994).

7.5.2 Epigenetic modification of p53

There are currently two examples of alterations in *p53* that do not involve direct alteration of the gene itself.

(i) The *MDM2* gene is amplified in human sarcomas (Oliner *et al.*, 1992). As described above, the MDM2 product inactivates the p53 transactivation function. One would therefore expect that those tumours with *MDM2* amplification would be devoid of *p53* mutations. This hypothesis was confirmed by several authors (Leach *et al.*, 1993; Cordon-Cardo *et al.*, 1994). Analysis of 24 human soft tissue sarcomas (11 malignant fibrous histiocytomas and 13 liposarcomas) showed that *p53* alteration could be found in 24 of the sarcomas and *MDM2* amplification was detected in another eight tumours, but no tumour contained an alteration in both genes (Leach *et al.*, 1993). This figure was also observed in brain tumours where *MDM2* amplification was observed in 10% of glioblastomas and astrocytomas (Reifenberger *et al.*, 1993). These results strongly suggest that *MDM2* amplification can be an alternative molecular mechanism by which *p53* is inactivated.

(ii) There is recent evidence associating specific HPVs with certain human anogenital cancers (Zur Hausen and Schneider, 1987; Howley, 1991), most notably cervical cancer. Recent studies have demonstrated that 84% of cervical carcinomas contain DNA from a high-risk HPV (mostly HPV16 and 18 and, to a lesser extent, HPV 31, 33, 35, 39, 45, 51, 52 and 56). The DNA is usually found to be integrated but there are some cases where it is apparently extrachromosomal. The finding that the E6 protein from high-risk HPV can induce the degradation of p53 either *in vitro* or *in vivo* has led to the proposal that such an inactivation pathway could be involved in the neoplastic process leading to a cervical cancer. This observation prompted some investigators to study the distribution of *p53* mutations in human primary cervical carcinoma (or cell lines) with and without HPV infection. In a first report, Crook *et al.* (1991) showed that six HPV-positive cervical cell lines expressed wild-type *p53*, whereas two apparently HPV-negative lines expressed mutant *p53*. Scheffner *et al.* (1991) reported that two other HPV-negative cervical cell lines expressed mutant *p53*. Analysis of tumour samples from 28 women with primary cancer of the cervix showed that 25 were HPV(16 or 18) positive but sequencing of the entire coding region of the *p53* gene failed to reveal any mutation (Crook *et al.*, 1992). By contrast, sequencing revealed point mutations in *p53* from the three HPV-negative tumours.

The fact that HPV-negative carcinomas have a worse prognosis than HPV-positive ones reinforces these results. Inactivation of p53 by E6 protein only leads to the loss of functional p53, whereas somatic mutation results in the expression of an altered p53 protein, which can elicit positive transforming activity (by interfering with wild-type p53).

This very attractive model is still subject to some controversy as several authors do not find such an inverse correlation between HPV-positive cancer and *p53* mutation (Borresen *et al*, 1992; Busby-Earle *et al.*, 1992 Mcgregor *et al.*, 1992). A recent report from Fujita *et al.* (1992) suggests that the number of HPV copies present in the transformed cell could be a more critical parameter.

7.5.3 Analysis of alterations in the p53 gene

Molecular analysis. PCR, followed by sequence analysis, allows direct study of the nature of the mutational event in the gene. In over 90% of cases, this event is a point mutation that alters only one of the 23 000 nucleotides in the gene. Unlike the *HRAS* gene, for which only three of the 189 codons are targets of oncogenic mutations, *p53* gene mutations may occur in 90 of the 393 codons required for the synthesis of the protein. This high degree of heterogeneity makes diagnosis more difficult because the region to be analysed extends over almost the entire gene. Thus, the molecular analysis of the *p53* gene is somewhat tricky, and unsuited for routine diagnostic analysis. It is, nonetheless, essential for molecular epidemiological analyses in which it is important to determine the relationship between the type of cancer and the type of mutational event. Semi-direct detection methods such as single-stranded conformation polymorphism (SSCP) or denaturant gradient gel electrophoresis (DGGE) may enable selection of the region of the gene to be analysed. However, these methods are not 100% reliable (see Moyret *et al.* (1994) for comparison between SSCP and DGGE). In the not-so-distant future, it is possible to imagine that new technical advances in sequencing methods (such as sequencing by hybridization on solid support) or in screening point mutations (such as ligase chain reaction) and associated techniques will permit routine molecular analysis.

Immunocytochemical analysis. One of the significant properties of mutant p53 proteins is their extended half-life. In normal cells, p53 is undetectable because it has an extremely short half-life (15–20 min). In transformed cells, the mutant protein is much more stable, with a half-life of 4–12 h, and it accumulates in the nucleus. It is therefore possible to perform an immunocytochemical diagnosis (coupled with a histological analysis) in tumour tissue to visualize this nuclear accumulation directly. This approach has been used in many different types of cancer with a generally good correlation between the molecular analysis (presence of a mutation) and the immunohistochemical analysis (overexpression of the mutant protein) (Dowell *et al.*, 1994; Hall and Lane, 1994). The advantage of this approach is that it can be routinely used in histology laboratories but there are also a number of disadvantages, including the fact that any mutations that abolish *p53* expression (splicing signal mutations, nonsense mutations, insertions or deletions) do not produce the protein and therefore give a negative result. These types of mutations are found in 5–10% of cases. However, it is now known that tumours can overexpress *p53* in the complete absence of mutations in any part of the gene. Since this overexpression is specific to tumour cells and does not affect normal tissue, this result is generally considered to be a consequence of an alteration of *p53* by unknown mechanisms. The problems of evaluating *p53* overexpression in tumours have been addressed in several recent reviews (Wynford-Thomas, 1992; Dowell *et al.*, 1994; Hall and Lane, 1994).

In the past 2 years, new monoclonal antibodies against human p53 have been produced by several laboratories (Midgley *et al.*, 1992; Vojtesek *et al.*, 1992; Legros *et al.*, 1993, 1994a; Tenaud *et al.*, 1994). Their advantage is that they can be used for immunohistochemical diagnosis in highly varied conditions, such as detection of p53 in paraffin-embedded sections after fixation in formalin or Bouin's solution (Tenaud *et al.*, 1994).

Serological analysis.   Ten per cent of breast cancer patients have anti-p53 antibodies in their serum (Crawford *et al.*, 1982). This value reaches 20% in children with B-cell lymphomas, while it is zero in patients with T-cell lymphomas (Caron de Fromentel *et al.*, 1987). These studies, conducted during a lull in scientific interest in *p53*, were reviewed more recently in light of new knowledge on *p53* inactivation and stabilization. Anti-p53 antibodies have been found in most human cancers (Davidoff *et al.*, 1992; Hassapoglidoi and Diamandis, 1992; Schlichtholz *et al.*, 1992; Winter *et al.*, 1992; Angelopoulou *et al.*, 1994). There is generally a good correlation between their frequency and that of *p53* gene alterations. In lung cancer, which has a high rate of *p53* mutation, the frequency of these anti-p53 antibodies is high (24%) (Schlichtholz *et al.*, 1994). In prostate cancer, where the *p53* mutation rate is low, or in mesotheliomas, where it is nil, the incidence of seropositivity is very low. Several multifactorial studies show a very good correlation between the presence of anti-p53 antibodies, overexpression of the mutant protein in the tumour and the presence of a mutation in the gene (Davidoff *et al.*, 1992; Winter *et al.*, 1992; Lubin *et al.*, 1993). Detailed analysis of these antibodies indicates that they recognize both wild-type and mutant p53 (Schlichtholz *et al.*, 1992, 1994; Labrecque *et al.*, 1993; Lubin *et al.*, 1993). The epitopes are mainly located in the amino and carboxy terminal regions of the protein, regions which are not in the hot-spot areas (Schlichtholz *et al.*, 1992, 1994; Lubin *et al.*, 1993). These immunodominant epitopes have also been detected in the serum of mice hyperimmunized with wild-type p53 (Legros *et al.*, 1994a). Taken together, these studies show that overexpression of the p53 protein in tumour cells is responsible for the appearance of auto-antibodies. Serological analysis has the following advantages: (i) simplicity of analysis (enzyme-linked immunosorbent assay); (ii) no need for tumour tissue; (iii) the possibility of following the fate of the antibodies during treatment of the patient.

7.5.4 p53 *gene alteration and clinical parameters*

Is it possible to correlate *p53* gene alterations with diagnostic or prognostic clinical parameters, and how may this information contribute to the treatment choice? In fact, since Cattoreti's study in 1988 the analysis of *p53* overexpression in breast cancer shows that it occurs mainly in patients with a poor prognosis (absence of oestrogen receptors and high-grade tumours). These findings in breast cancer have since been confirmed by either a molecular, immunohistochemical or serological approach (Callahan, 1992; Schlichtholz *et al.*, 1992;

Allred *et al.*, 1993; Andersen *et al.*, 1993; Thor and Yandell, 1993). Thor *et al.* (Thor *et al.*, 1992; Barnes *et al.*, 1993) reported that *p53* gene alteration could be considered as a new, independent marker associated with lower patient survival. The most promising result is described by Allred *et al.* (1993), who analysed a series of 700 N0 breast cancer patients (no lymph node involvement). Fifty-two per cent (362/700) of patients had a *p53* alteration (measured mainly by immuno-histochemistry). These same patients had a much shorter relapse-free survival than those without p53 accumulation. If confirmed, this result would be very important since it is currently quite difficult to evaluate the prognosis of N0 breast cancer and, in particular, the use of adjuvant therapy following surgery. *p53* analysis in these specific cases could be of major importance.

Many similar studies have been conducted in other types of cancer, with results that are not so clear-cut as in breast cancer. Overall, tumours with p53 accumulation are generally high grade and more aggressive. For lung cancer, the results diverge as to whether p53 accumulation is related to a poor prognosis (cf. letters by Mitsudomi (1994) and Passlick *et al.* (1994)). On the one hand, it is important to establish some level of standardization so that studies of p53 accu-mulation can be comparable from one series to another. On the other hand, again taking the case of lung cancer, it is now clear that *p53* gene alterations are an early event (Sundaresan *et al.*, 1992; Bennett *et al.*, 1993). In a study of lung tis-sues containing preinvasive squamous neoplasms from patients with or without lung cancer, Bennett *et al.* (1993) did not observe p53 protein accumulation in normal mucosas, but did in 6.7% of squamous metaplasias, 29.5% of mild dys-plasias, 26.9% of moderate dysplasias and 59.7% of severe dysplasias. More recently, Mao *et al.* (1994) were able to detect *p53* mutations in patient sputum specimens obtained 1 year prior to clinical diagnosis of lung cancer. Finally, Schlichtholz *et al.* (1994) were able to detect anti-p53 antibodies in high-risk patients (heavy smokers) more than 1 year prior to clinical signs of cancer. These results suggest that, in this type of cancer, p53 could be used as an early marker of the presence of a cancerous lesion containing an alteration of the *p53* gene.

7.5.5 p53 *in the clinical setting – what does the future hold?*

As discussed above, the detection of *p53* gene alterations in human cancers is now possible (Soussi *et al.*, 1994). The use of molecular epidemiological studies to analyse the role of chemical carcinogens in the neoplastic process will cer-tainly become a major research theme.

The application of our knowledge of *p53* in the clinical setting will most like-ly take longer, all the more so since, apart from breast cancer, the relationships between clinical parameters and *p53* gene alterations remain cloudy. However, Lowe *et al.* (1993) have reported a very interesting new finding concerning *p53*. It has long been known that some antitumour agents such as fluorouracil and ionizing radiation act by inducing apoptosis in tumour cells. These investigators showed that cells expressing mutant *p53* are totally resistant to apoptosis, whereas cells expressing wild-type *p53* are sensitive to these therapeutic agents. This observation is of major interest for several reasons. First, knowledge of the state of the *p53* gene in a tumour makes it possible to make a suitable treatment choice. Moreover, it suggests that the reintroduction of wild-type *p53* in these tumour cells or the conversion of mutant *p53* to a wild-type conformation could induce apoptosis in these cells. In fact, this hypothesis was recently proved correct, as Fujiwara *et al.* (1994) showed that infection of a cisplatin-resistant tumour with a recombinant adenovirus expressing wild-type *p53* induces a return to chemosensitivity of the tumour, which is destroyed by apoptosis. It is therefore not illusory to predict that after getting off to a rather poor start in the world of oncogenes and tumour suppressor genes, *p53* may have a very promising future.

References

Agoff SN, Hou J, Linzer DIH, Wu B. (1993) Regulation of the human hsp70 promoter by p53. *Science* **259**: 84–87.

Aguilar F, Hussain SP, Cerutti P. (1993) Aflatoxin-B(1) induces the transversion of G→T in codon 249 of the p53 tumor suppressor gene in human hepatocytes. *Proc. Natl Acad. Sci. USA* **90**: 8586–8590.

Aguilar F, Harris CC, Sun T, Hollstein M, Cerutti P. (1994) Geographic variation of p53 mutational profile in nonmalignant human liver. *Science* **264**: 1317–1319.

Allred DC, Clark GM, Elledge R, Fuqua SAW, Brown RW, Chamness GC, Osborne CK, Mcguire WL. (1993) Association of p53 protein expression with tumor cell proliferation rate and clinical outcome in node-negative breast cancer. *J. Natl Cancer Inst.* **85**: 200–206.

Andersen TI, Holm R, Nesland JM, Heimdal KR, Ottestad L, Borresen AL. (1993) Prognostic significance of TP53 alterations in breast carcinoma. *Br. J. Cancer* **68**: 540–548.

Angelopoulou K, Diamandis EP, Sutherland DJA, Kellen JA, Bunting PS. (1994) Prevalence of serum antibodies against the p53 tumor suppressor gene protein in various cancers. *Int. J. Cancer* **58**: 480–487.

Arai N, Nomura D, Yokota K, Wolf D, Brill E, Shohat O, Rotter V. (1986) Immunologically distinct p53 molecules generated by alternative splicing. *Mol. Cell. Biol.* **6**: 3232–3239.

Baker SJ, Fearon ER, Nigro J, Hamilton S, Preisinger AC, Jessup JM, van Tuinen P, Ledbetter DH, Barker DF, Nakamura Y, Whyte R, Vogelstein B. (1989) Chromosome 17 deletions and p53 gene mutations in colorectal carcinomas. *Science* **244**: 217–221.

Baker SJ, Markowitz S, Fearon ER, Willson JKV, Vogelstein B. (1990) Suppression of human colorectal carcinoma cell growth by wild-type-P53. *Science* **249**: 912–915.

Barak Y, Oren M. (1992) Enhanced binding of a 95-kDa protein to p53 in cells undergoing p53-mediated growth arrest. *EMBO J.* **11**: 2115–2121.

Barak Y, Juven T, Haffner R, Oren M. (1993) mdm2 expression is induced by wild type-p53 activity. *EMBO J.* **12**: 461–468.

Bargonetti J, Friedman PN, Kern SE, Vogelstein B, Prives C. (1991) Wild-type but not mutant p53 immunopurified proteins bind to sequences adjacent to the SV40 origin of replication. *Cell* **65**: 1083–1091.

Bargonetti J, Manfredi JJ, Chen XB, Marshak DR, Prives C. (1993) A proteolytic fragment from the central region of p53 has marked sequence-specific DNA-binding activity when generated from wild-type but not from oncogenic mutant p53-protein. *Gene Devel.* **7**: 2565–2574.

Barnes DM, Dublin EA, Fisher CJ, Levison DA, Millis RR. (1993) Immunohistochemical detection of p53 protein in mammary carcinoma: an important new independent indicator of prognosis? *Hum. Pathol.* **24**: 469–476.

Baudier J, Delphin C, Grunwald D, Khochbin S, Lawrence JJ. (1992) Characterization of the tumor suppressor protein-p53 as a protein kinase-C substrate and a S100b-binding protein. *Proc. Natl Acad. Sci. USA* **89**: 11627–11631.

Ben–Dori R, Resnitzky D, Kimchi A. (1987) Changes in p53 mRNA expression during terminal differentiation of murine erythroleukemia cells. *Virology* **161**: 607–611.

Bennett WP, Colby TV, Travis WD, Borkowski A, Jones RT, Lane DP, Metcalf RA, Samet JM, Takeshima Y, Gu JR, Vahakangas KH, Soini Y, Paakko P, Welsh JA, Trump BF, Harris CC. (1993) p53 protein accumulates frequently in early bronchial neoplasia. *Cancer Res.* **53**: 4817–4822.

Bienz B, Zakut–Houri R, Givol D, Oren M. (1984) Analysis of the gene coding for the murine cellular tumour antigen p53. *EMBO J.* **3**: 2179–2183.

Bischoff JR, Friedman PN, Marshak DR, Prives C, Beach D. (1990) Human p53 is phosphorylated by P60-Cdc2 and cyclin-B-Cdc2. *Proc. Natl Acad. Sci. USA* **87**: 4766–4770.

Borresen AL, Helland A, Nesland J, Holm R, Trope C, Kaern J. (1992) Papillomaviruses, p53, and cervical cancer. *Lancet* **339**: 1350–1351.

Brash DE, Rudolph JA, Simon JA, Lin A, McKenna GJ, Baden HP, Halperin AJ, Ponten J. (1991) A role for sunlight in skin cancer – UV-induced p53 mutations in squamous cell carcinoma. *Proc. Natl Acad. Sci. USA* **88**: 10124–10128.

Bressac B, Kew M, Wands J, Ozturk M. (1991) Selective G-mutation to T-mutation of p53 gene in hepatocellular carcinoma from southern africa. *Nature* **350**: 429–431.

Busby-Earle RMC, Steel CM, Williams ARW, Cohen B, Bird CC. (1992) Papillomaviruses, p53, and cervical cancer. *Lancet* **339**: 1350.

Callahan R. (1992) p53 mutations, another breast cancer prognostic factor. *J. Natl Cancer Inst.* **84**: 826–827.

Caron de Fromentel C, Soussi T. (1992) TP53 tumor suppressor gene: a model for investigating human mutagenesis. *Genes Chrom. Cancer* **4**: 1–15.

Caron de Fromentel C, May–Levin F, Mouriesse H, Lemerle J, Chandrasekaran K, May P. (1987) Presence of circulating antibodies against cellular protein p53 in a notable proportion of children with B-cell lymphoma. *Int. J. Cancer* **39**: 185–189.

Caron de Fromentel C, Pakdel F, Chapus A, Baney C, May P, Soussi T. (1992) Rainbow trout p53 – cDNA cloning and biochemical characterization. *Gene* **112**: 241–245.

Cattoretti G, Rilke F, Andrealo S, D'Amato L, Delia D. (1988) p53 expression in breast cancer. *Int. J. Cancer* **41**: 178–183.

Chandrasekaran K, McFarland V, Simmons D, Dziadek M, Gurney E, Mora P. (1981) Quantitation and characterization of a species-specific and embryo stage dependent 55-kilodalton phosphoprotein also present in cells transformed by simian virus 40. *Proc. Natl Acad. Sci. USA* **78**: 6953–6957.

166

Chandrasekaran K, Mora PT, Nagarajan L, Anderson WB. (1982) The amount of a specific cellular protein (p53) is a correlate of differentiation in embryonal carcinoma cells. *J. Cell Physiol.* **113**: 134–140.

Chen JD, Marechal V, Levine AJ. (1993) Mapping of the p53 and mdm-2 interaction domains. *Mol. Cell Biol.* **13**: 4107–4114.

Chen PL, Chen YM, Bookstein R, Lee WH. (1990) Genetic mechanisms of tumor suppression by the human p53 gene. *Science* **250**: 1576–1580.

Chiba I, Takahashi T, Nau MM, d'Amico D, Curiel DT, Mitsudomi T, Buchhagen DL, Carbon D, Piantadosi S, Koga H, Reissman PT, Slamon DJ, Holmes EC, Minna JD. (1990) Mutations in the p53 gene are frequent in primary, resected non-small-cell lung cancer. *Oncogene* **5**: 1603–1610.

Chin KV, Ueda K, Pastan I, Gottesman MM. (1992) Modulation of activity of the promoter of the human MDR1 gene by Ras and p53. *Science* **255**: 459–462.

Cho YJ, Gorina S, Jeffrey PD, Pavletich NP. (1994) Crystal structure of a p53 tumor suppressor DNA complex: understanding tumorigenic mutations. *Science* **265**: 346–355.

Cordon-Cardo C, Latres E, Drobnjak M, Oliva MR, Pollack D, Woodruff JM, Marechal V, Chen JD, Brennan MF, Levine AJ. (1994) Molecular abnormalities of mdm2 and p53 genes in adult soft tissue sarcomas. *Cancer Res.* **54**: 794–799.

Crawford L. (1983) The 53,000-dalton cellular protein and its role in transformation. *Int. Rev. Exp. Pathol.* **25**: 1–50.

Crawford LV, Pim DC, Bulbrook RD. (1982) Detection of antibodies against the cellular protein p53 in sera from patients with breast cancer. *Int. J. Cancer* **30**: 403–408.

Crook T, Wrede D, Vousden KH. (1991) p53 point mutation in HPV negative human cervical carcinoma cell lines. *Oncogene* **6**: 873–875.

Crook T, Wrede D, Tidy JA, Mason WP, Evans DJ, Vousden KH. (1992) Clonal p53 mutation in primary cervical cancer – association with human-papillomavirus-negative tumours. *Lancet* **339**: 1070–1073.

D'Amico D, Carbone D, Mitsudomi T, Nau M, Fedorko J, Russell E, Johnson B, Buchhagen D, Bodner S, Phelps R, Gazdar A, Minna JD. (1992) High frequency of somatically acquired p53 mutations in small-cell lung cancer cell lines and tumors. *Oncogene* **7**: 339–346.

Dang CV, Lee WMF. (1989) Nuclear and nucleolar targeting sequences of C-Erb-A, C-Myb, N-myc, p53, Hsp70, and HIV TAT proteins. *J. Biol. Chem.* **264**: 18019–18023.

Davidoff AM, Iglehart JD, Marks JR. (1992) Immune response to p53 is dependent upon p53/HSP70 complexes in breast cancers. *Proc. Natl Acad. Sci. USA* **89**: 3439–3442.

Deb S, Jackson CT, Subler MA, Martin DW. (1992) Modulation of cellular and viral promoters by mutant human p53-proteins found in tumor cells. *J. Virol.* **66**: 6164–6170.

Deb SP, Munoz RM, Brown DR, Subler MA, Deb S. (1994) Wild-type human p53 activates the human epidermal growth factor receptor promoter. *Oncogene* **9**: 1341–1349.

Deppert W, Buschhausendenker G, Patschinsky T, Steinmeyer K. (1990) Cell cycle control of p53 in normal (3T3) and chemically transformed (Meth-A) mouse cells. 2. requirement for cell cycle progression. *Oncogene* **5**: 1701–1706.

Diller L, Kassel J, Nelson CE, Gryka MA, Litwak G, Gebhardt M, Bressac B, Ozturk M, Baker SJ, Vogelstein B, Friend SH. (1990) p53 functions as a cell cycle control protein in osteosarcomas. *Mol. Cell Biol.* **10**: 5772–5781.

Donehower LA, Harvey M, Slagle BL, McArthur MJ, Montgomery CA, Butel JS, Bradley A. (1992) Mice deficient for p53 are developmentally normal but susceptible to spontaneous tumours. *Nature* **356**: 215–221.

Dony C, Kessel M, Gruss P. (1985) Post-transcriptional control of c-myc and p53 expression during differentiation of the embryonal carcinoma cell line F9. *Nature* **317**: 636–639.

Dowell SP, Wilson POG, Derias NW, Lane DP, Hall PA. (1994) Clinical utility of the immunocytochemical detection of p53 protein in cytological specimens. *Cancer Res.* **54**: 2914–2918.

Dulic V, Kaufmann WK, Wilson SJ, Tlsty TD, Lees E, Harper JW, Elledge SJ, Reed SI. (1994) p53-Dependent inhibition of cyclin-dependent kinase activities in human fibroblasts during radiation-induced gl arrest. *Cell* **76**: 1013–1023.

Dumaz N, Drougard C, Sarasin A, Dayagrosjean L. (1993) Specific UV-induced mutation spectrum in the p53 gene of skin tumors from DNA-repair-deficient xeroderma-pigmentosum patients. *Proc. Natl Acad. Sci. USA* **90**: 10529–10533.

Dumaz N, Stary A, Soussi T, Dayagrosjean L, Sarasin A. (1994) Can we predict solar ultraviolet radiation as the causal event in human tumors by analysing the mutation spectra of the p53 gene? *Mutat. Res.* **307**: 375–386.

Dutta A, Ruppert JM, Aster JC, Winchester E. (1993) Inhibition of DNA replication factor RPA by p53. *Nature* **365**: 79–82.

El–Deiry WS, Kern SE, Pietenpol JA, Kinzler KW, Vogelstein B. (1992) Definition of a consensus binding site for p53. *Nature Genet.* **1**: 45–49.

El–Deiry WS, Tokino T, Velculescu VE, Levy DB, Parsons R, Trent JM, Lin D, Mercer WE, Kinzler KW, Vogelstein B. (1993) WAF1, a potential mediator of p53 tumor suppression. *Cell* **75**: 817–825.

El–Deiry WS, Harper JW, O'Connor PM, Velculescu VE, Canman CE, Jackman J, Pietenpol JA, Burrell M, Hill DE, Wang YS, Wiman KG, Mercer WE, Kastan MB, Kohn KW, Elledge SJ, Kinzler KW, Vogelstein B. (1994) Waf1/Cip1 is induced in p53-mediated G1 arrest and apoptosis. *Cancer Res.* **54**: 1169–1174.

Eliyahu D, Michalovitz D, Eliyahu S, Pinhasikimhi O, Oren M. (1989) Wild-type p53 can inhibit oncogene-mediated focus formation. *Proc. Natl Acad. Sci. USA* **86**: 8763–8767.

Fakharzadeh S, Trusko RS, George D. (1991) Tumorigenetic potential associated with enhanced expression of a gene that is amplified in a mouse tumor cell line. *EMBO J.* **2**: 495–501.

Feinstein E, Gale RP, Reed J, Canaani E. (1992) Expression of the normal p53 gene induces differentiation of k562 cells. *Oncogene* **7**: 1853–1857.

Feitelson MA, Zhu M, Duan LX, London WT. (1993) Hepatitis-B x-antigen and p53 are associated in vitro and in liver tissues from patients with primary hepatocellular carcinoma. *Oncogene* **8**: 1109–1117.

Fields S, Jang SK. (1990) Presence of a potent transcription activating sequence in the p53 protein. *Science* **249**: 1046–1049.

Finlay CA, Hinds PW, Tan TH, Eliyahu D, Oren M, Levine AJ. (1988) Activating mutations for transformation by p53 produce a gene product that forms an hsc70–p53 complex with an altered half life. *J. Virol.* **8**: 531–539.

Finlay CA, Hinds PW, Levine AJ. (1989) The p53 proto-oncogene can act as a suppressor of transformation. *Cell* **57**: 1083–1093.

Fiscella M, Ullrich SJ, Zambrano N, Shields MT, Lin D, Leesmiller SP, Anderson CW, Mercer WE, Appella E. (1993) Mutation of the serine 15 phosphorylation site of human p53 reduces the ability of p53 to inhibit cell cycle progression. *Oncogene* **8**: 1519–1528.

Fritsche M, Haessler C, Brandner G. (1993) Induction of nuclear accumulation of the tumor-suppressor protein p53 by DNA-damaging agents. *Oncogene* **8**: 307–318.

Fujita M, Inoue M, Tanizawa O, Iwamoto S, Enomoto T. (1992) Alterations of the p53 gene in human primary cervical carcinoma with and without human papillomavirus infection. *Cancer Res*. **52**: 5323–5328.

Fujiwara T, Grimm EA, Mukhopadhyay T, Zhang WW, Owenschaub LB, Roth JA. (1994) Induction of chemosensitivity in human lung cancer cells in vivo by adenovirus-mediated transfer of the wild-type p53 gene. *Cancer Res*. **54**: 2287–2291.

Funk WD, Pak DT, Karas RH, Wright WE, Shay JW. (1992) A transcriptionally active DNA-binding site for human p53 protein complexes. *Mol. Cell Biol*. **12**: 2866–2871.

Gannon JV, Greaves R, Iggo R, Lane DP. (1990) Activating mutations in p53 produce a common conformational effect – a monoclonal antibody specific for the mutant form. *EMBO J*. **9**: 1595–1602.

Ginsberg D, Mechta F, Yaniv M, Oren M. (1991) Wild-type p53 can down-modulate the activity of various promoters. *Proc. Natl Acad. Sci. USA* **88**: 9979–9983.

Gu Y, Turck CW, Morgan DO. (1993) Inhibition of CDK2 activity in vivo by an associated 20K regulatory subunit. *Nature* **366**: 707–710.

Hainaut P, Milner J. (1992) Interaction of heat-shock protein-70 with p53 translated in vitro – evidence for interaction with dimeric p53 and for a role in the regulation of p53 conformation. *EMBO J*. **11**: 3513–3520.

Hainaut P, Milner J. (1993a) Redox modulation of p53 conformation and sequence-specific DNA binding in vitro. *Cancer Res*. **53**: 4469–4473.

Hainaut P, Milner J. (1993b) A structural role for metal ions in the wild-type conformation of the tumor suppressor protein-p53. *Cancer Res*. **53**: 1739–1742.

Halazonetis TD, Davis LJ, Kandil AN. (1993) Wild-type p53 adopts a mutant-like conformation when bound to DNA. *EMBO J*. **12**: 1021–1028.

Halevy O, Michalovitz D, Oren M. (1990) Different tumor-derived p53 mutants exhibit distinct biological activities. *Science* **250**: 113–116.

Hall PA, Lane DP. (1994) P53 in tumour pathology – can we trust immunohistochemistry – revisited. *J. Pathol*. **172**: 1–4.

Harper JW, Adami GR, Wei N, Keyomarsi K, Elledge SJ. (1993) The p21 CdK-interacting protein CiP1 is a potent inhibitor of G1 cyclin-dependent kinases. *Cell* **75**: 805–816.

Harvey M, McArthur MJ, Montgomery CA, Bradley A, Donehower LA. (1993) Genetic background alters the spectrum of tumors that develop in p53-deficient mice. *FASEB J*. **7**: 938–943.

Hassapoglidoi S, Diamandis EP. (1992) Antibodies to the p53 tumor suppressor gene product quantified in cancer patients serum with a time-resolved immunofluorometry technique. *Clin. Biochem*. **25**: 445–449.

Herrmann CPE, Kraiss S, Montenarh M. (1991) Association of case in kinase-II with immunopurified p53. *Oncogene* **6**: 877–884.

Hinds PW, Finlay CA, Frey AB, Levine AJ. (1987) Immunological evidence for the association of p53 with a heat shock protein, hsc70, in p53-plus-ras-transformed cell lines. *Mol. Cell Biol*. **7**: 2863–2869.

Hinds PW, Finlay CA, Quartin RS, Baker SJ, Fearon ER, Vogelstein B, Levine AJ. (1990) Mutant p53 DNA clones from human colon carcinomas cooperate with ras in transforming primary rat cells: a comparison of the 'Hot Spot' mutant phenotypes.

Cell Growth Differ. **1**: 571–580.

Hollstein M, Sidransky D, Vogelstein B, Harris CC. (1991) p53 mutations in human cancers. *Science* **253**: 49–53.

Howley PM. (1991) Role of human papillomaviruses in human cancer. *Cancer Res.* **51**: 5019s–5022s.

Hsu IC, Metcalf RA, Sun T, Welsh JA, Wang NJ, Harris CC. (1991) Mutational hotspot in the p53 gene in human hepatocellular carcinomas. *Nature* **350**: 427–428.

Huibregtse JM, Scheffner M, Howley PM. (1991) A cellular protein mediates association of p53 with the E6 oncoprotein of human papillomavirus type-16 or type-18. *EMBO J.* **10**: 4129–4135.

Hupp TR, Meek DW, Midgley CA, Lane DP. (1992) Regulation of the specific DNA binding function of p53. *Cell* **71**: 875–886.

Hupp TR, Meek DW, Midgley CA, Lane DP. (1993) Activation of the cryptic DNA binding function of mutant forms of p53. *Nucleic Acids Res.* **21**: 3167–3174.

Iggo R, Gatter K, Bartek J, Lane D, Harris AL. (1990) Increased expression of mutant forms of p53 oncogene in primary lung cancer. *Lancet* **335**: 675–679.

Iwabuchi K, Bartel PL, Li B, Marraccino R, Fields S. (1994) Two cellular proteins that bind to wild-type but not mutant p53. *Proc. Natl Acad. Sci. USA* **91**: 6098–6102.

Jenkins JR, Rudge K, Redmond S, Wade-Evans A. (1984) Cloning and expression analysis of full length mouse cDNA encoding the transformation associated protein p53. *Nucleic Acids Res.* **12**: 5609–5626.

Jenkins JR, Chumakov P, Addison C, Stürzbecher HW, Wade–Evans A. (1988) Two distinct regions of the murine p53 primary amino acid sequence are implicated in stable complex formation with simian virus 40 T antigen. *J. Virol.* **62**: 3902–3906.

Kastan MB, Onyekwere O, Sidransky D, Vogelstein B, Craig RW. (1991) Participation of p53 protein in the cellular response to DNA damage. *Cancer Res.* **51**: 6304–6311.

Kay HD, Mountjoy CP, Wu GQ, Cornish KG, Smith LJ. (1994) Sequence of a cDNA encoding the p53 protein in rhesus monkey (*Macaca mulatta*). *Gene* **138**: 223–226.

Kern SE, Kinzler KW, Bruskin A, Jarosz D, Friedman P, Prives C, Vogelstein B. (1991) Identification of p53 as a sequence-specific DNA-binding protein. *Science* **252**: 1708–1711.

Khochbin S, Principaud E, Chabanas A, Lawrence JJ. (1988) Early events in murine erythroleukemia cells induced to differentiate. Accumulation and gene expression of the transformation associated cellular protein p53. *J. Mol. Biol.* **200**: 55–64.

Kley N, Chung RY, Fay S, Loeffler JP, Seizinger BR. (1992) Repression of the basal c-fos promoter by wild-type p53. *Nucleic Acids Res.* **20**: 4083–4087.

Kraiss S, Quaiser A, Oren M, Montenarh M. (1988) Oligomerization of oncoprotein p53. *J. Virol.* **62**: 4737–4744.

Kress S, Sutter C, Strickland PT, Mukhtar H, Schweizer J, Schwarz M. (1992) Carcinogen-specific mutational pattern in the p53 gene in ultraviolet-B radiation-induced squamous cell carcinomas of mouse skin. *Cancer Res.* **52**: 6400–6403.

Kuerbitz SJ, Plunkett BS, Walsh WV, Kastan MB. (1992) Wild-type p53 is a cell cycle checkpoint determinant following irradiation. *Proc. Natl Acad. Sci. USA* **89**: 7491–7495.

Labrecque S, Naor N, Thomson D, Matlashewsky G. (1993) Analysis of the anti-p53 antibody response in cancer patients. *Cancer Res.* **53**: 3468–3471.

Lam KT, Calderwood SK. (1992) HSP70 binds specifically to a peptide derived from the highly conserved domain (I) region of p53. *Biochem. Biophys. Res. Commun.* **184**:

167–174.

Lane D. (1992) p53, guardian of the genome. *Nature* **358**: 15–16.

Lane DP, Crawford LV. (1979) T antigen is bound to a host protein in SV40-transformed cells. *Nature* **278**: 261–263.

Lavigueur A, Maltby V, Mock D, Rossant J, Pawson T, Bernstein A. (1989) High incidence of lung, bone, and lymphoid tumors in transgenic mice overexpressing mutant alleles of the p53 oncogene. *Mol. Cell. Biol.* **9**: 3982–3991.

Leach FS, Tokino T, Meltzer P, Burrell M, Oliner JD, Smith S, Hill DE, Sidransky D, Kinzler KW, Vogelstein B. (1993) p53 mutation and MDM2 amplification in human soft tissue sarcomas. *Cancer Res.* **53**: 2231–2234.

Lees-Miller SP, Sakaguchi K, Ullrich SJ, Appella E, Anderson CW. (1992) Human DNA-activated protein kinase phosphorylates serine-15 and serine-37 in the amino-terminal transactivation domain of human p53. *Mol. Cell. Biol.* **12**: 5041–5049.

Legros Y, McIntyre P, Soussi T. (1992) The cDNA cloning and immunological characterization of hamster p53. *Gene* **112**: 247–250.

Legros Y, Lacabanne V, D'Agay M, Larsen C, Pla M, Soussi T. (1993) Production of human p53 specific monoclonal antibodies and their use in immunohistochemical studies of tumor cells. *Bull. Cancer* **80**: 102–110.

Legros Y, Lafon C, Soussi T. (1994a) Linear antigenic sites defined by the B-cell response to human p53 are localized predominantly in the amino and carboxy-termini of the protein. *Oncogene* **9**: 2071–2076.

Legros Y, Meyer A, Ory K, Soussi T. (1994b) Mutations in the p53 produce a common conformational effect that can be detected with a panel of monoclonal antibodies directed toward the central part of the p53 protein. *Oncogene* **9**: 3689–3694.

Levine AJ, Momand J. (1990) Tumor suppressor genes – the p53 and retinoblastoma sensitivity genes and gene products. *Biochim. Biophys. Acta* **1032**: 119–136.

Lin D, Shields MT, Ullrich SJ, Appella E, Mercer WE. (1992) Growth arrest induced by wild-type p53 protein blocks cells prior to or near the restriction point in late G1 phase. *Proc. Natl Acad. Sci. USA* **89**: 9210–9214.

Lin JY, Chen JD, Elenbaas B, Levine AJ. (1994) Several hydrophobic amino acids in the p53 amino-terminal domain are required for transcriptional activation, binding to mdm-2 and the adenovirus 5 E1B 55-kD protein. *Gene Devel.* **8**: 1235–1246.

Linzer DIH, Levine AJ. (1979) Characterization of a 54 K dalton cellular SV40 tumor antigen present in SV40-transformed cells and in infected embryonal carcinoma cells. *Cell* **1**: 43–52.

Livingstone LR, White A, Sprouse J, Livanos E, Jacks T, Tlsty TD. (1992) Altered cell cycle arrest and gene amplification potential accompany loss of wild-type p53. *Cell* **70**: 923–935.

Louis JM, McFarland VW, May P, Mora PT. (1988) The phosphoprotein p53 is down regulated post transcriptionally during embryogenesis in vertebrates. *Biochim. Biophys. Acta, Gene Struct. Expr.* **950**: 395–402.

Lowe SW, Ruley HE, Jacks T, Housman DE. (1993) p53-dependent apoptosis modulates the cytotoxicity of anticancer agents. *Cell* **74**: 957–967.

Lu X, Lane DP. (1993) Differential induction of transcriptionally active p53 following UV or ionizing radiation – defects in chromosome instability syndromes? *Cell* **75**: 765–778.

Lubin R, Schlichtholz B, Bengoufa D, Zalcman G, Tredaniel J, Hirsch A, Caron de Fromentel C, Preudhomme C, Fenaux P, Fournier G, Mangin P, Laurentpuig P, Pelletier G, Schlumberger M, Desgrandchamps F, Leduc A, Peyrat JP, Janin N,

Bressac B, Soussi T. (1993) Analysis of p53 antibodies in patients with various cancers define B-cell epitopes of human p53 – distribution on primary structure and exposure on protein surface. *Cancer Res.* **53**: 5872–5876.

Mack DH, Vartikar J, Pipas JM, Laimins LA. (1993) Specific repression of TATA-mediated but not initiator-mediated transcription by wild-type-p53. *Nature* **363**: 281–283.

Magewu AN, Jones PA. (1994) Ubiquitous and tenacious methylation of the CpG site in codon 248 of the p53 gene may explain its frequent appearance as a mutational hot spot in human cancer. *Mol. Cell Biol.* **14**: 4225–4232.

Maheswaran S, Park S, Bernard A, Morris JF, Rauscher FJ, Hill DE, Haber DA. (1993) Physical and functional interaction between WT1 and p53 proteins. *Proc. Natl Acad. Sci. USA* **90**: 5100–5104.

Malkin D, Li FP, Strong LC, Fraumeni JF, Nelson CE, Kim DH, Kassel J, Gryka MA, Bischoff FZ, Tainsky MA, Friend SH. (1990) Germ line p53 mutations in a familial syndrome of breast cancer, sarcomas, and other neoplasms. *Science* **250**: 1233–1238.

Maltzman W, Czyzyk L. (1984) UV irradiation stimulates levels of p53 cellular tumor antigen in nontransformed mouse cells. *Mol. Cell Biol.* **54**: 1689–1694.

Mao L, Hruban RH, Boyle JO, Tockman M, Sidransky D. (1994) Detection of oncogene mutations in sputum precedes diagnosis of lung cancer. *Cancer Res.* **54**: 1634–1637.

Martinez J, Georgoff I, Martinez J, Levine AJ. (1991) Cellular localization and cell cycle regulation by a temperature-sensitive p53-protein. *Gene Devel.* **5**: 151–159.

Matlashewski G, Lamb P, Pim D, Peacok J, Crawford L, Benchimol S. (1984) Isolation and characterization of a human p53 cDNA clone: expression of the human p53. *EMBO J.* **3**: 3257–3262.

McCormick F, Harlow E. (1980) Association of a murine 53,000 dalton phosphoprotein with simian virus 40 large T antigen in transformed cells. *J. Virol.* **34**: 213–224.

Mcgregor JM, Levison DA, Macdonald DM, Yu CC. (1992) Papillomaviruses, p53, and cervical cancer. *Lancet* **339**: 1351.

Meek DW. (1994) Post-translational modification of p53. *Semin. Cancer Biol.* **5**: 203–210.

Meek DW, Simon S, Kikkawa U, Eckhart W. (1990) The p53 tumour suppressor protein is phosphorylated at serine-389 by casein kinase-II. *EMBO J.* **9**: 3253–3260.

Mercer WE, Nelson D, DeLeo AB, Old J, Baserga R. (1982) Microinjection of monoclonal antibody to protein p53 inhibits serum-induced DNA synthesis in 3T3 cells. *Proc. Natl Acad. Sci. USA* **79**: 6309–6312.

Mercer WE, Shields MT, Amin M, Sauve GJ, Appella E, Romano JW, Ullrich SJ. (1990) Negative growth regulation in a glioblastoma tumor cell line that conditionally expresses human wild-type p53. *Proc. Natl Acad. Sci. USA* **87**: 6166–6170.

Mercer WE, Shields MT, Lin D, Appella E, Ulrich SJ. (1991) Growth suppression induced by wild-type p53 protein is accompanied by selective down-regulation of proliferating-cell nuclear antigen expression. *Proc. Natl Acad. Sci. USA* **88**: 1958–1962.

Michalovitz D, Halevy O, Oren M. (1990) Conditional inhibition of transformation and of cell proliferation by a temperature-sensitive mutant of p53. *Cell* **62**: 671–680.

Michalovitz D, Halevy O, Oren M. (1991) p53 mutations – gains or losses. *J. Cell*

Biochem. **45**: 22–29.

Midgley CA, Fisher CJ, Bartek J, Vojtesek B, Lane D, Barnes D. (1992) Analysis of p53 expression in human tumors: an antibody raised against human p53 expressed in *E. coli. J. Cell Science* **101**: 183–189.

Milne DM, Palmer RH, Campbell DG, Meek DW. (1992) Phosphorylation of the p53 tumour-suppressor protein at 3 N-terminal sites by a novel casein kinase I-like enzyme. *Oncogene* **7**: 1361–1369.

Milne DM, Campbell DG, Caudwell FB, Meek DW. (1994) Phosphorylation of the tumor suppressor protein p53 by mitogen-activated protein kinases. *J. Biol. Chem.* **269**: 9253–9260.

Milner J. (1984) Different forms of p53 detected by monoclonal antibodies in non-dividing and dividing lymphocytes. *Nature* **310**: 143–145.

Milner J. (1991) A conformation hypothesis for the suppressor and promoter functions of p53 in cell growth control and in cancer. *Proc. R. Soc. Lond. (Biol)* **245**: 139–145.

Milner J. (1994) Forms and functions of p53. *Semin. Cancer Biol.* **5**: 211–219.

Milner J, Medcalf EA. (1990) Temperature-dependent switching between wild-type and mutant forms of p53-Val135. *J. Mol. Biol.* **216**: 481–484.

Milner J, Medcalf EA. (1991) Cotranslation of activated mutant p53 with wild type drives the wild-type p53 protein into the mutant conformation. *Cell* **65**: 765–774.

Milner J, Medcalf EA, Cook AC. (1991) Tumor suppressor p53 – analysis of wild-type and mutant p53 complexes. *Mol. Cell Biol.* **11**: 12–19.

Mitsudomi T. (1994) P53 in non-small-cell lung cancer – response. *J. Natl Cancer Inst.* **86**: 802–803.

Mitsudomi T, Steinberg SM, Nau MM, Carbone D, Damico D, Bodner S, Oie HK, Linnoila RI, Mulshine JL, Minna JD, Gazdar AF. (1992) p53 gene mutations in non-small-cell lung cancer cell lines and their correlation with the presence of ras mutations and clinical features. *Oncogene* **7**: 171–180.

Momand J, Zambetti GP, Olson DC, George D, Levine AJ. (1992) The mdm-2 oncogene product forms a complex with the p53 protein and inhibits p53-mediated transactivation. *Cell* **69**: 1237–1245.

Mowat M, Cheng A, Kimura N, Bernstein A, Benchimol S. (1985) Rearrangements of the cellular p53 gene in erythroleukaemic cells transformed by Friend virus. *Nature* **314**: 633–636.

Moyret C, Theillet C, Puig PL, Moles JP, Thomas G, Hamelin R. (1994) Relative efficiency of denaturing gradient gel electrophoresis and single strand conformation polymorphism in the detection of mutations in exons 5 to 8 of the p53 gene. *Oncogene* **9**: 1739–1743.

Munroe DG, Rovinski B, Bernstein A, Benchimol S. (1988) Loss of highly conserved domain on p53 as a result of gene deletion during friend virus-induced erythroleukemia. *Oncogene* **2**: 621–624.

Okuda M, Umeda A, Sakai T, Ohashi T, Momoi Y, Youn HY, Watari T, Goitsuka R, Tsujimoto H, Hasegawa A. (1994) Cloning of feline p53 tumor-suppressor gene and its aberration in hematopoietic tumors. *Int. J. Cancer* **58**: 602–607.

Oliner JD, Kinzler KW, Meltzer PS, Georges DL, Vogelstein B. (1992) Amplification of a gene encoding a p53 associated protein in human sarcomas. *Nature* **358**: 80–83.

Oliner JD, Pietenpol JA, Thiagalingam S, Gvuris J, Kinzler KW, Vogelstein B. (1993) Oncoprotein MDM2 conceals the activation domain of tumour suppressor-p53. *Nature* **362**: 857–860.

O'Reilly D, Miller LK. (1988) Expression and complex formation of simian virus 40 large T antigen and mouse p53 in insect cells. *J. Virol.* **62**: 3109–3119.

Oren M, Levine A. (1983) Molecular cloning of a cDNA specific for the murine p53 cellular tumor antigen. *Proc. Natl Acad. Sci. USA* **80**: 56–59.

Oren M, Maltzman W, Levine AJ. (1981) Post-translational regulation of the 53 K cellular tumor antigen in normal and transformed cells. *Mol. Cell Biol.* **1**: 101–110.

Ory K, Legros Y, Auguin C, Soussi T. (1994) Analysis of the most representative tumor-derived p53 mutants reveals that changes in protein conformation are not correlated with loss of transactivation or inhibition of cell proliferation. *EMBO J.* **13**: 3496–3504.

Otto A, Deppert W. (1993) Upregulation of mdm-2 expression in meth a tumor cells tolerating Wild-Type p53. *Oncogene* **8**: 2591–2603.

Ozturk M, et al. (1991) p53 mutation in hepatocellular carcinoma after aflatoxin exposure. *Lancet* **338**: 1356–1359.

Passlick B, Izbicki JR, Riethmuller G, Pantel K. (1994) P53 in non-small-cell lung cancer. *J. Natl Cancer Inst.* **86**: 801–802.

Pavletich NP, Chambers KA, Pabo CO. (1993) The DNA-binding domain of p53 contains the 4 conserved regions and the major mutation hot spots. *Gene Devel.* **7**: 2556–2564.

Pennica D, Goeddel DV, Hayflick JJ, Reich NC, Anderson CW, Levine AJ. (1984) The amino acid sequence of murine p53 determined from a c-DNA clone. *Virology* **134**: 477–782.

Picksley SM, Vojtesek B, Sparks A, Lane DP. (1994) Immunochemical analysis of the interaction of p53 with MDM2; fine mapping of the MDM2 binding site on p53 using synthetic peptides. *Oncogene* **9**: 2523–2529.

Pietenpol JA, Vogelstein B. (1993) No room at the p53 inn. *Nature* **365**: 17–18.

Pietenpol JA, Tokino T, Thiagalingam S, Eldeiry WS, Kinzler KW, Vogelstein B. (1994) Sequence-specific transcriptional activation is essential for growth suppression by p53. *Proc. Natl Acad. Sci. USA* **91**: 1998–2002.

Pinhasi–Kimhi O, Michalovitz D, Ben–Zeev A, Oren M. (1986) Specific interaction between the p53 cellular tumour antigen and major heat shock proteins. *Nature* **320**: 182–184.

Ponchel F, Puisieux A, Tabone E, Michot JP, Froschl G, Morel AP, Frebourg T, Fontaniere B, Oberhammer F, Ozturk M. (1994) Hepatocarcinoma-specific mutant p53–249Ser induces mitotic activity but has no effect on transforming growth factor beta 1-mediated apoptosis. *Cancer Res.* **54**: 2064–2068.

Puisieux A, Lim S, Groopman J, Ozturk M. (1991) Selective targeting of p53 gene mutational hotspots in human cancers by etiologically defined carcinogens. *Cancer Res.* **51**: 6185–6189.

Rady P, Scinicariello F, Wagner RF, Tyring SK. (1992) p53 mutations in basal cell carcinomas. *Cancer Res.* **52**: 3804–3806.

Raycroft L, Wu H, Lozano G. (1990) Transcriptional activation by wild-type but not transforming mutants of the p53 anti-oncogene. *Science* **249**: 1049–1051.

Reich NC, Levine AJ. (1984) Growth regulation of a cellular tumour antigen, p53, in non transformed cells. *Nature* **308**: 199–201.

Reifenberger G, Liu L, Ichimura K, Schmidt EE, Collins VP. (1993) Amplification and overexpression of the MDM2 gene in a subset of human malignant gliomas without p53 mutations. *Cancer Res.* **53**: 2736–2739.

Rideout WM, Coetzee GA, Olumi AF, Jones PA. (1990) 5-Methylcytosine as an endogenous mutagen in the human LDL receptor and p53 gene. *Science* **249**: 1288–1290.

Rigaudy P, Eckhart W. (1989) Nucleotide sequence of a cDNA encoding the monkey cellular phosphoprotein-p53. *Nucleic Acids Res.* **17**: 8375.

Rogel A, Popliker M, Webb CG, Oren M. (1985) p53 cellular tumour antigen: analysis of mRNA levels in normal adult tissues, embryos and tumours. *Mol. Cell Biol.* **5**: 2851–2855.

Samad A, Carroll RB. (1991) The tumor suppressor p53 is bound to RNA by a stable covalent linkage. *Mol. Cell Biol.* **11**: 1598–1606.

Santhanam U, Ray A, Sehgal PB. (1991) Repression of the interleukin-6 gene promoter by p53 and the retinoblastoma susceptibility gene product. *Proc. Natl Acad. Sci. USA* **88**: 7605–7609.

Sarnow P, Ho YS, Williams J, Levine AJ. (1982) Adenovirus E1B-58Kd tumor antigen and SV40 large tumor antigen physically associated with the same 54 Kd cellular protein in transformed cells. *Cell* **28**: 387–394.

Sato M, Nishigori C, Zghal M, Yagi T, Takebe H. (1993) Ultraviolet-specific mutations in p53 gene in skin tumors in xeroderma-pigmentosum patients. *Cancer Res.* **53**: 2944–2946.

Scheffner M, Werness BA, Huibregtse JM, Levine AJ, Howley PM. (1990) The E6 oncoprotein encoded by human papillomavirus type-16 and type-18 promotes the degradation of p53. *Cell* **63**: 1129–1136.

Scheffner M, Munger K, Byrne JC, Howley PM. (1991) The state of the p53 and retinoblastoma genes in human cervical carcinoma cell lines. *Proc. Natl Acad. Sci. USA* **88**: 5523–5527.

Schlichtholz B, Legros Y, Gillet D, Gaillard C, Marty M, Lane D, Calvo F, Soussi T. (1992) The immune response to p53 in breast cancer patients is directed against immunodominant epitopes unrelated to the mutational hot spot. *Cancer Res.* **52**: 6380–6384.

Schlichtholz B, Tredaniel J, Lubin R, Zalcman G, Hirsch A, Soussi T. (1994) Analyses of p53 antibodies in sera of patients with lung carcinoma define immunodominant regions in the p53 protein. *Br. J. Cancer* **69**: 809–816.

Schmid P, Lorenz A, Hameister H, Montenarh M. (1991) Expression of p53 during mouse embryogenesis. *Development* **113**: 857–865.

Schmieg FI, Simmons DT. (1988) Characterization of the in vitro interaction between SV40 T antigen p53: mapping the p53 binding site. *Virology* **164**: 132–140.

Schwartz D, Goldfinger N, Rotter V. (1993) Expression of p53 protein in spermatogenesis is confined to the tetraploid pachytene primary spermatocytes. *Oncogene* **8**: 1487–1494.

Seto E, Usheva A, Zambetti GP, Momand J, Horikoshi N, Weinmann R, Levine AJ, Shenk T. (1992) Wild-type p53 binds to the TATA-binding protein and represses transcription. *Proc. Natl Acad. Sci. USA* **89**: 12028–12032.

Shaulsky G, Benzeev A, Rotter V. (1990a) Subcellular distribution of the p53 protein during the cell cycle of Balb/c 3T3 cells. *Oncogene* **5**: 1707–1711.

Shaulsky G, Goldfinger N, Benzeev A, Rotter V. (1990b) Nuclear accumulation of p53 protein is mediated by several nuclear localization signals and plays a role in tumorigenesis. *Mol. Cell. Biol.* **10**: 6565–6577.

Shaulsky G, Goldfinger N, Peled A, Rotter V. (1991) Involvement of wild-type p53 in pre-B-cell differentiation in vitro. *Proc. Natl Acad. Sci. USA* **88**: 8982–8986.

Shaw P, Bovey R, Tardy S, Sahli R, Sordat B, Costa J. (1992) Induction of apoptosis by wild-type p53 in a human colon tumor-derived cell line. *Proc. Natl Acad. Sci. USA* **89**: 4495–4499.

Shiio Y, Yamamoto T, Yamaguchi N. (1992) Negative regulation of Rb expression by the p53 gene product. *Proc. Natl Acad. Sci. USA* **89**: 5206–5210.

Shohat O, Greenberg M, Reisman D, Oren M, Rotter V. (1987) Inhibition of cell growth mediated by plasmids encoding p53 anti-sense. *Oncogene* **1**: 277–283.

Soussi T, Caron de Fromentel C, Méchali M, May P, Kress M. (1987) Cloning and characterization of a cDNA from *Xenopus laevis* coding for a protein homologous to human and murine p53. *Oncogene* **1**: 71–78.

Soussi T, Bègue A, Stehelin D, May P. (1988a) Nucleotide sequence of a cDNA encoding the chicken p53 nuclear oncoprotein. *Nucleic Acids Res.* **16**: 11383.

Soussi T, Caron de Fromentel C, Breugnot C, May E. (1988b) Nucleotide sequence of a cDNA encoding the rat p53 nuclear oncoprotein. *Nucleic Acids Res.* **16**: 11384.

Soussi T, Caron de Fromentel C, May P. (1990) Structural aspects of the p53 protein in relation to gene evolution. *Oncogene* **5**: 945–952.

Soussi T, Legros Y, Lubin R, Ory K, Schlichtholz B. (1994) Multifactorial analysis of p53 alteration in human cancer – a review. *Int. J. Cancer* **57**: 1–9.

Srivastava S, Zou ZQ, Pirollo K, Blattner W, Chang EH. (1990) Germ-line transmission of a mutated p53 gene in a cancer-prone family with Li–Fraumeni syndrome. *Nature* **348**: 747–749.

Stenger JE, Mayr GA, Mann K, Tegtmeyer P. (1992) Formation of stable p53 homotetramers and multiples of tetramers. *Mol. Carcinogen.* **5**: 102–106.

Stürzbecher HW, Chumakov P, Welch WJ, Jenkins JR. (1987) Mutant p53 proteins bind hsp72/73 cellular heat shock-related proteins in SV 40-transformed monkey cells. *Oncogene* **1**: 201–211.

Stürzbecher HW, Maimets T, Chumakov P, Brain R, Addison C, Simanis V, Rudge K, Philp R, Grimaldi M, Court W, Jenkins JR. (1990) p53 interacts with p34cdc2 in mammalian cells – implications for cell cycle control and oncogenesis. *Oncogene* **5**: 795–801.

Stürzbecher HW, Brain R, Addison C, Rudge K, Remm M, Grimaldi M, Keenan E, Jenkins JR. (1992) A C-terminal alpha-helix plus basic region motif is the major structural determinant of p53 tetramerization. *Oncogene* **7**: 1513–1523.

Subler MA, Martin DW, Deb S. (1992) Inhibition of viral and cellular promoters by human wild-type p53. *J. Virol.* **66**: 4757–4762.

Subler MA, Martin DW, Deb S. (1994) Activation of the human immunodeficiency virus type 1 long terminal repeat by transforming mutants of human p53. *J. Virol.* **68**: 103–110.

Sundaresan V, Ganly P, Hasleton P, Rudd R, Sinha G, Bleehen NM, Rabbits P. (1992) p53 and chromosome 3 abnormalities, characteristic of malignant lung tumours, are detectable in preinvasive lesions of the bronchus. *Oncogene* **7**: 1989–1997.

Szekely L, Selivanova G, Magnusson KP, Klein G, Wiman KG. (1993) EBNA-5, an Epstein–Barr virus-encoded nuclear antigen, binds to the retinoblastoma and p53 proteins. *Proc. Natl Acad. Sci. USA* **90**: 5455–5459.

Takahashi T, Nau MM, Chiba I, Birrer MJ, Rosenberg RK, Vinocour M, Levitt M, Pass H, Gazdar AF, Minna JD. (1989) p53 – a frequent target for genetic abnormalities in lung cancer. *Science* **246**: 491–494.

Takahashi T, Takahashi T, Suzuki H, Hida T, Sekido Y, Ariyoshi Y, Ueda R. (1991) The p53 gene is very frequently mutated in small-cell lung cancer with a distinct nucleotide substitution pattern. *Oncogene* **6**: 1775–1778.

Tan TH, Wallis J, Levine AJ. (1986) Identification of the protein p53 domain involved in formation of the simian virus 40 large T-antigen–p53 protein complex. *J. Virol.* **59**: 574–583.

Taylor JA, Watson MA, Devereux TR, Michels RY, Saccomanno G, Anderson M. (1994) P53 mutation hotspot in radon-associated lung cancer. *Lancet* **343**: 86–87.

Tchang F, Gusse M, Soussi T, Mechali M. (1993) Stabilization and expression of high levels of p53 during early development in *Xenopus laevis. Devel. Biol.* **159**: 163–172.

Tenaud C, Negoescu A, Labat-Moleur F, Legros Y, Soussi T, Brambilla E. (1994) Methods in pathology – p53 immunolabeling in archival paraffin-embedded tissues: optimal protocol based on microwave heating for eight antibodies on lung carcinomas. *Modern Pathol.* **7**: 853–859.

Thor AD, Yandell DW. (1993) Prognostic significance of p53 overexpression in node-negative breast carcinoma – preliminary studies support cautious optimism. *J. Natl Cancer Inst.* **85**: 176–177.

Thor AD, Moore DH, Edgerton SM, Kawasaki ES, Reihsaus E, Lynch HT, Marcus JN, Schwartz L, Chen LC, Mayall BH, Smith HS. (1992) Accumulation of p53 tumor suppressor gene protein – an independent marker of prognosis in breast cancers. *J. Natl Cancer Inst.* **84**: 845–855.

Tornaletti S, Pfeifer GP. (1994) Slow repair of pyrimidine dimers at p53 mutation hotspots in skin cancer. *Science* **263**: 1436–1438.

Venitt S, Biggs PJ. (1994) Radon, mycotoxins, p53, and uranium mining. *Lancet* **343**: 795.

Vogelstein B, Fearon ER, Hamilton SR, Kern SE, Preisinger AC, Leppert M, Nacamura V, White R, Smits AM, Bos JL. (1988) Genetic alterations during colorectal-tumor development. *N. Engl. J. Med.* **319**: 525–532.

Vojtesek B, Bartek J, Midgley CA, Lane DP. (1992) An immunochemical analysis of the human nuclear phosphoprotein-p53 – new monoclonal antibodies and epitope mapping using recombinant-p53. *J. Immunol. Methods* **151**: 237–244.

Wang XW, Forrester K, Yeh H, Feitelson MA, Gu JR, Harris CC. (1994) Hepatitis B virus X protein inhibits p53 sequence-specific DNA binding, transcriptional activity, and association with transcription factor ERCC3. *Proc. Natl Acad. Sci. USA* **91**: 2230–2234.

Wang Y, Eckhart W. (1992) Phosphorylation sites in the amino-terminal region of mouse p53. *Proc. Natl Acad. Sci. USA* **89**: 4231–4235.

Wang Y, Reed M, Wang P, Stenger JE, Mayr G, Anderson ME, Schwedes JF, Tegtmeyer P. (1993) p53 domains – identification and characterization of 2 autonomous DNA-binding regions. *Gene Devel.* **7**: 2575–2586.

Weintraub H, Hauschka S, Tapscott SJ. (1991) The MCK enhancer contains a p53 responsive element. *Proc. Natl Acad. Sci. USA* **88**: 4570–4571.

Werness BA, Levine AJ, Howley PM. (1990) Association of human papillomavirus type-16 and type-18 E6 proteins with p53. *Science* **248**: 76–79.

Winter SF, Minna JD, Johnson BE, Takahashi T, Gazdar AF, Carbone DP. (1992) Development of antibodies against p53 in lung cancer patients appears to be dependent on the type of p53 mutation. *Cancer Res.* **52**: 4168–4174.

Wu XW, Bayle JH, Olson D, Levine AJ. (1993) The p53 mdm-2 autoregulatory feedback loop. *Gene Devel.* **7**: 1126–1132.

Wynford-Thomas D. (1992) p53 in tumour pathology – can we trust immunocytochemistry. *J. Pathol.* **166**: 329–330.

Xiong Y, Hannon GJ, Zhang H, Casso D, Kobayashi R, Beach D. (1993) p21 is a universal inhibitor of cyclin kinases. *Nature* **366**: 701–704.

Yin YX, Tainsky MA, Bischoff FZ, Strong LC, Wahl GM. (1992) Wild-type p53 restores cell cycle control and inhibits gene amplification in cells with mutant p53 alleles. *Cell* **70**: 937–948.

Yonish-Rouach E, Resnitzky D, Lotem J, Sachs L, Kimchi A, Oren M. (1991) Wild-type p53 induces apoptosis of myeloid leukaemic cells that is inhibited by interleukin-6. *Nature* **352**: 345–347.

Zhang Q, Gutsch D, Kenney S. (1994) Functional and physical interaction between p53 and BZLF1 – implications for Epstein–Barr virus latency. *Mol. Cell Biol.* **14**: 1929–1938.

Ziegler A, Leffell DJ, Kunala S, Sharma HW, Gailani M, Simon JA, Halperin AJ, Baden HP, Shapiro PE, Bale AE, Brash DE. (1993) Mutation hotspots due to sunlight in the p53 gene of nonmelanoma skin cancers. *Proc. Natl Acad. Sci. USA* **90**: 4216–4220.

Zur Hausen H, Schneider A. (1987) The role of papillomaviruses in human anogenital cancer. *Papovaviridae* **2**: 245–263.

Molecular biology of lung cancer

Frederic J. Kaye, Young Whan Kim and Gregory A. Otterson

8.1 Introduction

Our understanding of the molecular basis of human cancers has been revolutionized by the identification of a large (but finite) family of genes that both participate in normal cell growth regulation and serve as targets for mutations in essentially all human malignancies. These findings have also demonstrated that, although important aetiological, clinical and biological differences exist among tumours of varying histological subtype, a central concept of mutational activation/inactivation underlies neoplastic transformation unifying research efforts by investigators studying diverse tumour models. For example, abnormalities in either DNA mismatch repair activity, activation of the *RAS* and *MYC* genes, or inactivation of the *RB, p53* and *CDKN2* genes are common mutational events in lung cancers as well as in many other sporadic or inherited malignancies. These observations offer hope that future technological advances in the prevention, diagnosis or therapy of cancer will be widely applicable.

For the past 10 years, genes targeted for mutations in cancer have been broadly characterized as either dominant oncogenes (activating mutations in one allele with 'gain of function' transforming activity) or as recessive oncogenes or tumour suppressor genes (inactivating mutations of both alleles with 'loss of wild-type function'). Although this classification has been useful for tabulating cancer genes, it is based largely on epidemiological or *in vitro* experimental data that does not address directly the roles these genes played in normal cell physiology. As more detailed information has become available about oncogene and tumour suppressor gene functions, new organizational groupings, such as cell cycle or growth factor/signal transduction pathways, have developed which incorporate both dominant and recessive oncogenes and which allow investigators, for the first time, to begin to understand relationships that underlie the multi-step genetic theory of tumour induction and progression. In the case of the molecular biology of lung cancer, recent advances in our understanding of cell cycle events have contributed to a working model to begin to explain the significance of the remarkably high incidence of *RB*, *p53* and *CDKN2* mutations associated with these tumours. More information, however, is still needed to integrate these findings with other common gene mutations (*RAS* and *MYC*) and with other cytogenetic abnormalities (chromosome 3p and many others) in order to ascertain which genes are rate limiting for the malignant phenotype, and whether individuals within the population can be identified with an inherited or acquired susceptibility for the development of lung cancer.

Ultimately, lung cancer is an iatrogenic (and fatal) disease of increasing incidence world-wide which results largely from the accumulated active and passive exposures to tobacco products (Hoffmann *et al.*, 1983; Boring *et al.*, 1994). Since strategies for either the early detection of lung tumours or for the effective treatment of established lung tumours are unlikely to be highly successful (given the present level of technology) a greater emphasis on prevention is essential. For example, the recognition that acquired genomic instability and/or defective DNA repair activity are early steps that may accelerate subtle genetic mutations has suggested novel potential targets for preventive therapies. In addition, the ability of molecular studies of tumour tissues to provide a fingerprint that is more characteristic of the carcinogen exposure rather than the histological tissue type is another example of how genetic analyses may provide tools to help identify and isolate carcinogenic compounds in the environment. Finally, public dissemination of the detailed (and unambiguous) knowledge of mutated genes in the lung tissues of patients with cancer may serve to encourage more individuals to take a stronger stand against unnecessary exposures to exogenous agents that may accelerate this process.

8.2 Histological types of lung cancers

8.2.1 Small cell lung cancer

Small cell lung cancer (SCLC) is a distinct histological, clinical, biochemical and biological type of lung cancer that is strongly linked to cigarette consumption and which comprises approximately 25% of all lung cancers (Hoffmann *et al.*, 1983; Boring *et al.*, 1994). The diagnosis of SCLC remains exclusively based on its histological features by light microscopy, although these patients also present with characteristic clinical and X-ray findings which show that SCLC is a highly aggressive tumour that disseminates early (Ihde, 1992). For this reason, as well as the observation that SCLC is initially responsive to standard chemotherapy, the surgical resection of tumour is rarely performed resulting in a general lack of primary tumour material for research investigations. Because of its unique biochemical properties and the ability to generate continuous cell lines in culture, however, SCLC tumour lines have been a popular model system for studying the expression of a wide range of neuro-endocrine peptides and bioenergetic enzymes (Carney *et al.*, 1985). For example, SCLC cell lines have been demonstrated to express large amounts of neuron-specific enolase, the brain isoenzyme of creatine kinase, dopa-decarboxylase, calcitonin, pro-opiomelanocortin peptide, adrenocorticotrophic hormone (ACTH), arginine vasopressin, atrial natriuretic factor, chromogranin A, insulin-like growth factor, and gastrin-releasing peptide or bombesin (Moody *et al.*, 1981; Carney *et al.*, 1985; Said *et al.*, 1985; Chretien *et al.*, 1986; Minuto *et al.*, 1986; Sausville *et al.*, 1986; Kaye *et al.*, 1987; Stewart *et al.*, 1989; White *et al.*, 1989; Bliss *et al.*, 1990). In most of these cases, the proteins synthesized are enzymatically active, and several markers have been implicated in clinical disease, such as ACTH for Cushing's syndrome (Lokich, 1982) and arginine vasopressin and atrial natriuretic factor for hyponatraemia (Bliss *et al.*, 1990). Other peptides, such as gastrin-releasing peptide and bombesin, have been hypothesized as participating in a positive growth feedback loop (autocrine hypothesis) (Cuttitta *et al.*, 1985), while the expression of neuron-specific enolase and the brain isoenzyme of creatine kinase remains unexplained. These observations have subsequently resulted in several clinical trials. For example, monoclonal antibodies directed against the gastrin-releasing peptide/bombesin molecule in patients with extensive SCLC refractory to conventional treatments is currently under investigation (Avis *et al.*, 1991). In addition, creatine analogues are being tested for growth inhibition *in vitro* and *in vivo* in tumours expressing high levels of creatine kinase enzyme activity (Martin *et al.*, 1994). Since several of these enzymes are associated with the neural amine precursor uptake and decarboxylation (APUD) system, these findings have suggested that SCLC may arise from rare clusters of neuro-endocrine APUD cells, referred to as Kulchitsky cells, which are located beneath the bronchial mucosa. The cell of origin for SCLC, however, has not been defined clearly and conflicting histological lineages have been reported based on cell surface markers (Ruff and Pert, 1984). Characterizing the cell of origin of SCLC and its patterns of gene expression will be essential to the study of premalignant lesions and toward understanding why, in SCLC, patterns of gene inactivation are strikingly distinct from those observed in non-small cell lung cancers (NSCLCs), despite similar carcinogen exposures.

8.2.2 Non-small cell lung cancer

NSCLC comprises the majority of lung cancers and represents a collection of several histologically distinct forms referred to as squamous cell carcinomas, adenocarcinomas, bronchioloalveolar carcinomas (a subtype of adenocarcinoma), large cell carcinomas and undifferentiated carcinomas. Occasionally, tumours of the pleural surface (the mesothelial envelope of cells surrounding the lungs) called mesotheliomas are also included in the classification of NSCLC. These tumours clinically resemble each other with a predisposition to local invasion within the chest, reduced responsiveness to standard chemotherapy or radiation treatments, and with surgical resection, if feasible, as the primary treatment modality (Ihde and Minna, 1991a,b; Ihde, 1992). As a consequence, academic centres and clinical cooperative groups have stored large numbers of NSCLC primary tumour samples which facilitates the study of genetic events without the confounding variable of cell culture selection bias. NSCLC cell lines, however, are readily established from biopsy samples and offer the advantage, for genetic studies, of a continuous supply of tumour cells free from contaminating normal stroma cells. Given the sensitivity of the polymerase chain reaction assay, which is used frequently for genomic DNA analyses, even small amounts of contaminating normal tissue are a potential source of confusion. Therefore, primary tumour and cell line data have served complementary roles in investigating the genetic aetiology of lung cancer.

The clinical course and histological features of NSCLC also predict that the early events underlying this tumour will be easier to define than for SCLC. For example, NSCLC appears to arise out of the surface of a preneoplastic (dysplastic) bronchial epithelium. In contrast, early lesions of SCLC are often detected as submucosal (beneath the bronchial epithelium) tumours without an obvious clinical counterpart to the early, mid and late stages of tumorigenesis as evident with carcinomas of the colon. In addition, primary NSCLC tumours occasionally demonstrate histological features which overlap with other types of lung cancer, including SCLC. This finding initially resulted in the proposal of a 'unitarian theory of lung cancer' with a multipotent lung cell as a target for malignancy; however, this hypothesis has been difficult to prove (Baylin and Gazdar, 1981). Nonetheless, approximately, 10–15% of NSCLC tumours exhibit features of neuro-endocrine differentiation that are characteristic of SCLC, and a similar proportion of SCLC tumours express no neuro-endocrine gene products (Linnoila et al., 1988). The clinical relevance of these observations is under investigation in addition to a search to identify biomarkers that may serve as reliable indicators for early disease, for response to radiation therapy or chemotherapy, and for patient survival.

8.3 Cytogenetics of lung cancer

Cytogenetic studies continue to play a central role in both encouraging and guiding investigations into the genetic aetiology of lung cancer. For example, early analyses using SCLC samples from cultured cell lines identified a consensus deletion involving the short arm of human chromosome 3 (Whang-Peng et al., 1982a, b). Although this finding was initially debated (Wurster-Hill et al., 1984), later studies using the more sensitive molecular technique of restriction fragment length polymorphism (RFLP) analysis confirmed that in almost 100% of SCLC cases and approximately 50% of NSCLC cases there is evidence for deletion of one chromosome 3p allele (Brauch et al., 1987; Naylor et al., 1987; Johnson et al., 1988b). Using the Knudson two-hit theory of recessive tumorigenesis as a model, these findings predicted that an essential gene for control of lung cell growth and differentiation resides within the deleted region of chromosome 3p. The 'first hit' would generally represent an inherited or acquired subtle mutation within one allele of the target gene and the 'second hit' would result in a greater loss of chromosomal material spanning the remaining wild-type allele, which may be visible either cytogenetically or by more sensitive molecular techniques.

Additional studies have been undertaken including an extensive series at the National Cancer Institute of Japan which analysed many genetic loci from primary lung tumours of all histological subtypes (Yokota *et al.*, 1987). Evidence was found for deletions within chromosomes 3p, 13q and 17p, the latter two subsequently shown to represent targeted mutations within the *RB* (Harbour *et al.*, 1988; Yokota *et al.*, 1988) and *p53* (Takahashi *et al.*, 1989) genes, respectively (see below). More recently, however, non-random chromosomal abnormalities have been noted at many different genetic loci (*Table 8.1*). The almost bewildering accumulation of cytogenetic and molecular evidence for DNA loss within SCLC and NSCLC samples raises the question as to whether each abnormality represents another important gene that awaits isolation and characterization. This issue is being addressed at a rapid pace. For example, the 9p21 deletions undoubtedly represent inactivation of tandem *p16* and *p15* genes which encode specific inhibitors for cyclin-dependent kinase (cdk) activity (Olopade *et al.*, 1993; Serrano *et al.*, 1993; Hannon and Beach, 1994; Kamb *et al.*, 1994; Nobori *et al.*, 1994). In addition, many of the other chromosomal abnormalities, such as 1p, 3p, 5q and 11p deletions, are also detected in a range of other human malignancies. As with the haematological malignancies and other solid tumours, classic and molecular cytogenetics will continue to play a major role in the ongoing identification of cancer genes important in the genesis or progression of human lung cancer.

Table 8.1. Selected cytogenetic abnormalities in lung cancer

SCLC	NSCLC	
1p	1p10–13	11p13–15
2q11–13; q33	1q11–12	11q12–q23
3p14–23	2q	13p
5q13–21; q33–35	3p14–25	13q
6p	3q	14p
6q	5p13	15p
9p	5q11–14; q21	16q2
9q	6p	17p11–13
13q14	6q15–27	17q11
17p13	7p	18q
18+	7q	19p
	8p21.3–23.1	19q13
	9p21	21p1
	9q32–34	22q

Data from Whang-Peng *et al.*, 1982a,b, 1991; Morstyn *et al.*, 1987; Yokota *et al.*, 1987; DeFusco *et al.*, 1989; Lukeis *et al.*, 1990, 1993; Sozzi *et al.*, 1991; Miura *et al.*, 1992; Testa *et al.*, 1992; Center *et al.*, 1993; Fujiwara *et al.*, 1994; Hosoe *et al.*, 1994; Johansson *et al.*, 1994; Kohno *et al.*, 1994; Merlo *et al.*, 1994a,b; Shiseki *et al.*, 1994; Wieland and Bohm, 1994.

8.4 G1/S transition of the cell cycle is a common target for mutations in lung cancer

Similar to a jigsaw puzzle where a starting framework is needed in order to begin to put together seemingly disparate pieces, a working model for the interconnecting processes of cell growth also needs a firm starting framework. For the moment, the machinery directly underlying the cell cycle appears to be a suitable reference point to piece together the multiple feedback loops involving external signalling, transcriptional activation and cell division. Furthermore, it is now clear that deregulation of cell cycle progression is a hallmark of neoplastic transformation in a wide range of tumour types, including lung cancer.

The eukaryotic cell cycle is governed by the sequential activation of specific cdk–cyclin complexes synchronized with distinct phases of the cell cycle (Hunter and Pines, 1994; King *et al.*, 1994). The activated cdk–cyclin complex, in turn, phosphorylates downstream protein substrates to coordinate the cascade of events required for progression through the cell cycle. Although simple eukaryotes such as yeast require only a single cdk molecule with a few accompanying cyclins to complete cell division, mammalian cells require a large family of related genes encoding cdk and cyclin, each of which appears to play a role designed for a specific stage of the cell cycle, such as G1, S, G2 and M (Sherr, 1993), or even in the differentiation process of non-mitotic cells (Tsai *et al.*, 1994). Theoretically, any phase of the cell division machinery could be vulnerable to mutational events leading to cell growth deregulation; however, mutations within genes essential for the M phase transition have not been reported. In contrast, genes involved in the G1/S transition are especially frequent targets for mutations in human cancer, including lung cancer (*Figure 8.1*). The protein products encoded by these genes include: (i) those which mediate signal transduction from external growth factors (i.e. RAS, MYC) (Wong *et al.*, 1986; Rodenhuis *et al.*, 1987; Johnson *et al.*, 1988a); (ii) those which 'sense' the presence of DNA damage to mediate G1 arrest (p53) (Takahashi *et al.*, 1989); (iii) the G1 cyclin proteins (cyclins D1, D2, A, E) (Brechot, 1993; Sherr, 1993) and their cdk inhibitors (p21, p16, p15, p27 and others) (Hunter and Pines, 1994); and (iv) the immediate downstream substrates for these G1 kinases (RB and perhaps others) (Weinberg, 1991).

8.5 Recessive genes commonly targeted in lung cancer

8.5.1 Lung cancer and p53

The recognition that the *p53* gene can function as a tumour suppressor gene by mediating G1 checkpoint control is one of the most important developments in the brief history of molecular oncology and is reviewed in detail in Chapter 7. The surprising observation that mice homozygous for a null *p53* allele were born phenotypically normal, albeit with an increased predisposition to malignant tumours (Donehower *et al.*, 1992), helped formulate the model in which the potent growth

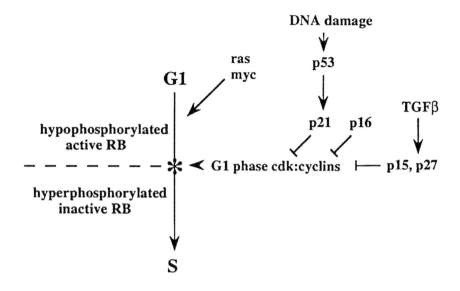

Figure 8.1. Genes involved in the G1/S transition of the cell cycle (*MYC*, *RAS*, *p53*, *RB*, *p16*, *p15*(?), G1 cyclins) are potential targets for mutations in lung cancer. TGF-β, transforming growth factor β.

suppressive activity of *p53* functions primarily as a 'guardian of the genome' (Lane, 1992). In this capacity, the p53 product responds to a variety of DNA-damaging stimuli, such as UV or γ irradiation, by enhancing its DNA binding and transcriptional activity. The induced expression of p53-regulated genes, such as *p21* (a general cdk inhibitor), *GADD45* and others, may then mediate a G1 pause in affected cells to allow for DNA repair and, if this is not feasible, to proceed with programmed cell death (Kastan *et al.*, 1992; el-Deiry *et al.*, 1993; el-Deiry *et al.*, 1994; Nelson and Kastan 1994; Zhan *et al.*, 1994). In the absence of p53 function, therefore, cells containing sublethal (and potentially transforming) DNA damage continue to replicate eventually producing a neoplastic cell clone. This hypothesis has also predicted a mechanism for tumour cell death in response to DNA damage from γ irradiation or from chemotherapeutic agents, and experiments using *in vitro* and *in vivo* models confirm that apoptotic cell death (with a prolonged survival of the host) is a consistent feature of tumour cells that retain wild-type p53 function (Lowe *et al.*, 1993a,b, 1994).

Following the initial observations that the *p53* gene is mutated in a wide range of human cancers (Hollstein *et al.*, 1991), analysis of lung cancer samples showed that 70–100% of SCLC lines and primary tumours (Takahashi *et al.*, 1989, 1991; Hensel *et al.*, 1991; Lehman *et al.*, 1991; D'Amico *et al.*, 1992; Miller *et al.*, 1992; Sameshima *et al.*, 1992) and 40–60% of NSCLC lines and primary tumours had

DNA or RNA evidence for mutational inactivation (Chiba *et al.*, 1990; Lehman *et al.*, 1991; Kishimoto *et al.*, 1992; Miller *et al.*, 1992; Mitsudomi *et al.*, 1992; Suzuki *et al.*, 1992). The high frequency of *p53* mutations in these tumours suggested that this event was an essential and perhaps early event. However, studies in other tumour types, such as colon and bladder cancer, suggested instead that *p53* inactivation occurs later in the course of tumour progression. To address the timing of *p53* mutations in lung cancer, several investigators have carefully micro-dissected and analysed dysplatic, and presumably premalignant, bronchial lesions. In these studies, *p53* mutations were observed in different cases of preneoplastic bronchial dysplasia (Sozzi *et al.*, 1992; Sundaresan *et al.*, 1992) suggesting that these mutations were early events in lung cancer, contrary to colon, ovarian and thyroid cancer. In addition, using enhanced immunohistochemical staining as a marker for mutant p53 (due to an abnormally increased protein half-life), another study observed that mutant p53 expression was detected in 0% of normal mucosas, 8.3% of squamous metaplasias, 37.5% of mild dysplasias, 12.5% of moderate dysplasias, 93.8% of severe dysplasias and 55% of carcinoma-*in-situ* lesions (Bennett *et al.*, 1993). These data suggest that *p53* may participate in pulmonary tumorigenesis, that *p53* alterations can occur before microinvasion and that the *p53* tumour suppressor gene may serve as a marker for early diagnosis.

Mutational analysis of the *p53* gene in lung cancer resembles data from other adult tumours with the preponderance of mutations occurring within the evolutionarily conserved DNA binding domains (Hollstein *et al.*, 1991). More interestingly, the types of mutation appear to be loosely correlated with specific types of carcinogen exposure. For example, G → T transversions of *p53* predominate in lung tumours, while G → A transitions are more frequent in colonic tumours, and mutations in thymidine dimers predominate in UV light-associated skin cancers (Puisieux *et al.*, 1991; Vogelstein and Kinzler, 1992; Harris and Hollstein, 1993). In addition, an AGG → ATG transversion at codon 259, which occurs rarely in tobacco-associated lung cancer, was observed in 31% (16/52) of cases of radon-associated lung cancer in uranium miners (Taylor *et al.*, 1994). Furthermore, lung cancers arising in atomic bomb survivors demonstrated predominantly a G → A transition pattern within the *p53* gene, which was distinct from the G → T transversion which the authors detected in their tobacco-associated lung cancers (Takeshima *et al.*, 1993). These findings validate the model of carcinogen-accelerated tumorigenesis, and have suggested objective assays to attempt to identify exogenous and endogenous carcinogens.

Similar to studies performed in breast cancer (Isola *et al.*, 1992), several investigators have reported an association between altered p53 protein expression and inferior survival in patients with NSCLC (Quinlan *et al.*, 1992; Horio *et al.*, 1993; Marchetti *et al.*, 1993). Although the identification of a prognostic factor is of uncertain value in a disease without effective non-surgical treatment, these findings also support growing experimental data that p53(+) and p53(–/mutant) tumours are indeed biologically distinct entities. For example, p53(+) tumour cells *in vitro* or p53(+) tumours explanted to nude mice are highly susceptible to programmed cell death following a variety of noxious stimuli (such as radiation therapy or chemotherapy) as discussed above (Kastan *et al.*, 1992; el-Deiry *et al.*, 1993, 1994; Lowe *et al.*, 1994; Nelson and Kastan 1994; Zhan *et al.*, 1994). Conversely, the absence of wild-type p53 function may facilitate the escape of neo-plastic cell clones with additional genetic mutations favouring enhanced cell proliferation and tissue invasion. In summary, although these interpretations are necessarily over-simplified, they are important in that they begin to address the molecular basis for the clinical therapy of human cancer.

The potent *in vitro* growth suppressor activity of the *p53* gene in lung cancer cell lines (Takahashi *et al.*, 1992) and the high frequency of mutational events in lung cancer has recently stimulated clinical trials in human lung cancer using two divergent approaches. In one approach, the recognition that discrete epitopes of the mutant p53 product are presented on the surface of tumour cells where they may render these cells susceptible to immune destruction has led to an effort to pulse immune cells from lung cancer patients with the mutant p53 peptides pre-dicted to be expressed on the surface of their tumour cells (Winter *et al.*, 1993; Wiedenfeld *et al.*, 1994). In the second approach, the biological effect of a *p53* gene delivered by a retroviral vector would be analysed in tumours from patients with unresectable lung cancer (Cai *et al.*, 1993). Although both trials are just beginning, the refinement of techniques to make non-immunogenic cells immunogenic and the development of new gene delivery vectors are major priorities at many acad-emic centres at this time, and suggest that many more clinical studies will be ongoing soon.

8.5.2 Lung cancer and the RB gene

The cloning of the *RB* gene in 1986 (Friend *et al.*, 1986) validated the two-hit the-ory of recessive tumorigenesis and has served as the paradigm for the concept of a tumour suppressor gene as detailed in Chapter 1. Interest in studying the *RB* gene in lung cancer resulted from RFLP data suggesting 13q allele loss in lung cancer (Yokota *et al.*, 1987) and from a review of cytogenetic data from SCLC that showed many examples of deletions and reciprocal and non-reciprocal transloca-tions involving the *RB* locus at chromosomal band 13q14 (Harbour *et al.*, 1988). In addition, although SCLC and retinoblastoma tumours have different clinical features, they share several characteristics in common, such as the presence of neural differentiation, fastidious growth in cell culture as non-adherent clusters of cells, and the frequent amplification and overexpression of members of the *MYC* gene family. Using nucleic acid (Harbour *et al.*, 1988) and protein (Yokota *et al.*, 1988) analysis, investigators showed that the majority of SCLC cell lines and approximately 10% of NSCLC lines had targeted mutations within the *RB* gene. Later protein analyses, which are more sensitive for detecting subtle point muta-tions that are not visible by Southern and Northern blotting techniques, were performed on a larger group of lung tumour samples and showed that 90% of SCLC and 15% of NSCLC had either absent or mutant RB protein expression (Horowitz *et al.*, 1990; Shimizu *et al.*, 1994). At least five different stable RB mutant proteins have been isolated from SCLC tumours including the products of aberrant splicing of exons 16, 21 and 22 (Horowitz *et al.*, 1990; Shew *et al.*, 1990) and point mutations resulting in single missense mutations within exons 20 and 21 (Bignon *et al.*, 1990; Kaye *et al.*, 1990; Kratzke and Kaye, unpublished data). In addition, the identical exon 21 and exon 22 splicing defect has been detected in cDNA isolated from independent SCLC and NSCLC cell lines and from a primary NSCLC tumour (Kaye *et al.*, unpublished data). Each of these stable *in vivo* mutants encodes the same properties of defective phosphorylation in response to cell cycle signals and absent protein binding ('pocket') activity (Kaye *et al.*, 1990). The location of these *in vivo* mutations also corresponds to two non-contiguous domains referred to as domains A and B which were shown by *in vitro* deletional analysis to be required for protein binding activity (Hu *et al.*, 1990; Huang *et al.*, 1990; Kaelin *et al.*, 1990). These findings suggested that mutations within the *RB* gene in lung cancer have specifically affected the ability of RB to achieve a tertiary structure essential for protein binding activity and have not targeted the phosphorylation sites themselves, most of which are located C-terminally to domain B. In view of the importance of the RB protein-binding 'pocket' for tumour suppressor activity (Qin *et al.*, 1992), the identification and characterization of the cellular proteins that reversibly bind to the 'pocket' will be essential for understanding RB-mediated tumour suppressor pathways.

To address the clinical implication of *RB* mutations in lung cancer, primary NSCLC tumours, obtained through the cooperative Lung Cancer Study Group, were subjected to immunostaining with a monoclonal α-RB antibody and scored as wild-type or abnormal. It was observed that approximately 30% of the samples had an abnormal RB staining pattern; however, no correlation of RB status with either time to relapse or overall survival was detected in the matched patient samples (Reissmann *et al.*, 1993). This resembled the findings of an analysis of 171 SCLC and NSCLC cell lines where RB status was scored as either wild-type, mutant or absent by immunoblotting techniques. In this study, no significant associations were observed between RB status and age, sex, extent of disease, clinical performance status, smoking history, previous treatment, best clinical response or overall survival (Shimizu *et al.*, 1994). This type of analysis, however, was weakened by the poor overall prognosis of the lung cancer patients and by the small number of SCLC samples with wild-type RB and, conversely, by the small number of NSCLC samples with absent or mutant RB. Alternatively, other investigators have reported that advanced stage NSCLC had a higher frequency of altered RB immunohistochemical staining than lower staged tumours (Xu *et al.*, 1991) and, more recently, that altered RB immune staining was associated with decreased survival in early stage NSCLC (Xu *et al.*, 1994). Therefore, the issue of the clinical utility of ascertaining the status of the *RB* gene product in primary lung tumours is still unclear. Nonetheless, it is still likely that an important biological difference would be evident between RB(+) and RB(−) tumours which is underscored by the large difference in the incidence of *RB* inactivation observed between SCLC and NSCLC tumours.

The high incidence of *RB* inactivation in SCLC, as well as the observation that all naturally occurring *RB* mutations had selectively targeted the RB 'pocket' domain which is essential for tumour suppressor activity, provides strong evidence that the *RB* gene functions as a tumour suppressor gene in lung cancer. Final evidence, however, resides with functional experiments that test the ability of RB to suppress parameters of tumorigenicity following the reintroduction of the gene into RB(−) lung cells. Using plasmid-based vectors and selectable markers, the tumorigenicity in nude mice and the ability of tumour cells to clone in soft agar were markedly suppressed in both NSCLC and SCLC tumour cell lines by the stable expression of an exogenous RB product (Kratzke *et al.*, 1993; Ookawa *et al.*, 1993). This suppression was incomplete, however, as demonstrated by the occasional appearance of smaller xenograft tumours arising from transfected cells that continued to express detectable levels of RB protein. In addition, the inclusion of a small amount of an extract enriched in extracellular matrix (Matrigel) at the site of tumour inoculation in the nude mouse was consistently associated with the loss of RB-mediated tumour suppressor activity (Kratzke *et al.*, 1993). Although these findings are poorly explained, there is evidence to support an 'RB threshold' level of RB binding activity that is required for growth arrest, and multiple mechanisms may overcome this threshold (Kratzke *et al.*, 1994). For example, since the hypophosphorylated form of RB is the predominant growth suppressor molecule, either the overexpression of members of the G1 cdk or cyclin family (Sherr, 1993) or the inactivation of specific cdk inhibitors (Otterson *et al.*, 1994) may result in growth deregulation by reducing the effective levels of hypophosphorylated RB. Regardless of the mechanism, however, the observation that RB has been less effective than the *p53* gene product in inducing growth arrest or programmed cell death in tumour cells *in vitro* has lessened its potential as an effective agent for clinical gene therapy trials.

8.5.3 CDKN2/*p16^INK4* gene product

The recent identification of the *CDKN2* gene as a target for mutations in NSCLC was a controversial but extremely important finding that highlighted many recurrent themes in the genetic aetiology of lung cancer. First, the gene was initially predicted on the basis of cytogenetic data implicating loss of DNA on chromosomal band 9p21 in approximately 85% of NSCLC samples (Lukeis *et al.*, 1990; Coleman *et al.*, 1994; Mead *et al.*, 1994; Merlo *et al.*, 1994a). Second, the gene was originally identified as the locus conferring a familial predisposition to melanoma (Kamb *et al.*, 1994; Nobori *et al.*, 1994), reinforcing the observation that histologically distinct tumours share genetic aetiologies. Finally, the *CDKN2*/p16^INK4 gene product plays an important role in regulating the kinase activity of the cdk enzymes that control the transit through the G1/S boundary (Serrano *et al.*, 1993), confirming the key role of cell cycle elements in neoplastic transformation. The controversy surrounding the initial report of *CDKN2* as a tumour suppressor gene involved in multiple different malignancies focused on the recognition that the frequency of homozygous deletions of *CDKN2* was higher in cultured tumour cell lines than in primary tumours (Bonetta, 1994; Cairns *et al.*, 1994; Marx, 1994; Spruck *et al.*, 1994; Zhang *et al.*, 1994). However, germline mutations of *CDKN2* have been detected in families with familial melanoma and analysis of sporadic primary tumours from the bladder, oesophagus, and lung have shown a 20–85% incidence of small intragenic deletion, nonsense and missense mutations (Hayashi *et al.*, 1994; Mori *et al.*, 1994; Spruck *et al.*, 1994). In addition, reintro-duction of the *CDKN2* gene in tumour cells was associated with decreased foci formation (Guan *et al.*, 1994; Okamoto *et al.*, 1994; Otterson *et al.*, unpublished data). These findings confirm the importance of the *CDKN2* gene in human can-cers; however, the identification of an adjacent highly homologous gene on chromosome 9p, designated *MTS2* (multiple tumour suppressor), has raised the hypothesis that the homozygous deletions observed in many cultured tumour cell lines may be the result of selective growth pressures to delete the functions of both genes (Hannon and Beach, 1994).

Since the *CDKN2* gene product (p16^{INK4}) specifically blocks the phosphoryla-tion *in vitro* of the *RB* gene product by inhibiting the kinase activity of the cdk4 and cdk6 molecules (Serrano *et al.*, 1993), a recent study analysed the status of p16^{INK4} in lung cancer cell lines that expressed either a wild-type or an absent/mutant RB product (Otterson *et al.*, 1994). A striking inverse correlation was observed with only 6/55 SCLC (11%) showing the p16^{INK4} protein to be absent, and all six belonged to the rare subset of SCLC with wild-type RB. Conversely, the p16^{INK4} protein was detectable in all 48 SCLC samples with absent or mutant RB. This inverse correlation was also demonstrated in NSCLC samples, while a similar relationship was not observed with p16^{INK4} and p53 or RAS. These find-ings, therefore, suggested a common p16/RB tumour suppressor pathway which could be inactivated by mutations in either *CDKN2* or *RB*, but mutations in both were not required. The propensity for SCLC to target the *RB* gene while *CDKN2* mutations predominate in NSCLC samples, however, is still an unanswered issue, as is the role of the adjacent *MTS2/p15* gene in the genetic aetiology of lung can-cer.

8.5.4 Other predicted recessive oncogenes

The identification of a predicted tumour suppressor gene on chromosome 3p for lung cancer has remained surprisingly elusive, reflecting the large size of human chromosome 3 and the inability to perform linkage analysis due to the absence of a clear inherited pattern for this disease. Efforts to implicate previously identified genes that mapped within the consensus deletion region of chromosome 3p have been largely unsuccessful. These have included the thyroid hormone receptor (*ERBC*-β), a retinoic acid receptor (*RAR*β), a transcribed gene of unknown function at the *DNF15S2* locus, a tyrosine phosphatase (*PTP*-γ), and the *VHL* gene which confers susceptibility to selective malignancies associated with the von Hippel–Lindau syndrome (Drabkin *et al.*, 1988; Mattei *et al.*, 1988; Leduc *et al.*, 1989; Naylor *et al.*, 1989; Harbour *et al.*, 1990; Gebert *et al.*, 1991; LaForgia *et al.*, 1991; Latif *et al.*, 1993; Sekido *et al.*, 1994). An alternate strategy has been to search for homozygous deletions using large numbers of mapped 3p probes that have become available through the international genome mapping efforts. These studies have identified a homozygous deletion near 3p12–14 in one cell line (Rabbitts *et al.*, 1990) and near chromosome 3p21 in several others (Hibi *et al.*, 1992; Daly *et al.*, 1993; Yamakawa *et al.*, 1993). Using the precedent of the cloning of other tumour suppressor genes (such as the *RB* gene), the identification of lung tumour cell lines that contain small homozygous deletions will provide a valuable reagent for defining the pertinent breakpoints and isolating the intervening transcribed genes. It is, therefore, hopeful that in the near future the elusive 3p gene(s) will be identified.

In addition, cytogenetic studies have suggested many different loci as non-random targets for deletions in lung cancer. Accordingly, several previously described tumour suppressor genes, such as *DCC* (deleted in colorectal cancer), *APC* (adenomatous polyposis coli), *MCC* (mutated in colon cancer), *WT1* (Wilms' tumour) and others, have been studied in lung cancer; however, their roles in the aetiology of lung cancer are still uncertain (Testa and Graziano, 1993; Bepler and Garcia-Blanco, 1994; Hosoe *et al.*, 1994; Wieland and Bohm, 1994). More recently, 45% of primary SCLC tumours were identified with alterations of microsatellite loci defined by deletions or expansions of $(CA)_n$ dinucleotide repeats (Merlo *et al.*, 1994b). This instability of microsatellite loci is most frequently due to mutations within the *hMSH2* gene and can result in widespread genetic damage. In order to determine whether mutations within members of the DNA repair gene family are an important aetiological event, however, further studies must be awaited, especially since lung cancer is not a recognized feature in patients with hereditary non-polyposis colorectal cancer or the Lynch syndrome who carry germline mutations of these genes.

8.6 Dominant genes commonly targeted in lung cancer

8.6.1 MYC

Homogeneously staining regions and double minute chromosome fragments observed in the cytogenetic study of several SCLC cell lines were subsequently determined to represent regions of DNA amplification of the *MYC*, *MYCN* and *MYCL* oncogenes (Little *et al.*, 1983; Nau *et al.*, 1985, 1986; Kaye *et al.*, 1988).

Although approximately 25–30% of the SCLC cell lines had amplification of one member of the *MYC* family, the incidence in primary tumour samples was lower (5–18%) suggesting that this was a late event in lung tumorigenicity (Wong *et al.*, 1986; Johnson *et al.*, 1988a; Brennan *et al.*, 1991; Richardson and Johnson, 1993). Since amplification of the *MYCN* gene had been reported to be associated with a more aggressive clinical behaviour and worse prognosis in patients with neuro-blastoma (Brodeur *et al.*, 1984), several reports investigated the association between *MYC* amplification and survival in lung cancer. These studies showed a small decrease in overall survival in patients with tumours with *MYC* amplifica-tion, but no correlation was observed with *MYCN* or *MYCL* (Johnson *et al.*, 1988a; Brennan *et al.*, 1991).

Although *MYC* gene deregulation is closely linked with Burkitt's lymphoma and several other solid tumours (including lung cancer), the causal relationship is poorly understood. Over the past decade, the prevailing hypothesis has been that the aberrant expression of a normal *MYC* gene product disrupted growth and/or differentiation signals in these tumour cells (Lee, 1989). However, recent data have suggested that, in Burkitt's lymphoma tumours, missense mutations are pre-sent within an N-terminal MYC homology box which may affect its normal transcriptional activity (Bhatia *et al.*, 1993; Gu *et al.*, 1994). In this model, the RB-related 'pocket' protein, p107, binds *in vivo* to the wild-type MYC product to repress its transcriptional activity, analogous to the ability of RB to suppress E2F1 transcriptional activity (Nevins, 1992). In the presence of the N-terminal MYC missense mutations, however, p107 has lost the ability to repress gene transcrip-tion (Gu *et al.*, 1994). If these findings are confirmed, it will be important to ascertain whether similar acquired mutations are present within the *MYC* genes isolated from lung tumour samples with and without *MYC* gene amplification. In addition, there may be other mechanisms which may affect *MYC* gene function. For example, although chromosomal rearrangements of *MYC* genes are not gen-erally observed in human carcinomas, two independent SCLC samples were identified with an intrachromosomal rearrangement fusing an amplified *MYCL* gene with a novel transcribed gene designated *rlf* (Makela *et al.*, 1991). The role of this chimeric mRNA in the development or progression of these tumours is still under investigation.

8.6.2 RAS

The *RAS* oncogenes are members of a large gene family that were initially identi-fied by their homology to the transforming elements of certain RNA tumour viruses and by their ability to transform animal cells in culture (Barbacid, 1987). Normally, RAS functions as a cytoplasmic guanine nucleotide binding protein that mediates information from external signals towards the nucleus. Acquired mis-sense mutations, however, appear to lock the protein in a conformation that promotes continued growth stimulation. Fortunately, analysis of the activation of the three mammalian *RAS* genes, *HRAS*, *KRAS* and *NRAS*, in human tumour samples has been relatively easy to perform since these mutations have been generally restricted to codons 12, 13 and 61 which are located in or near the guanine nucleotide binding domain (Tong *et al.*, 1989). The likelihood of finding a *RAS* mutation, however, varies greatly among tumour types, and the striking observation in lung cancer was that, while mutations were detected in 20–30% of NSCLC primary tumours and culture cell lines, 0% of SCLC samples had evidence for mutational activation (Rodenhuis *et al.*, 1987; Rodenhuis and Slebos, 1990; Slebos *et al.*, 1990; Mitsudomi *et al.*, 1991a, b; Slebos *et al.*, 1991). Similarly, *RAS* mutations are found rarely in breast cancer, while approximately 50% of colon cancers and 90% of exocrine pancreatic cancers carry these mutations. The role of these *RAS* mutations in the aetiology of these tumours is still uncertain; however, a recent report studying the molecular determinants of dysplasia in colon cancers observed that the timing of *RAS* mutations may have a critical impact on outcome. For example, *RAS* mutations preceding *APC* mutations appear to result in small non-dysplastic lesions that do not progress to cancer. In contrast, *APC* mutations are found in essentially all dysplastic lesions, some of which may also harbour *RAS* mutations, which are believed to occur as a secondary event (Jen *et al.*, 1994). In this model, therefore, mutant *RAS* functions later in the course of tumorigenesis to help drive dysplastic tissues toward a more invasive phenotype. This hypothesis is in agreement with the observation that *RAS* mutations also appear to occur later in the course of development of invasive NSCLC (Sugio *et al.*, 1994) and that *RAS* mutations are associated with both tobacco exposure and with a shortened survival (Slebos *et al.*, 1990; Mitsudomi *et al.*, 1991a; Slebos *et al.*, 1991).

An exciting development in the application of this information is the effort to manipulate the biological properties of the tumour cell *in vivo* by modulating the expression of mutant genes, such as the *RAS* oncogene. For example, pharmacological agents screened for their ability to block enzymatic reactions (farnesyltransferase) required for *RAS* activity are under active investigation (Kohl *et al.*, 1993). In addition, the use of an antisense *KRAS* construct delivered in a retroviral vector has shown the ability to inhibit growth *in vitro* and *in vivo* in animal models, and efforts are underway to test this strategy in a pilot clinical trial (Georges *et al.*, 1993; Zhang *et al.*, 1993).

8.6.3 Other dominant oncogenes

Activation of numerous other oncogenes has been reported in isolated studies. This includes amplification and/or overexpression of the *MYB*, *CSF1R*, *EGFR*, *ERBB2* (*neu*), *RAF1*, *KIT*, and *nm23/NDP* kinase genes (Cline and Battifora, 1987; Kiefer *et al.*, 1987; Sekido *et al.*, 1991; Hida *et al.*, 1994; Ozeki *et al.*, 1994; Yu *et al.*, 1994). The biological importance of these isolated findings, however, is still under investigation.

8.7 Conclusion

Although the amount of information concerning the molecular genetics of lung cancer is expanding rapidly, many questions still remain. What is the cell of origin of SCLC? Is there a defined temporal sequence of mutations that results in an orderly progression from dysplasia to lung cancer? Are there genetically inherited susceptibilities to the development of lung cancer that can be unambiguously identified in the population? Can individual carcinogen molecules in the pulmonary epithelium be identified and eliminated? Will a lung tumour with multiple different genetic abnormalities be amenable to pharmacological therapies (drugs or genes) targeted toward a single pathway? We are already seeing the first wave of genetic biomarkers that are being developed for the early diagnosis of lung cancer (Mao *et al.*, 1994). New strategies for the prevention and treatment of this disease will be greater challenges.

References

Avis IL, Kovacs TO, Kasprzyk PG, Treston AM, Bartholomew R, Walsh JH, Cuttitta F, Mulshine JL. (1991) Preclinical evaluation of an anti-autocrine growth factor monoclonal antibody for treatment of patients with small-cell lung cancer. *J. Natl Cancer Inst.* **83**: 1470–1476.

Barbacid M. (1987) ras genes. *Annu. Rev. Biochem.* **56**: 779–827.

Baylin SB, Gazdar AF. (1981) Endocrine biochemistry in the spectrum of human lung cancer: implications for the cellular origin of small cell carcinoma. In: *Small Cell Lung Cancer* (ed. F Greco). Grune and Stratton, New York, pp. 123–145.

Bennett WP, Colby TV, Travis WD, Borkowski A, Jones RT, Lane DP, Metcalf RA, Samet JM, Takeshima Y, Gu JR. (1993) p53 protein accumulates frequently in early bronchial neoplasia. *Cancer Res.* **53**: 4817–4822.

Bepler G, Garcia–Blanco MA. (1994) Three tumor-suppressor regions on chromosome 11p identified by high-resolution deletion mapping in human non-small-cell lung cancer. *Proc. Natl Acad. Sci. USA* **91**: 5513–5517.

Bhatia K, Huppi K, Spangler G, Siwarski D, Iyer R, Magrath I. (1993) Point mutations in the c-Myc transactivation domain are common in Burkitt's lymphoma and mouse plasmacytomas. *Nature Genet.* **5**: 56–61.

Bignon YJ, Shew JY, Rappolee D, Naylor SL, Lee EY, Schnier J, Lee WH. (1990) A single Cys706 to Phe substitution in the retinoblastoma protein causes the loss of binding to SV40 T antigen. *Cell Growth Differ.* **1**: 647–651.

Bliss D Jr, Battey JF, Linnoila RI, Birrer MJ, Gazdar AF, Johnson BE. (1990) Expression of the atrial natriuretic factor gene in small cell lung cancer tumors and tumor cell lines. *J. Natl Cancer Inst.* **82**: 305–310.

Bonetta L. (1994) Tumor-suppressor genes. Open questions on p16. *Nature* **370**: 180.

Boring CC, Squires TS, Tong T, Montgomery S. (1994) Cancer statistics, 1994. *Cancer J. Clin.* **44**: 7–26.

Brauch H, Johnson B, Hovis J, Yano T, Gazdar A, Pettengill OS, Graziano S, Sorenson GD, Poiesz BJ, Minna J, *et al.* (1987) Molecular analysis of the short arm of chromosome 3 in small-cell and non-small-cell carcinoma of the lung. *N. Engl. J. Med.* **317**: 1109–13.

Brechot C. (1993) Oncogenic activation of cyclin A. *Curr. Opin. Genet. Devel.* **3**: 11–18.

Brennan J, O'Connor T, Makuch RW, Simmons AM, Russell E, Linnoila RI, Phelps RM, Gazdar AF, Ihde DC, Johnson BE. (1991) myc family DNA amplification in 107 tumors and tumor cell lines from patients with small cell lung cancer treated with different combination chemotherapy regimens. *Cancer Res.* **51**: 1708–1712.

Brodeur GM, Seeger RC, Schwab M, Varmus HE, Bishop JM. (1984) Amplification of N-myc in untreated human neuroblastomas correlates with advanced disease stage. *Science* **224**: 1121–1124.

Cai DW, Mukhopadhyay T, Liu Y, Fujiwara T, Roth JA. (1993) Stable expression of the wild-type p53 gene in human lung cancer cells after retrovirus-mediated gene transfer. *Hum. Gene Ther.* **4**: 617–624.

Cairns P, Mao L, Merlo A, Lee DJ, Schwab D, Eby Y, Tokino K, van der Riet P, Blaugrund JE, Sidransky D. (1994) Rates of p16 (MTS1) mutations in primary tumors with 9p loss. *Science* **265**: 415–417.

Carney DN, Gazdar AF, Bepler G, Guccion JG, Marangos PJ, Moody TW, Zweig MH, Minna JD. (1985) Establishment and identification of small cell lung cancer cell lines having classic and variant features. *Cancer Res.* **45**: 2913–2923.

Center R, Lukeis R, Vrazas V, Garson M. (1993) Y chromosome loss and rearrangement in non-small cell lung cancer. *Int. J. Cancer.* **55**: 390–393.

Chiba I, Takahashi T, Nau MM, D'Amico D, Curiel DT, Mitsudomi T, Buchhagen DL, Carbone D, Piantadosi S, Koga H, *et al.* (1990) Mutations in the p53 gene are frequent in primary, resected non-small cell lung cancer. Lung Cancer Study Group. *Oncogene* **5**: 1603–1610.

Chretien MF, Pouplard–Barthelaix A, Dubois MP, Simard C, Rebel A. (1986) Somatostatin and adrenocorticotrophic hormone like immunoreactivity in small cell carcinoma of the lung. *J. Clin. Pathol.* **39**: 418–422.

Cline MJ, Battifora H. (1987) Abnormalities of protooncogenes in non-small cell lung cancer. Correlations with tumor type and clinical characteristics. *Cancer* **60**: 2669–2674. [Published erratum appears in *Cancer* **61**: 1064, 1988.]

Coleman A, Fountain JW, Nobori T, Olopade OI, Robertson G, Housman DE, Lugo TG. (1994) Distinct deletions of chromosome 9p associated with melanoma versus glioma, lung cancer, and leukemia. *Cancer Res.* **54**: 344–348.

Cuttitta F, Carney DN, Mulshine J, Moody TW, Fedorko J, Fischler A, Minna JD. (1985) Bombesin-like peptides can function as autocrine growth factors in human small-cell lung cancer. *Nature* **316**: 823–826.

Daly MC, Xiang RH, Buchhagen D, Hensel CH, Garcia DK, Killary AM, Minna JD, Naylor SL. (1993) A homozygous deletion on chromosome 3 in a small cell lung cancer cell line correlates with a region of tumor suppressor activity. *Oncogene* **8**: 1721–1729.

D'Amico D, Carbone D, Mitsudomi T, Nau M, Fedorko J, Russell E, Johnson B, Buchhagen D, Bodner S, Phelps R, *et al.* (1992) High frequency of somatically acquired p53 mutations in small-cell lung cancer cell lines and tumors. *Oncogene* **7**: 339–346.

DeFusco P, Frytak S, Dahl R, Weiland L, Unni K, Dewald G. (1989) Cytogenetic studies in 11 patients with small cell carcinoma of the lung. *Mayo Clin. Proc.* **64**: 168–176.

Donehower LA, Harvey M, Slagle BL, McArthur MJ, Montgomery CA, Butel JA, Bradley A. (1992) Mice deficient for p53 are developmentally normal but susceptible to spontaneous tumours. *Nature* 356: 215–221.

Drabkin H, Kao FT, Hartz J, Hart L, Gazdar A, Weinberger C, Evans R, Gerber M. (1988) Localization of human ERBA2 to the 3p22–3p24.1 region of chromosome 3 and variable deletion in small cell lung cancer. *Proc. Natl Acad. Sci. USA* 85: 9258–9262.

el-Deiry WS, Tokino T, Velculescu VE, Levy DB, Parsons R, Trent JM, Lin D, Mercer WE, Kinzler KW, Vogelstein B. (1993) WAF1, a potential mediator of p53 tumor suppression. *Cell* 75: 817–825.

el-Deiry WS, Harper JW, O'Connor PM, Velculescu VE, Canman CE, Jackman J, Pietenpol JA, Burrell M, Hill DE, Wang Y, *et al*. (1994) WAF1/CIP1 is induced in p53-mediated G1 arrest and apoptosis. *Cancer Res.* 54: 1169–1174.

Friend SH, Bernards R, Rogelj S, Weinberg RA, Rapaport JM, Albert DM, Dryja TP. (1986) A human DNA segment with properties of the gene that predisposes to retinoblastoma and osteosarcoma. *Nature* 323: 643–646.

Fujiwara Y, Ohata H, Emi M, Okui K, Koyoma K, Tsuchiya E, Nakajima T, Monden M, Mori T, Nakamura Y. (1994) A 3-Mb physical map of the chromosome region 8p21.3–p22, including a 600 kb region commonly deleted in human hepatocellular carcinoma, colorectal cancer, and non-small cell lung cancer. *Genes Chrom. Cancer* 10: 7–14.

Gebert JF, Moghal N, Frangioni JV, Sugarbaker DJ, Neel BG. (1991) High frequency of retinoic acid receptor beta abnormalities in human lung cancer. *Oncogene* 6: 1859–1868.

Georges RN, Mukhopadhyay T, Zhang Y, Yen N, Roth JA. (1993) Prevention of orthotopic human lung cancer growth by intratracheal instillation of a retroviral antisense K-ras construct. *Cancer Res.* 53: 1743–1746.

Gu W, Bhatia K, Magrath IT, Dang CV, Dalla–Favera R. (1994) Binding and suppression of the Myc transcriptional activation domain by p107. *Science* 264: 251–254.

Guan KL, Jenkins CW, Li Y, Nichols MA, Wu X, O'Keefe CL, Matera AG, Xiong Y. (1994) Growth suppression by p18, a p16INK4/MTS1- and p14INK4B/MTS2-related CDK6 inhibitor, correlates with wildtype pRB function. *Genes Devel.* 8: 2939–2952.

Hannon GJ, Beach D. (1994) p15INK4B is a potential effector of TGF-beta-induced cell cycle arrest [see comments]. *Nature* 371: 257–261.

Harbour JW, Lai SL, Whang–Peng J, Gazdar AF, Minna JD, Kaye FJ. (1988) Abnormalities in structure and expression of the human retinoblastoma gene in SCLC. *Science* 241: 353–357.

Harbour JW, Lai SL, Gazdar AF, Minna JD, Kaye FJ. (1990) Expression in lung cancer of a transcribed sequence at the DNF15S2 locus at chromosome 3P21. *Anticancer Res.* 10: 23–27.

Harris CC, Hollstein M. (1993) Clinical implications of the p53 tumor-suppressor gene. *N. Engl. J. Med.* 329: 1318–1327.

Hayashi N, Sugimoto Y, Tsuchiya E, Ogawa M, Nakamura Y. (1994) Somatic mutations of the MTS (multiple tumor suppressor) 1/CDK41 (cyclin-dependent kinase-4 inhibitor) gene in human primary non-small cell lung carcinomas. *Biochem. Biophys. Res. Commun.* 202: 1426–1430.

Hensel CH, Xiang RH, Sakaguchi AY, Naylor SL. (1991) Use of the single strand conformation polymorphism technique and PCR to detect p53 gene mutations in small cell lung cancer. *Oncogene* **6**: 1067–1071.

Hibi K, Takahashi T, Yamakawa K, Ueda R, Sekido Y, Ariyoshi Y, Suyama M, Takagi H, Nakamura Y, Takahashi T. (1992) Three distinct regions involved in 3p deletion in human lung cancer. *Oncogene* **7**: 445–449.

Hida T, Ueda R, Sekido Y, Hibi K, Matsuda R, Ariyoshi Y, Sugiura T, Takahashi T, Takahashi T. (1994) Ectopic expression of c-kit in small-cell lung cancer. *Int. J. Cancer Suppl.* **8**: 108–109.

Hoffmann D, Hecht SS, Wynder EL. (1983) Tumor promoters and cocarcinogens in tobacco carcinogenesis. *Environ. Health Perspect.* **50**: 247–257.

Hollstein M, Sidransky D, Vogelstein B, Harris CC. (1991) p53 mutations in human cancers. *Science* **253**: 49–53.

Horio Y, Takahashi T, Kuroishi T, Hibi K, Suyama M, Niimi T, Shimokata K, Yamakawa K, Nakamura Y, Ueda R, *et al.* (1993) Prognostic significance of p53 mutations and 3p deletions in primary resected non-small cell lung cancer. *Cancer Res.* **53**: 1–4.

Horowitz JM, Park SH, Bogenmann E, Cheng JC, Yandell DW, Kaye FJ, Minna JD, Dryja TP, Weinberg RA. (1990) Frequent inactivation of the retinoblastoma anti-oncogene is restricted to a subset of human tumor cells. *Proc. Natl Acad. Sci. USA* **87**: 2775–2779.

Hosoe S, Ueno K, Shigedo Y, Tachibana I, Osaki T, Kumagai T, Tanio Y, Kawase I, Nakamura Y, Kishimoto T. (1994) A frequent deletion of chromosome 5q21 in advanced small cell and non-small cell carcinoma of the lung. *Cancer Res.* **54**: 1787–1790.

Hu QJ, Dyson N, Harlow E. (1990) The regions of the retinoblastoma protein needed for binding to adenovirus E1A or SV40 large T antigen are common sites for mutations. *EMBO J.* **9**: 1147–1155.

Huang S, Wang NP, Tseng BY, Lee WH, Lee EH. (1990) Two distinct and frequently mutated regions of retinoblastoma protein are required for binding to SV40 T antigen. *EMBO J.* **9**: 1815–1822.

Hunter T, Pines J. (1994) Cyclins and cancer II: cyclin D and CDK inhibitors come of age. *Cell* **79**: 573–582.

Ihde DC. (1992) Chemotherapy of lung cancer. *N. Engl. J. Med.* **327**: 1434–1441.

Ihde DC, Minna JD. (1991a) Non-small cell lung cancer. Part I: Biology, diagnosis, and staging. *Curr. Probl. Cancer* **15**: 61–104.

Ihde DC, Minna JD. (1991b) Non-small cell lung cancer. Part II: Treatment. *Curr. Probl. Cancer* **15**: 105–154.

Isola J, Visakorpi T, Holli K, Kallioniemi OP. (1992) Association of overexpression of tumor suppressor protein p53 with rapid cell proliferation and poor prognosis in node-negative breast cancer patients. *J. Natl Cancer Inst.* **84**: 1109–1114.

Jen J, Powell SM, Papadopoulos N, Smith KJ, Hamilton SR, Vogelstein B, Kinzler KW. (1994) Molecular determinants of dysplasia in colorectal lesions. *Cancer Res.* **54**: 5523–5526.

Johansson M, Dietrich C, Mandahl N, Hambraeus G, Johansson L, Clausen P, Mitelman F, Heim S. (1994) Karyotypic characterization of bronchial large cell carcinoma. *Int. J. Cancer* **57**: 463–467.

Johnson BE, Makuch RW, Simmons AD, Gazdar AF, Burch D, Cashell AW. (1988a) myc family DNA amplification in small cell lung cancer patients' tumors and corresponding cell lines. *Cancer Res.* **48**: 5163–5166.

Johnson BE, Sakaguchi AY, Gazdar AF, Minna JD, Burch D, Marshall A, Naylor SL. (1988b) Restriction fragment length polymorphism studies show consistent loss of chromosome 3p alleles in small cell lung cancer patients' tumors. *J. Clin. Invest.* **82**: 502–507.

Kaelin W Jr, Ewen ME, Livingston DM. (1990) Definition of the minimal simian virus 40 large T antigen- and adenovirus E1A-binding domain in the retinoblastoma gene product. *Mol. Cell Biol.* **10**: 3761–3769.

Kamb A, Gruis NA, Weaver-Feldhaus J, Liu Q, Harshman K, Tavtigian SV, Stockert E, Day R, Johnson BE, Skolnick MH. (1994) A cell cycle regulator potentially involved in genesis of many tumor types. *Science* **264**: 436–440.

Kastan MB, Zhan Q, el-Deiry WS, Carrier F, Jacks T, Walsh WV, Plunkett BS, Vogelstein B, Fornace A Jr. (1992) A mammalian cell cycle checkpoint pathway utilizing p53 and GADD45 is defective in ataxia-telangiectasia. *Cell* **71**: 587–597.

Kaye FJ, McBride OW, Battey JF, Gazdar AF, Sausville EA. (1987) Human creatine kinase-B complementary DNA. Nucleotide sequence, gene expression in lung cancer, and chromosomal assignment to two distinct loci. *J. Clin. Invest.* **79**: 1412–1420.

Kaye FJ, Battey J, Nau M, Brooks B, Seifter E, De Greve J, Birrer M, Sausville E, Minna J. (1988) Structure and expression of the human L-myc gene reveal a complex pattern of alternative mRNA processing. *Mol. Cell Biol.* **8**: 186–195.

Kaye FJ, Kratzke RA, Gerster JL, Horowitz JM. (1990) A single amino acid substitution results in a retinoblastoma protein defective in phosphorylation and oncoprotein binding. *Proc. Natl Acad. Sci. USA* **87**: 6922–6926.

Kiefer PE, Bepler G, Kubasch M, Havemann K. (1987) Amplification and expression of protooncogenes in human small cell lung cancer cell lines. *Cancer Res.* **47**: 6236–6242.

King RW, Jackson PK, Kirschner MW. (1994) Mitosis in transition. *Cell* **79**: 563–571.

Kishimoto Y, Murakami Y, Shiraishi M, Hayashi K, Sekiya T. (1992) Aberrations of the p53 tumor suppressor gene in human non-small cell carcinomas of the lung. *Cancer Res.* **52**: 4799–4804.

Kohl N, Mosser S, de Solms S, Giuliani E, Pompliano D, Graham S, Smith R, Scolnick E, Oliff A, Gibbs J. (1993) Selective inhibition of ras-dependent transformation by a farnesyltransferase inhibitor. *Science* **260**: 1934–1937.

Kohno T, Morishita K, Takano H, Shapiro D, Yokota J. (1994) Homozygous deletion at chromosome 2q33 in human small-cell lung carcinoma identified by arbitrarily primed PCR genomic fingerprinting. *Oncogene* **9**: 103–108.

Kratzke RA, Shimizu E, Geradts J, Gerster JL, Segal S, Otterson GA, Kaye FJ. (1993) RB-mediated tumor suppression of a lung cancer cell line is abrogated by an extract enriched in extracellular matrix. *Cell Growth Differ.* **4**: 629–635.

Kratzke RA, Otterson GA, Hogg A, Coxon AB, Geradts J, Cowell JK, Kaye FJ. (1994) Partial inactivation of the RB product in a family with incomplete penetrance of familial retinoblastoma and benign retinal tumours. *Oncogene* **9**: 1321–1362.

LaForgia S, Morse B, Levy J, Barnea G, Cannizzaro LA, Li F, Nowell PC, Boghosian-Sell L, Glick J, Weston A. (1991) Receptor protein-tyrosine phosphatase gamma is a candidate tumor suppressor gene at human chromosome region 3p21. *Proc. Natl Acad. Sci. USA* **88**: 5036–5040.

Lane DP. (1992) Cancer. p53, guardian of the genome. *Nature* **358**: 15–16.

Latif F, Tory K, Gnarra J, Yao M, Duh FM, Orcutt ML, Stackhouse T, Kuzmin I, Modi W, Geil L, et al. (1993) Identification of the von Hippel–Lindau disease tumor suppressor gene. *Science* **260**: 1317–1320.

Leduc F, Brauch H, Hajj C, Dobrovic A, Kaye F, Gazdar A, Harbour JW, Pettengill OS, Sorenson GD, van den Berg A, *et al.* (1989) Loss of heterozygosity in a gene coding for a thyroid hormone receptor in lung cancers. *Am. J. Hum. Genet.* **44**: 282–287.

Lee EY, Chang CY, Hu N, Wang YC, Lai CC, Herrup K, Lee WH, Bradley A. (1992) Mice deficient for Rb are nonviable and show defects in neurogenesis and haematopoiesis. *Nature* **359**: 288–294.

Lehman TA, Bennett WP, Metcalf RA, Welsh JA, Ecker J, Modali RV, Ullrich S, Romano JW, Appella E, Testa JR, *et al.* (1991) p53 mutations, ras mutations, and p53–heat shock 70 protein complexes in human lung carcinoma cell lines. *Cancer Res.* **51**: 4090–4096.

Linnoila RI, Mulshine JL, Steinberg SM, Funa K, Matthews MJ, Cotelingam JD, Gazdar AF. (1988) Neuroendocrine differentiation in endocrine and nonendocrine lung carcinomas. *Am. J. Clin. Pathol.* **90**: 641–652.

Little CD, Nau MM, Carney DN, Gazdar AF, Minna JD. (1983) Amplification and expression of the c-myc oncogene in human lung cancer cell lines. *Nature* **306**: 194–196.

Lokich JJ. (1982) The frequency and clinical biology of the ectopic hormone syndromes of small cell carcinoma. *Cancer* **50**: 2111–2114.

Lowe SW, Ruley HE, Jacks T, Housman DE. (1993a) p53-dependent apoptosis modulates the cytotoxicity of anticancer agents. *Cell* **74**: 957–967.

Lowe SW, Schmitt EM, Smith SW, Osborne BA, Jacks T. (1993b) p53 is required for radiation-induced apoptosis in mouse thymocytes. *Nature* **362**: 847–849.

Lowe SW, Bodis S, McClatchey A, Remington L, Ruley HE, Fisher DE, Housman DE, Jacks T. (1994) p53 status and the efficacy of cancer therapy in vivo. *Science* **266**: 807–810.

Lukeis R, Irving L, Garson M, Hasthorpe S. (1990) Cytogenetics of non-small cell lung cancer: analysis of consistent non-random abnormalities. *Genes Chrom. Cancer* **2**: 116–124.

Lukeis R, Ball D, Garson M, Hasthorpe S. (1993) Chromosome abnormalities in non-small cell lung cancer pleural effusions: cytogenetic indicators of disease subgroups. *Genes Chrom. Cancer* **8**: 262–269.

Makela TP, Kere J, Winqvist R, Alitalo K. (1991) Intrachromosomal rearrangements fusing L-myc and rlf in small-cell lung cancer. *Mol. Cell Biol.* **11**: 4015–4021.

Mao L, Hruban R, Boyle J, Tockman M, Sidransky D. (1994) Detection of oncogene mutations in sputum precedes diagnosis of lung cancer. *Cancer Res.* **54**: 1634–1637.

Marchetti A, Buttitta F, Merlo G, Diella F, Pellegrini S, Pepe S, Macchiarini P, Chella A, Angeletti CA, Callahan R. *et al.* (1993) p53 alterations in non-small metastatic involvement of hilar and mediastinal lymph nodes. *Cancer Res.* **53**: 2846–2851.

Martin KJ, Winslow ER, Kaddurah–Daouk R. (1994) Cell cycle studies of cyclocreatine, a new anticancer agent. *Cancer Res.* **54**: 5160–5165.

Marx J. (1994) A challenge to p16 gene as a major tumor suppressor. *Science* **264**: 1846.

Mattei MG, de The H, Mattei JF, Marchio A, Tiollais P, Dejean A. (1988) Assignment of the human hap retinoic acid receptor RAR beta gene to the p24 band of chromosome 3. *Hum. Genet.* **80**: 189–190.

Mead LJ, Gillespie MT, Irving LB, Campbell LJ. (1994) Homozygous and hemizygous deletions of 9p centromeric to the interferon genes in lung cancer. *Cancer Res.* **54**: 2307–2309.

Merlo A, Gabrielson E, Askin F, Sidransky D. (1994a) Frequent loss of chromosome 9 in human primary non-small cell lung cancer. *Cancer Res.* **54**: 640–642.

Merlo A, Mabry M, Gabrielson E, Vollmer R, Baylin SB, Sidransky D. (1994b) Frequent microsatellite instability in primary small cell lung cancer. *Cancer Res.* **54**: 2098–2101.

Miller CW, Simon K, Aslo A, Kok K, Yokota J, Buys CH, Terada M, Koeffler HP. (1992) p53 mutations in human lung tumors. *Cancer Res.* **52**: 1695–1698.

Minuto F, Del Monte P, Barreca A, Fortini P, Cariola G, Catrambone G, Giordano G. (1986) Evidence for an increased somatomedin-C/insulin-like growth factor I content in primary human lung tumors. *Cancer Res.* **46**: 985–988.

Mitsudomi T, Steinberg SM, Oie HK, Mulshine JL, Phelps R, Viallet J, Pass H, Minna JD, Gazdar AF. (1991a) ras gene mutations in non-small cell lung cancers are associated with shortened survival irrespective of treatment intent. *Cancer Res.* **51**: 4999–5002.

Mitsudomi T, Viallet J, Mulshine JL, Linnoila RI, Minna JD, Gazdar AF. (1991b) Mutations of ras genes distinguish a subset of non-small-cell lung cancer cell lines from small-cell lung cancer cell lines. *Oncogene* **6**: 1353–1362.

Mitsudomi T, Steinberg SM, Nau MM, Carbone D, D'Amico D, Bodner S, Oie HK, Linnoila RI, Mulshine JL, Minna JD. (1992) p53 gene mutations in non-small-cell lung cancer cell lines and their correlation with the presence of ras mutations and clinical features. *Oncogene* **7**: 171–180.

Miura I, Graziano S, Cheng J, Doyle A, Testa J. (1992) Chromosome alterations in human small cell lung cancer: frequent involvement of 5q. *Cancer Res.* **52**: 1322–1328.

Moody TW, Pert CB, Gazdar AF, Carney DN, Minna JD (1981) High levels of intracellular bombesin characterize human small-cell lung carcinoma. *Science* **214**: 1246–1248.

Mori T, Miura K, Aoki T, Nishihira T, Mori S, Nakamura Y. (1994) Frequent somatic mutation of the MTS1/CDK4I (multiple tumor suppressor/cyclin-dependent kinase 4 inhibitor) gene in esophageal squamous cell carcinoma. *Cancer Res.* **54**: 3396–3397.

Morstyn G, Brown J, Novak U, Gardner J, Bishop J, Garson M. (1987) Heterogeneous cytogenetic abnormalities in small cell lung cancer cell lines. *Cancer Res.* **47**: 3322–3327.

Nau MM, Brooks BJ, Battey J, Sausville E, Gazdar AF, Kirsch IR, McBride OW, Bertness V, Hollis GF, Minna JD. (1985) L-myc, a new myc-related gene amplified and expressed in human small cell lung cancer. *Nature* **318**: 69–73.

Nau MM, Brooks BJ, Carney DN, Gazdar AF, Battey JF, Sausville EA, Minna JD. (1986) Human small-cell lung cancers show amplification and expression of the N-myc gene. *Proc. Natl Acad. Sci. USA* **83**: 1092–1096.

Naylor SL, Johnson BE, Minna JD, Sakaguchi AY. (1987) Loss of heterozygosity of chromosome 3p markers in small-cell lung cancer. *Nature* **329**: 451–454.

Naylor SL, Marshall A, Hensel C, Martinez PF, Holley B, Sakaguchi AY. (1989) The DNF15S2 locus at 3p21 is transcribed in normal lung and small cell lung cancer. *Genomics* **4**: 355–361.

Nelson WG, Kastan MB. (1994) DNA strand breaks: the DNA template alterations that trigger p53-dependent DNA damage response pathways. *Mol. Cell Biol.* **14**: 1815–1823.

Nevins JR. (1992) E2F: a link between the Rb tumor suppressor protein and viral oncoproteins. *Science* **258**: 424–429.

Nobori T, Miura K, Wu DJ, Lois A, Takabayashi K, Carson DA. (1994) Deletions of the cyclin-dependent kinase-4 inhibitor gene in multiple human cancers. *Nature* **368**: 753–756.

Okamoto A, Demetrick DJ, Spillare EA, Hagiwara K, Hussain SP, Bennett WP, Forrester K, Gerwin B, Serrano M, Beach D, Harriss CC. (1994) Mutations and altered expression of p16INK4 in human cancer. *Proc. Natl Acad. Sci. USA* **91**: 11045–11049.

Olopade OI, Buchhagen DL, Malik K, Sherman J, Nobori T, Bader S, Nau MM, Gazdar AF, Minna JD, Diaz MO. (1993) Homozygous loss of the interferon genes defines the critical region on 9p that is deleted in lung cancers. *Cancer Res.* **53**: 2410–2415.

Ookawa K, Shiseki M, Takahashi R, Yoshida Y, Terada M, Yokota J. (1993) Reconstitution of the RB gene suppresses the growth of small-cell lung carcinoma cells carrying multiple genetic alterations. *Oncogene* **8**: 2175–2181.

Otterson GA, Kratzke RA, Coxon A, Kim YW, Kaye FJ. (1994) Absence of p16INK4 protein is restricted to the subset of lung cancer lines that retains wildtype RB. *Oncogene* **9**: 3375–3378.

Ozeki Y, Takishima K, Mamiya G. (1994) Immunohistochemical analysis of nm23/NDP kinase expression in human lung adenocarcinoma: association with tumor progression in Clara cell type. *Jpn J. Cancer Res.* **85**: 840–846.

Puisieux A, Lim S, Groopman J, Ozturk M. (1991) Selective targeting of p53 gene mutational hotspots in human cancers by etiologically defined carcinogens. *Cancer Res.* **51**: 6185–6189.

Qin XQ, Chittenden T, Livingston DM, Kaelin W Jr. (1992) Identification of a growth suppression domain within the retinoblastoma gene product. *Genes Devel.* **6**: 953–964.

Quinlan DC, Davidson AG, Summers CL, Warden HE, Doshi HM. (1992) Accumulation of p53 protein correlates with a poor prognosis in human lung cancer. *Cancer Res.* **52**: 4828–4831.

Rabbitts P, Bergh J, Douglas J, Collins F, Waters J. (1990) A submicroscopic homozygous deletion at the D3S3 locus in a cell line isolated from a small cell lung carcinoma. *Genes Chrom. Cancer* **2**: 231–238.

Reissmann PT, Koga H, Takahashi R, Figlin RA, Holmes EC, Piantadosi S, Cordon-Cardo C, Slamon DJ. (1993) Inactivation of the retinoblastoma susceptibility gene in non-small-cell lung cancer. The Lung Cancer Study Group. *Oncogene* **8**: 1913–1919.

Richardson GE, Johnson BE. (1993) The biology of lung cancer. *Semin. Oncol.* **20**: 105–127.

Rodenhuis S, Slebos, RJ. (1990) The ras oncogenes in human lung cancer. *Am. Rev. Respir. Dis.* **142**: S27–30.

Rodenhuis S, van de Wetering ML, Mooi WJ, Evers SG, van Zandwijk N, Bos JL. (1987) Mutational activation of the K-ras oncogene. A possible pathogenetic factor in adenocarcinoma of the lung. *N. Engl. J. Med.* **317**: 929–935.

Ruff MR, Pert CB. (1984) Small cell carcinoma of the lung: macrophage-specific antigens suggest hemopoietic stem cell origin. *Science* **225**: 1034–1036.

Said JW, Vimadalal S, Nash G, Shintaku IP, Heusser RC, Sassoon AF, Lloyd RV. (1985) Immunoreactive neuron-specific enolase, bombesin, and chromogranin as markers for neuroendocrine lung tumors. *Hum. Pathol.* **16**: 236–240.

Sameshima Y, Matsuno Y, Hirohashi S, Shimosato Y, Mizoguchi H, Sugimura T, Terada M, Yokota J. (1992) Alterations of the p53 gene are common and critical events for the maintenance of malignant phenotypes in small-cell lung carcinoma. *Oncogene* **7**: 451–457.

Sausville EA, Lebacq–Verheyden AM, Spindel ER, Cuttitta F, Gazdar AF, Battey JF. (1986) Expression of the gastrin-releasing peptide gene in human small cell lung cancer. Evidence for alternative processing resulting in three distinct mRNAs. *J. Biol. Chem.* **261**: 2451–2457.

Sekido Y, Obata Y, Ueda R, Hida T, Suyama M, Shimokata K, Ariyoshi Y, Takahashi T. (1991) Preferential expression of c-kit protooncogene transcripts in small cell lung cancer. *Cancer Res.* **51**: 2416–2419.

Sekido Y, Bader S, Latif F, Gnarra JR, Gazdar AF, Linehan WM, Zbar B, Lerman MI, Minna JD. (1994) Molecular analysis of the von Hippel–Lindau disease tumor suppressor gene in human lung cancer cell lines. *Oncogene* **9**: 1599–1604.

Serrano M, Hannon GJ, Beach D. (1993) A new regulatory motif in cell-cycle control causing specific inhibition of cyclin D/CDK4. *Nature* **366**: 704–707.

Sherr CJ. (1993) Mammalian G1 cyclins. *Cell* **73**: 1059–1065.

Shew JY, Chen PL, Bookstein R, Lee EY, Lee WH. (1990) Deletion of a splice donor site ablates expression of the following exon and produces an unphosphorylated RB protein unable to bind SV40 T antigen. *Cell Growth Differ.* **1**: 17–25.

Shimizu E, Coxon A, Otterson GA, Steinberg SM, Kratzke RA, Kim YW, Fedorko J, Oie H, Johnson BE, Mulshine JL, Minna JD, Gazdar AF, Kaye FJ. (1994) RB protein status and clinical correlation from 171 cell lines respresenting lung cancer, extrapulmonary small cell carcinoma, and mesothelioma. *Oncogene* **9**: 2441–2448.

Shiseki M, Kohno T, Nishikawa R, Sameshima Y, Mizoguchi H, Yokota J. (1994) Frequent allelic losses on chromosomes 2q, 18q and 22q in advanced non-small cell lung carcinoma. *Cancer Res.* **54**: 5643–5648.

Slebos RJ, Kibbelaar RE, Dalesio O, Kooistra A, Stam J, Meijer CJ, Wagenaar SS, Vanderschueren RG, van Zandwijk N, Mooi WJ, *et al.* (1990) K-ras oncogene activation as a prognostic marker in adenocarcinoma of the lung. *N. Engl. J. Med.* **323**: 561–565.

Slebos RJ, Hruban RH, Dalesio O, Mooi WJ, Offerhaus GJ, Rodenhuis S. (1991) Relationship between K-ras oncogene activation and smoking in adenocarcinoma of the human lung. *J. Natl Cancer Inst.* **83**: 1024–1027.

Sozzi G, Miozzo M, Tagliabue E, Calderone C, Lombardi L, Pilotti-Pastorino U, Pierotti M, Della Porta G. (1991) Cytogenetic abnormalities and overexpression of receptors for growth factors in normal bronchial epithelium and tumor samples of lung cancer patients. *Cancer Res.* **51**: 400–404.

Sozzi G, Miozzo M, Donghi R, Pilotti S, Cariani CT, Pastorino U, Della Porta G, Pierotti MA. (1992) Deletions of 17p and p53 mutations in preneoplastic lesions of the lung. *Cancer Res.* **52**: 6079–6082.

Spruck C, Gonzalez–Zulueta M, Shibata A, Simoneau AR, Lin MF, Gonzales F, Tsai YC, Jones PA. (1994) p16 gene in uncultured tumours [letter] [see comments]. *Nature* **370**: 183–184.

Stewart MF, Crosby SR, Gibson S, Twentyman PR, White A. (1989) Small cell lung cancer cell lines secrete predominantly ACTH precursor peptides not ACTH. *Br. J. Cancer* **60**: 20–24.

Sugio K, Kishimoto Y, Virmani AK, Hung JY, Gazdar AF. (1994) K-ras mutations are a relatively late event in the pathogenesis of lung carcinomas. *Cancer Res.* 54: 5811–5815.

Sundaresan V, Ganly P, Hasleton P, Rudd R, Sinha G, Bleehen NM, Rabbitts P. (1992) p53 and chromosome 3 abnormalities, characteristic of malignant lung tumours, are detectable in preinvasive lesions of the bronchus. *Oncogene* 7: 1989–1997.

Suzuki H, Takahashi T, Kuroishi T, Suyama M, Ariyoshi Y, Takahashi T, Ueda R. (1992) p53 mutations in non-small cell lung cancer in Japan: association between mutations and smoking. *Cancer Res.* 52: 734–736.

Takahashi T, Nau MM, Chiba I, Birrer MJ, Rosenberg RK, Vinocour M, Levitt M, Pass H, Gazdar AF, Minna JD. (1989) p53: a frequent target for genetic abnormalities in lung cancer. *Science* 246: 491–494.

Takahashi T, Takahashi T, Suzuki H, Hida T, Sekido Y, Ariyoshi Y, Ueda R. (1991) The p53 gene is very frequently mutated in small-cell lung cancer with a distinct nucleotide substitution pattern. *Oncogene* 6: 1775–1778.

Takahashi T, Carbone D, Takahashi T, Nau MM, Hida T, Linnoila I, Ueda R, Minna JD. (1992) Wild-type but not mutant p53 suppresses the growth of human lung cancer cells bearing multiple genetic lesions. *Cancer Res.* 52: 2340–2343.

Takeshima Y, Seyama T, Bennett WP, Akiyama M, Tokuoka S, Inai K, Mabuchi K, Land CE, Harris CC. (1993) p53 mutations in lung cancers from non-smoking atomic-bomb survivors. *Lancet* 342: 1520–1521.

Taylor JA, Watson MA, Devereux TR, Michels RY, Saccomanno G, Anderson M. (1994) p53 mutation hotspot in radon-associated lung cancer. *Lancet* 343: 86–87.

Testa JR, Graziano SL. (1993) Molecular implications of recurrent cytogenetic alterations in human small cell lung cancer. *Cancer Detect. Prevent.* 17: 267–277.

Testa J, Siegfried J. (1992) Chromosome abnormalities in human non-small cell lung cancer. *Cancer Res.* 523: 2702s–2706s.

Tong LA, de Vos AM, Milburn MV, Jancarik J, Noguchi S, Nishimura S, Miura K, Ohtsuka E, Kim SH. (1989) Structural differences between a ras oncogene protein and the normal protein. *Nature* 337: 90–93.

Tsai LH, Delalle I, Caviness V Jr, Chae T, Harlow E. (1994) p53 is a neural-specific regulatory subunit of cyclin-dependent kinase 5. *Nature* 371: 419–423.

Vogelstein B, Kinzler KW. (1992) Carcinogens leave fingerprints [news]. *Nature* 355: 209–210.

Weinberg RA. (1991) Tumor suppressor genes. *Science* 254: 1138–1146.

Whang–Peng J, Bunn P Jr, Kao–Shan CS, Lee EC, Carney DN, Gazdar A, Minna JD. (1982a) A nonrandom chromosomal abnormality, del 3p(14–23), in human small cell lung cancer (SCLC). *Cancer Genet. Cytogenet.* 6: 119–134.

Whang–Peng J, Kao–Shan CS, Lee EC, Bunn PA, Carney DN, Gazdar AF, Minna JD. (1982b) Specific chromosome defect associated with human small-cell lung cancer; deletion 3p(14–23). *Science* 215: 181–182.

Whang–Peng J, Gazdar A, Steinberg S, Oie H, Linoilla I, Mulshine J, Nau M, Minna J. (1991) Nonrandom structural and numerical chromosome changes in non-small cell lung cancer. *Genes Chrom. Cancer* 3: 168–188.

White A, Stewart MF, Farrell WE, Crosby SR, Lavender PM, Twentyman PR, Rees LH, Clark AJ. (1989) Pro-opiomelanocortin gene expression and peptide secretion in human small-cell lung cancer cell lines. *J. Mol. Endocrinol.* 3: 65–70.

Wiedenfeld EA, Fernandez–Vina M, Berzofsky JA, Carbone DP. (1994) Evidence for selection against human lung cancers bearing p53 missense mutations which occur within the HLA A *0201 peptide consensus motif. *Cancer Res.* **54**: 1175–1177.

Wieland I, Bohm M. (1994) Frequent allelic deletion at a novel locus on chromosome 5 in human lung cancer. *Cancer Res.* **54**: 1772–1774.

Winter SF, Sekido Y, Minna JD, McIntire D, Johnson BE, Gazdar AF, Carbone DP. (1993) Antibodies against autologous tumor cell proteins in patients with small-cell lung cancer: association with improved survival [see comments]. *J. Natl Cancer Inst.* **85**: 2012–2018.

Wong AJ, Ruppert JM, Eggleston J, Hamilton SR, Baylin SB, Vogelstein B. (1986) Gene amplification of c-myc and N-myc in small cell carcinoma of the lung. *Science* **233**: 461–446.

Wurster–Hill DH, Cannizzaro LA, Pettengill OS, Sorenson GD, Cate CC, Maurer LH. (1984) Cytogenetics of small cell carcinoma of the lung. *Cancer Genet. Cytogenet.* **13**: 303–330.

Xu HJ, Hu SX, Cagle PT, Moore GE, Benedict WF. (1991) Absence of retinoblastoma protein expression in primary non-small cell lung carcinomas. *Cancer Res.* **51**: 2735–2739.

Xu HJ, Quinlan DC, Davidson AG, Hu SX, Summers CL, Li J, Benedict WF. (1994) Altered retinoblastoma protein expression and prognosis in early-stage non-small-cell lung carcinoma. *J. Natl Cancer Inst.* **86**: 695–699.

Yamakawa K, Takahashi T, Horio Y, Murata Y, Takahashi E, Hibi K, Yokoyama S, Ueda R, Takahashi T, Nakamura Y. (1993) Frequent homozygous deletions in lung cancer cell lines detected by a DNA marker located at 3p21.3–p22. *Oncogene* **8**: 327–330.

Yokota J, Wada M, Shimosato Y, Terada M, Sugimura T. (1987) Loss of heterozygosity on chromosomes 3, 13, and 17 in small-cell carcinoma and on chromosome 3 in adenocarcinoma of the lung. *Proc. Natl Acad. Sci. USA* **84**: 9252–9256.

Yokota J, Akiyama T, Fung YK, Benedict WF, Namba Y, Hanaoka M, Wada M, Terasaki T, Shimosato Y, Sugimura T, Terada M. (1988) Altered expression of the retinoblastoma (RB) gene in small-cell carcinoma of the lung. *Oncogene* **3**: 471–475.

Yu D, Wang SS, Dulski KM, Tsai CM, Nicolson GL, Hung MC. (1994) c-erbB-2/neu overexpression enhances metastatic potential of human lung cancer cells by induction of metastasis-associated properties. *Cancer Res.* **54**: 3260–3266.

Zhan Q, Bae I, Kastan MB, Fornace A Jr. (1994) The p53-dependent gamma-ray response of GADD45. *Cancer Res.* **54**: 2755–2760.

Zhang Y, Mukhopadhyay T, Donehower LA, Georges RN, Roth JA. (1993) Retroviral vector-mediated transduction of K-ras antisense RNA into human lung cancer cells inhibits expression of the malignant phenotype. *Hum. Gene Ther.* **4**: 451–460.

Zhang SY, Klein–Szanto AJ, Sauter ER, Shafarenko M, Mitsunaga S, Nobori T, Carson DA, Ridge JA, Goodrow TL (1994) Higher frequency of alterations in the p16/CDKN2 gene in squamous cell carcinoma cell lines than in primary tumors of the head and neck. *Cancer Res.* **54**: 5050–5053.

Rhabdomyosarcoma

David N. Shapiro

9.1 Introduction

Rhabdomyosarcoma, a malignant tumour of skeletal muscle, is the most common soft tissue sarcoma in persons younger than 21 years of age, and accounts for between 5% and 8% of all cases of childhood cancer. The histological classification scheme for rhabdomyosarcomas is generally based upon their resemblance to normal fetal skeletal muscle prior to nervous system innervation (Horn and Enterline, 1958). These tumours are classified into two broad histotypes; embryonal and alveolar, each with characteristic pathological, cytogenetic and clinical features (Raney *et al.*, 1989). Embryonal rhabdomyosarcomas occur in younger children and account for approximately 60% of cases; the primary tumour is usually located in specific anatomical sites including the head and neck region, genito-urinary tract and orbit. Histologically, the tumour is characterized by variable numbers of malignant spindle and primitive round cells that may contain cross-striations typical of skeletal muscle. In contrast, alveolar rhabdomyosarcomas often occur during adolescence as extremity or axial primary tumours. This histological variant is characterized by the presence of fibrovascular septa that form alveolar-like spaces filled with primitive, poorly cohesive, monomorphous, malignant cells. Patients with tumours of alveolar histology are generally considered to have a worse clinical prognosis as compared to those with embryonal rhabdomyosarcoma (Crist *et al.*, 1990).

Consistent with the fact that they originate from skeletal muscle precursors, rhabdomyosarcomas share patterns of expression of muscle-specific genes with the normal cognate tissue, fetal skeletal muscle. For example, rhabdomyosarcomas have consistently been shown to express transcripts for the MyoD family of muscle-specific regulatory factors (Davis *et al.*, 1987; Olson, 1990; Weintraub *et al.*, 1991; Li and Olson, 1992; Olson and Klein, 1994). Members of the MyoD family [MyoD (*MYF3*), myogenin (*MYF4*), *MYF5*, herculin (*MYF6*)] share homology within a conserved basic helix–loop–helix (HLH) motif that mediates DNA binding, protein dimerization and transcriptional activity (Weintraub *et al.*, 1991). These muscle-specific transcription factors are sufficient to orchestrate the coordinated expression of the skeletal myogenic programme and directly regulate the expression of essential skeletal muscle-specific structural genes (Tapscott and Weintraub, 1991). Furthermore, the expression of MyoD family members is generally accepted to be consistent with commitment to the myogenic lineage and therefore serves as a useful marker for identifying skeletal muscle precursors. Numerous studies have now demonstrated the expression of MyoD transcripts in all rhabdomyosarcomas that have been examined; less frequent expression of myogenin, *MYF5*, and herculin has also been observed (Scrable *et al.*, 1989b; Clark *et al.*, 1991; Tonin *et al.*, 1991; Hosoi *et al.*, 1992). Interestingly, recent data has suggested that the endogenous rhabdomyosarcoma MyoD protein is relatively non-functional as a transcriptional activator, perhaps due to a deficiency in the tumour cells of a cooperating co-factor (Tapscott *et al.*, 1993). Although no correlation between tumour histology and the expression pattern of these transcription factors has been reported, the restriction of their expression to cells of myogenic lineage has proven a useful adjunct in the pathological diagnosis of rhabdomyosarcoma and in their differentiation from other primitive paediatric neoplasms (Dias *et al.*, 1990b).

9.2 Cytogenetics

Cytogenetic analysis has become increasingly important for the characterization of childhood tumours and has provided a reliable means of grouping seemingly dissimilar forms of the same disease. Furthermore, the identification of non-random translocations has served as the starting point for the positional cloning of important genes located at several of these tumour-specific translocation breakpoints. Cytogenetic analysis of rhabdomyosarcomas has demonstrated chromosomal abnormalities in both embryonal and alveolar histologies. Detailed cytogenetic studies have been reported for over 60 cases of rhabdomyosarcoma; approximately two thirds of these are of alveolar histology (Seidal *et al.*, 1982; Potluri and Gilbert, 1985; Trent *et al.*, 1985; Turc-Carel *et al.*, 1986; Douglass *et al.*, 1987, 1993; Lizard-Nacol *et al.*, 1987; Rowe *et al.*, 1987; Engel *et al.*, 1988; Nojima *et al.*, 1990; Dal Cin *et al.*, 1991; Whang-Peng *et al.*, 1992). Importantly, several chromosomal regions appear to show non-random involvement in specific structural rearrangements.

A characteristic chromosomal translocation, t(2;13)(q35;q14), was first described by Seidal in 1982 in alveolar rhabdomyosarcoma (Seidal *et al.*, 1982). Subsequent studies have now confirmed the presence of this translocation in over 70% of successfully karyotyped alveolar rhabdomyosarcomas and it is now considered specific for tumours of this histology (Whang-Peng *et al.*, 1992). Additionally, a less frequent variant abnormality, t(1;13)(p36;q14), has been described, which occurs in approximately 10–15% of tumours of alveolar histology (Biegel *et al.*, 1991; Douglass *et al.*, 1991; Whang-Peng *et al.*, 1992). This translocation involves a cytogenetically indistinguishable region on chromosome 13q14; furthermore, the pathological and clinical features of patients with the t(1;13) are similar to those with the more common t(2;13).

Approximately two-thirds of alveolar tumours have been noted to have near-tetraploid DNA content as determined by flow cytometry; the remaining cases are usually diploid (Molenaar *et al.*, 1988; Shapiro *et al.*, 1991; Pappo *et al.*, 1993; Wijnaendts *et al.*, 1993; Mathieu *et al.*, 1994; Niggli *et al.*, 1994). This result is not unexpected because karyotypic evidence supports the development of tetraploidy by endoreduplication of a primary diploid tumour line (Douglass *et al.*, 1987). Interestingly, near-tetraploidy is virtually never observed in embryonal rhabdomyosarcomas and is apparently pathognomonic for tumours of alveolar histology (Shapiro *et al.*, 1991; Pappo *et al.*, 1993).

A variety of chromosomal abnormalities have been reported for embryonal rhabdomyosarcoma, which include deletion of chromosome 1p with hyperdiploidy, trisomy 2 and ring chromosome 13 among others (Potluri and Gilbert, 1985; Hayashi *et al.*, 1990; Kubo *et al.*, 1991; Magnani *et al.*, 1991; Voullaire *et al.*, 1991; Olegard *et al.*, 1992; Whang-Peng *et al.*, 1992). The complexity and variability of these cytogenetic abnormalities argue against the presence of consistent, karyotypically apparent structural rearrangements. As with alveolar rhabdomyosarcomas, however, reproducible ploidy patterns have been noted with embryonal tumours. These studies have generally shown that about two-thirds of embryonal rhabdomyosarcomas have hyperdiploid DNA content while the remaining tumours are usually diploid (Shapiro *et al.*, 1991; Pappo *et al.*, 1993; Wijnaendts *et al.*, 1993; Mathieu *et al.*, 1994; Niggli *et al.*, 1994). Interestingly, these investigations have also shown that diploid embryonal rhabdomyosarcomas frequently respond less favourably to standard chemotherapeutic regimens.

9.3 Molecular cytogenetics of alveolar rhabdomyosarcoma

Studies that were designed to identify the genetic loci disrupted by the t(2;13) sought initially to refine the location of the chromosome 2 and 13 breakpoints on physical maps of these chromosomes. Isotopic *in situ* hybridization to metaphase chromosomes from rhabdomyosarcoma cell lines was used to position the breakpoint proximal to the anonymous marker D13S10, esterase D (*ESD*) and *RB1*, and distal to the anonymous markers D13S6 and D13S1; these studies also eliminated *RB1* as a potential target for disruption by this translocation (Valentine *et al.*, 1989). Further refinement of the chromosome 13 and 2 breakpoint regions was possible by mapping studies, using somatic cell hybrids retaining either the der(2) or der(13) chromosomes, and by the analysis of long-range physical maps of these regions prepared using pulsed-field gel electrophoresis (Barr *et al.*, 1991a, b, 1992; Mitchell *et al.*, 1991; Shapiro *et al.*, 1992). Examination of 14 candidate genes that had been positioned in these regions failed to detect any rearrangements in several alveolar rhabdomyosarcoma cell lines. Included among these candidate genes were such potential targets as the *FLT1* and *FLT2* receptor tyrosine kinases. Nevertheless, these studies were able to delimit the chromosome 13 breakpoint region to an area between the anonymous marker D13S29 and *TUBBP2*, an interval estimated to comprise approximately two megabases (Barr *et al.*, 1991b; Shapiro *et al.*, 1992). Analogous studies of the chromosome 2 breakpoint region showed that it was flanked proximally by inhibin-α (*INHA*) and distally by intestinal alkaline phosphatase (*ALP1*), a genetic distance of at least 5 cM (Barr *et al.*, 1992).

Subsequent to the refinement of the t(2;13) breakpoints on physical maps of chromosomes 2 and 13, Waardenburg syndrome, which is characterized by deafness and pigmentary disturbances and had previously been linked to a region on distal chromosome 2q, was shown to be associated with mutations of the *PAX3* gene (Baldwin *et al.*, 1992; Farrer *et al.*, 1992; Tassabehji *et al.*, 1992; Hoth *et al.*, 1993). Other studies had established that *PAX3* is a member of a large superfamily of developmental control genes which encode transcription factors containing a characteristic DNA binding domain termed the paired box (Gruss and Walther, 1992). The paired box was first found in three *Drosophila* segmentation genes and subsequently detected in the genomes of mouse, human, nematode, zebrafish and chick (Bopp *et al.*, 1986, 1989; Burri *et al.*, 1989; Noll, 1993). Cloning and analysis of the murine homologue, *Pax-3*, revealed the presence of both a 128-amino acid paired box and a 78-amino acid paired-type homeodomain in the 5' half of the gene; together, these regions are associated with coordinate, sequence-specific DNA binding activity (Goulding *et al.*, 1991). Importantly, deletions within the *Pax-3* paired box domain have been identified in the murine mutant *splotch*, which is associated with neural tube defects and shares some phenotypic features with Waardenburg syndrome (Epstein *et al.*, 1991, 1993; Walther *et al.*, 1991). Furthermore, *Pax-3* was mapped by linkage analysis to the region between the inhibin-α and intestinal alkaline phosphatase loci on proximal mouse chromosome 1, a segment which is syntenically conserved with the distal human 2q region (Schurr, 1990).

Using a chromosome 2 somatic cell hybrid mapping panel, human *PAX3* was positioned within the same physical interval on chromosome 2q as the alveolar rhabdomyosarcoma-specific t(2;13) breakpoint and was shown to be rearranged in tumour DNA from cell lines containing this translocation (Barr *et al.*, 1993). *PAX3* structural rearrangements were then confirmed to be a direct consequence of the translocation and to result in disruption of the gene 3' of the paired box and homeodomain. Fine-mapping showed that all rearrangements occur within different positions of the same 20-kb intron located between the last two *PAX3* exons. In all cases, these rearrangements result in the translocation of the 5' region of *PAX3* to the tumour-derived der(13) chromosome and the 3' region of *PAX3* to the tumour-derived der(2) chromosome (*Figure 9.1*).

Isolation and characterization of fusion cDNA clones has shown that the *PAX3* rearrangement results in the creation of a novel chimeric gene composed

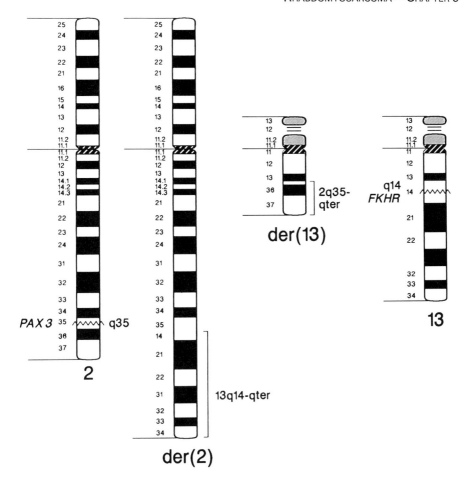

Figure 9.1. Ideograms of normal human chromosomes 2 and 13 along with the locations of the *PAX3* gene at 2q35 and the *FKHR* gene at 13q14. The tumour-specific der(2) and der(13) chromosomes are also indicated.

of 5′ *PAX3* sequences juxtaposed to 3′ sequences derived from a member of the forkhead family of transcription factors located on chromosome 13q14 (designated *ALV* or *FKHR*). This rearrangement creates an 836-amino acid fusion protein with a predicted molecular mass of 97 kDa, consisting of the PAX3 paired box and homeodomains and the carboxyl-terminal residues of the conserved forkhead DNA binding domain, together with the unique FKHR carboxyl-terminal region (*Figure 9.2*) (Galili *et al.*, 1993; Shapiro *et al.*, 1993). More than 40 forkhead genes have been identified from species ranging from yeast to human; all forkhead family members share a highly conserved 100-amino acid motif, termed the forkhead domain, which has sequence-specific

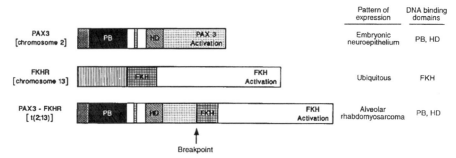

Figure 9.2. Schematic diagram depicting the results of the t(2;13)(q35;q14) in alveolar rhabdomyosarcoma. The creation of the active der(13) chromosome results in a chimeric fusion gene containing both *PAX3* DNA binding domains [the paired box (PB) and homeodomain (HD)], the bisected *FKHR* DNA binding domain and the *FKHR* activation domain. The distribution of the transcripts for *PAX3*, *FKHR* and the t(2;13) fusion gene is shown.

DNA binding activity (Weigel *et al.*, 1989; Weigel and Jackle, 1990; Lai *et al.*, 1991; Hacker *et al.*, 1992; Pierrou *et al.*, 1994). Structural analysis of this domain has identified two potential helix forming regions in the amino-terminal half and a basic region in the carboxyl-terminal half; features that are similar to the HLH family of transcription factors (Brennan, 1993; Clark *et al.*, 1993). In several of the forkhead family members, other regions of the protein have been demonstrated to confer potent transcriptional regulatory activity (Weigel *et al.*, 1990; Pani *et al.*, 1992; Clevidence *et al.*, 1993). Genetic and functional studies of forkhead genes have also indicated the contribution of these genes to control of embryonic development and adult tissue-specific gene expression (Lai *et al.*, 1991; Hromas *et al.*, 1993; Kaestner *et al.*, 1993; Miller *et al.*, 1993).

Based upon amino acid identity and sequence alignment within the DNA binding domain, *FKHR* is most closely related to a forkhead homology group containing the *Drosophila fkh* and *sloppy* (*slp-1* and *slp-2*) genes, and the rat and human brain factor 1 (*BF1*) genes (Weigel *et al.*, 1989; Grossniklaus *et al.*, 1992; Tao and Lai, 1992; Murphy *et al.*, 1994). Interestingly, *FKHR* shows the highest degree of identity with *AFX1*, a recently identified forkhead family member fused in-frame to *MLL* as the result of the t(X;11) in acute lymphoblastic leukaemia (Parry *et al.*, 1994). A 6.0-kb *FKHR* transcript, encoding a 655-amino acid protein with a predicted molecular mass of 72 kDa, is ubiquitously expressed in virtually all fetal and adult tissues.

Northern and reverse transcriptase polymerase chain reaction (RT-PCR) analyses of several independently derived alveolar rhabdomyosarcoma cell lines have indicated that the t(2;13) always results in chimeric transcripts from the der(13) and often from the der(2) chromosomes (Galili *et al.*, 1993; Shapiro *et al.*, 1993). The uniform size of these transcripts indicates that similar translocation breakpoints occur within the same introns of *PAX3* and *FKHR*; sequence analysis has confirmed that *PAX3* is always fused in-frame after Asp[391] to 3' *FKHR* sequences. The der(13) allele expressing the PAX3–FKHR fusion is most likely critical in the aetiology of alveolar rhabdomyosarcoma for several reasons. First, the reciprocal der(2) gene product would lack both the PAX3 paired box and homeodomains; together the paired box and homeodomains have been implicated in novel coordinate DNA binding specificity, which is different from that displayed by either domain alone (Treisman *et al.*, 1989; Chalepakis *et al.*, 1991; Czerny *et al.*, 1993). Furthermore, although both chimeric transcripts have been identified in many alveolar rhabdomyosarcoma cell lines, sensitive RT-PCR analysis has failed to detect expression of the der(2) transcript in some cell lines expressing the der(13) message. These observations are consistent with Northern analysis of tumour cell RNA which has indicated that, in those tumours expressing both chimeric transcripts, the der(13) transcript is significantly more abundant. Finally, only the der(13)-derived 97-kDa fusion protein has been identified by immunoprecipitation (Galili *et al.*, 1993). Taken together, these data indicate that the der(13) encodes the protein product involved in alveolar rhabdomyosarcoma tumorigenesis.

The predicted PAX3–FKHR protein has several important structural features that may be of functional importance. First, the recent identification of the avian retroviral *qin* oncogene and its human homologue as forkhead family members suggests that the mammalian forkhead genes, under certain circumstances, possess transforming activity (Li and Vogt, 1993; Kastury *et al.*, 1994). However, based on functional and structural studies demonstrating the essential nature of the amino terminal of the forkhead domain for DNA binding, it is unlikely that the disrupted carboxyl FKHR forkhead domain present in the t(2;13) fusion protein contributes to sequence-specific DNA interactions (Clark *et al.*, 1993; Clevidence *et al.*, 1993). Of potential importance, however, is the fact that the extreme carboxyl terminal of FKHR contains an acidic domain, similar to the transcriptional regulatory domains of other transcription factors such as AP-1 and *JUN* (Mitchell and Tjian, 1989). In contrast, retention of the intact PAX3 paired box and homeodomain suggests that these domains contribute to the DNA binding specificity of the fusion protein. Moreover, the reported *in vitro* transforming potential of *PAX* genes is dependent on the structural integrity of the paired domain and retention of its DNA binding capacity (Maulbecker and Gruss, 1993). The fusion protein may thus bind to normal *PAX3* genomic targets but aberrantly regulate transcription by either excessive activation or repression through novel *FKHR* 3' regulatory sequences. This hypothesis has been confirmed by recent experimental data demonstrating that both *in vitro* and *in vivo* t(2;13) fusion proteins are capable of binding to and transcriptionally activating model *PAX3* binding sites without demonstrable contribution from the bisected FKHR domain (*Figure 9.3*) (Sublett *et al.*, submitted for publication). Consistent with this model, a potent transactivation domain has been shown to be present in the 60 amino acids of

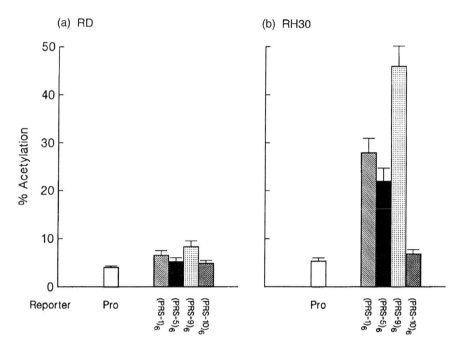

Figure 9.3. Transactivation studies using mutant and idealized *PAX3* reporter constructs transfected into the embryonal rhabdomyosarcoma cell line RD (panel a) and the t(2;13) alveolar rhabdomyosarcoma cell line RH30 (panel b). Only the alveolar rhabdomyosarcoma cell line containing the chimeric t(2;13) transcription factor is able to activate various *PAX3* reporter constructs containing optimized sixfold multimerized paired box and homeodomain binding sites (PRS-1, -5, -9). Specificity is demonstrated by the failure of the PRS-10 construct containing both paired box and homeodomain recognition mutations to be activated in RH30 cells. By contrast, none of the reporter constructs are activated when transfected into the embryonal rhabdomyosarcoma cell line RD which does not contain the t(2;13). (PRS-1, -5, -9, -10)$_6$ represent CAT reporter constructs containing sixfold multimerized *PAX* binding sites; Pro represents the control pTKCAT-Promoter vector without *PAX* binding sites.

the carboxyl terminal of the FKHR protein (Sublett *et al.*, submitted for publication).

Alternatively, the translocation could also activate the oncogenic potential of *PAX3* and/or *FKHR* by simply increasing protein expression levels. The t(2;13) may increase chimeric transcript expression by juxtaposing positive *FKHR* regulatory elements, such as an enhancer element or RNA stabilization sequences, accounting for the expression of chimeric *PAX3–FKHR* transcripts in cells lacking detectable expression of normal *PAX3* transcripts from their unrearranged chromosome 2 allele. Conversely, chimeric protein expression levels could be increased by removal of potential *PAX3* 3' negative regulatory elements by the translocation (Jackson, 1993). Presently, however, experimental data to support these models of augmented chimeric protein expression are not available.

Although the (2;13)(q35;q14) translocation has been found in the majority of cases of alveolar rhabdomyosarcoma, several cases with a variant (1;13)(p36;q14) translocation have been reported (Biegel *et al.*, 1991; Douglass *et al.*, 1991; Whang-Peng *et al.*, 1992). In a manner analogous to the t(2;13), the t(1;13) was recently shown to rearrange the *PAX7* gene on chromosome 1 and fuse it to identical *FKHR* sequences on chromosome 13 (Davis *et al.*, 1994). This fusion results in a chimeric transcript encoding the PAX7 paired box and homeodomain along with the carboxyl-terminal FKHR transactivation domain. The predicted amino acid sequences of the PAX7 paired box and homeodomain display 94% and 97% identity, respectively, with the corresponding sequences of the PAX3 domains (Jostes *et al.*, 1990). This very high sequence identity suggests that the PAX3 and PAX7 DNA binding domains recognize very similar target sites and thus may regulate a common set of target genes important for rhabdomyosarcoma tumorigenesis.

The involvement of *PAX3* and *PAX7* in the t(2;13) and t(1;13) of alveolar rhabdomyosarcoma is very similar to recent data concerning the t(11;22) and t(21;22) of Ewing's sarcoma and peripheral neuroepithelioma (Zucman *et al.*, 1993; Sorensen *et al.*, 1994). In the latter case, the *EWS* gene is fused to *FLI1* or *ERG*, two members of the ETS family of transcription factors. These translocations generate chimeric transcription factors composed of the amino-terminal region of EWS fused to the carboxyl-terminal region of FLI1 or ERG, which contain intact ETS DNA binding domains. The amino acid sequences of the FLI1 and ERG DNA binding domains are 98% identical, suggesting that these fusion proteins may also aberrantly regulate a common set of target genes important for tumorigenesis.

By contrast with Ewing's sarcoma, in which both chimeric transcripts are expressed from the *EWS* promoter, the chimeric transcripts in the alveolar rhabdomyosarcoma translocations are expressed from either the *PAX3* or *PAX7* promoter. *In situ* hybridization studies during mouse embryogenesis demonstrate that *Pax-3* and *Pax-7* are expressed with distinct but overlapping developmental patterns (Jostes *et al.*, 1990; Goulding *et al.*, 1991; Gruss and Walther, 1992; Chalepakis *et al.*, 1993; Strachan and Read, 1994; Tremblay and Gruss, 1994). In addition to expression in specific regions of the developing nervous system, both genes are expressed in somites around the time of dermomyotome formation. Whereas *Pax-3* expression occurs prior to myoblast migration and formation of the skeletal muscle precursors, *Pax-7* expression commences a few days later and persists during the differentiation of the trunk and limb musculature. The finding of *Pax-3* and *Pax-7* expression in skeletal muscle precursors is consistent with the activity of their respective promoters in the development of alveolar rhabdomyosarcoma.

Although the genes involved in both alveolar rhabdomyosarcoma translocations have been identified, progress in understanding their specific mechanisms of transformation will depend on finding their individual or shared downstream targets. In this regard, specific target genes for both *PAX3* and *PAX7* are unknown, and optimal DNA recognition sequences for most paired box and forkhead genes remain to be identified. Clearly, defining altered patterns of gene expression in these tumours will be essential in the elucidation of the pathways responsible for tumorigenesis in alveolar rhabdomyosarcoma.

9.4 Molecular cytogenetics of embryonal rhabdomyosarcoma

Unlike alveolar rhabdomyosarcoma, no consistent translocations or other characteristic karyotypic abnormalities have been reported for embryonal rhabdomyosarcoma which could be important mechanisms contributing to tumorigenesis (Potluri and Gilbert, 1985; Olegard *et al.*, 1992). However, other genetic mechanisms including mitotic non-disjunction and loss (with or without reduplication of the remaining homologue), localized gene conversion, point mutation, small deletion, and mitotic recombination have all been reported to be contributing factors for tumour development. The existence of mitotic recombination in initiated progenitor cells giving rise to a specific tumour has suggested the chromosomal location for numerous tumour suppressor loci, even in the absence of known cytogenetic aberrations, by delineating the smallest overlapping region of somatic homozygosity shared among tumours of a similar phenotype.

In this regard, restriction fragment length polymorphism analysis of embryonal rhabdomyosarcomas has shown consistent loss of heterozygosity (LOH) through mitotic recombination for loci on chromosome 11p thus implying the location of a tumour suppressor gene; this loss has not been observed for alveolar rhabdomyosarcomas (Scrable *et al.*, 1987, 1989b). The smallest region affected in these cases encompasses 11p15.5-pter and includes the loci for the haemoglobin β gene cluster (*HBBC*), tyrosine hydroxylase (*TH*), *H19*, insulin (*INS*), insulin-like growth factor 2 (*IGF2*) and the Harvey *RAS* sarcoma virus oncogene (*HRAS*) (Scrable *et al.*, 1990; Newsham *et al.*, 1991). Importantly, this region has also been implicated in the development of other embryonal tumours, including Wilm's tumour, adrenal carcinoma and hepatoblastoma, as well as tumours of the lung, bladder, ovary and breast (Koufos *et al.*, 1985). Furthermore, the Beckwith–Wiedemann syndrome, an autosomal dominant syndrome consisting of generalized somatic hyperplasia and a predisposition for the development of embryonal tumours (including Wilm's tumour and rhabdomyosarcoma), has been mapped to precisely the same region of LOH as that found in embryonal rhabdomyosarcoma (Henry *et al.*, 1991, 1993).

Functional evidence for the location of a rhabdomyosarcoma tumour suppressor gene at 11p15.5 has been provided by the transfer of an intact chromosome 11, using microcell hybridization, into an embryonal rhabdomyosarcoma having

LOH for this region (Loh Jr *et al.*, 1992). By selective retention of either the short or long arms of chromosome 11 in the microcell hybrids, dramatic loss of proliferative capacity was elicited without loss of tumorigenicity. Thus, in addition to the known suppressor locus at 11p15.5, these results indicate the existence of a tumour suppressor on 11q, unrecognized by previous molecular analyses (Weissman *et al.*, 1987).

Refinement of the precise location of the chromosome 11p15.5 suppressor locus has recently been established by microcell fusion with subchromosomal transferable fragments containing smaller segments of chromosome 11p (Koi *et al.*, 1993). These studies have localized the suppressor locus to an approximately 4500-kb region between the anonymous markers D11S719 and D11S724. Importantly, the gene appears to reside between, but is excluded from, two regions proposed to contain it, namely the centromeric region near the calcitonin gene and the more telomeric region near *IGF2*. Interestingly, this same region also lies directly between the two clusters of germline translocation breakpoints in some patients with the Beckwith–Wiedemann syndrome and in the region of uniparental disomy in this disorder (Henry *et al.*, 1993).

Although the evidence for a tumour suppressor locus at 11p15.5 appears incontrovertible, the mechanism by which this gene is inactivated has yet to be established. Studies investigating the parental origin of alleles in this genomic region in familial and sporadic cases of embryonal rhabdomyosarcoma have shown that isodisomic chromosome 11p alleles are consistently of paternal origin (Scrable *et al.*, 1989a). Thus, genomic imprinting of the paternal allele in embryonal rhabdomyosarcoma may be an alternative first step to inactivation by mutation for the attainment of nullizygosity at the 11p15.5 suppressor locus. Similar presumed epigenetic modifications with the preferential inheritance of parental alleles have been reported for a variety of other tumours including retinoblastoma, Wilm's tumour, and osteosarcoma (Sakai *et al.*, 1991; Rainier *et al.*, 1993; Tycko, 1994).

9.5 Conclusions

Molecular analyses have clearly demonstrated that rhabdomyosarcoma, which has traditionally been classified by histological criteria, can also be defined by specific and characteristic genetic lesions. Alveolar tumours harbour tumour-specific translocations, resulting in the formation of chimeric transcription factors. By contrast, embryonal tumours have LOH for a region on chromosome 11p15.5, implying the inactivation of a tumour suppressor gene. While other molecular defects have also been described in rhabdomyosarcomas, including mutations of the *p53* tumour suppressor gene, activating point mutations in the *NRAS* and *KRAS* cellular oncogenes, imprinting of *IGF2*, and occasional *NMYC* gene amplification, none are specific for these tumours and therefore are not presumed to be of initial aetiological importance (Stratton *et al.*, 1989; Mulligan *et al.*, 1990;

Dias *et al.*, 1990a; Shapiro *et al.*, 1994). Future understanding of the role of these molecular defects in rhabdomyosarcoma tumorigenesis will undoubtedly result in improved, targeted treatment strategies for this malignancy.

Acknowledgements

This work was supported in part by Grant CA-23099 and Cancer Center CORE Grant CA-21765, from the National Institutes of Health, and by the American Lebanese Associated Charities (ALSAC).

References

Baldwin CT, Hoth CF, Amos JA, da–Silva EO, Milunsky A. (1992) An exonic mutation in the *HuP2* paired domain gene causes Waardenburg's syndrome. *Nature* 355: 637–638.

Barr FG, Biegel JA, Sellinger B, Womer RB, Emanuel BS. (1991a) Molecular and cytogenetic analysis of chromosomal arms 2q and 13q in alveolar rhabdomyosarcoma. *Genes Chrom. Cancer* 3: 153–161.

Barr FG, Sellinger B, Emanuel BS. (1991b) Localization of the rhabdomyosarcoma t(2;13) breakpoint on a physical map of chromosome 13. *Genomics* 11: 941–947.

Barr FG, Holick J, Nycum L, Biegel JA, Emanuel BS. (1992) Localization of the t(2;13) breakpoint of alveolar rhabdomyosarcoma on a physical map of chromosome 2. *Genomics* 13: 1150–1156.

Barr FG, Galili N, Holick J, Biegel JA, Rovera G, Emanuel BS. (1993) Rearrangement of the PAX3 paired box gene in the paediatric solid tumour alveolar rhabdomyosarcoma. *Nature Genet.* 3: 113–117.

Biegel JA, Meek RS, Parmiter AH, Conard K, Emanuel BS. (1991) Chromosomal translocation t(1;13)(p36;q14) in a case of rhabdomyosarcoma. *Genes Chrom. Cancer* 3: 483–484.

Bopp D, Burri M, Baumgartner S, Frigerio G, Noll M. (1986) Conservation of a large protein domain in the segmentation gene *paired* and in functionally related genes of Drosophila. *Cell* 47: 1033–1040.

Bopp D, Jamet E, Baumgartner S, Burri M, Noll M. (1989) Isolation of two tissue-specific Drosophila paired box genes, Pox meso and Pox neuro. *EMBO J.* 8: 3447–3457.

Brennan RG. (1993) The winged-helix DNA-binding motif: another helix-turn-helix takeoff. *Cell* 74: 773–776.

Burri M, Tromvoukis Y, Bopp D, Frigerio G, Noll M. (1989) Conservation of the paired domain in metazoans and its structure in three isolated human genes. *EMBO J.* 8: 1183–1190.

Chalepakis G, Fritsch R, Fickenscher H, Deutsch U, Goulding M, Gruss P. (1991) The molecular basis of the undulated/Pax-1 mutation. *Cell* 66: 873–884.

Chalepakis G, Stoykova A, Wijnholds J, Tremblay P, Gruss P. (1993) Pax: gene regulators in the developing nervous system. *J. Neurobiol.* 24: 1367–1384.

Clark J, Rocques PJ, Braun T, Bober E, Arnold HH, Fisher C, Fletcher C, Brown K, Gusterson BA, Carter RL, *et al.* (1991) Expression of members of the myf gene family in human rhabdomyosarcomas. *Br. J. Cancer* **64**: 1039–1042.

Clark KL, Halay ED, Lai E, Burley SK. (1993) Co-crystal structure of the HNF-3/fork head DNA-recognition motif resembles histone H5. *Nature* **364**: 412–420.

Clevidence DE, Overdier DG, Tao W, Qian X, Pani L, Lai E, Costa RH. (1993) Identification of nine tissue-specific transcription factors of the hepatocyte nuclear factor 3/forkhead DNA-binding-domain family. *Proc. Natl Acad. Sci. USA* **90**: 3948–3952.

Crist WM, Garnsey L, Beltangady MS, Gehan E, Ruymann F, Webber B, Hays DM, Wharam M, Maurer HM. (1990) Prognosis in children with rhabdomyosarcoma: a report of the Intergroup Rhabdomyosarcoma Studies I and II. *J. Clin. Oncol.* **8**: 443–452.

Czerny T, Schaffner G, Busslinger M. (1993) DNA sequence recognition by Pax proteins: bipartite structure of the paired domain and its binding site. *Genes Devel.* **7**: 2048–2061.

Dal Cin P, Brock P, Aly MS, Casteels–Van Daele M, De Wever I, Van Damme B, Van den Berghe H. (1991) A variant (2;13) translocation in rhabdomyosarcoma. *Cancer Genet. Cytogenet.* **55**: 191–195.

Davis RJ, D'Cruz CM, Lovell MA, Biegel JA, Barr FG. (1994) Fusion of PAX7 to FKHR by the variant t(1;13)(p36;q14) translocation in alveolar rhabdomyosarcoma. *Cancer Res.* **54**: 2869–2872.

Davis RL, Weintraub H, Lassar AB. (1987) Expression of a single transfected cDNA converts fibroblasts to myoblasts. *Cell* **51**: 987–1000.

Dias P, Kumar P, Marsden HB, Gattamaneni HR, Heighway J, Kumar S. (1990a) N-myc gene is amplified in alveolar rhabdomyosarcomas (RMS) but not in embryonal RMS. *Int. J. Cancer* **45**: 593–596.

Dias P, Parham DM, Shapiro DN, Webber BL, Houghton PJ. (1990b) Myogenic regulatory protein (MyoD1) expression in childhood solid tumors: diagnostic utility in rhabdomyosarcoma. *Am. J. Pathol.* **137**: 1283–1291.

Douglass EC, Valentine M, Etcubanas E, Parham D, Webber BL, Houghton PJ, Houghton JA, Green AA. (1987) A specific chromosomal abnormality in rhabdomyosarcoma. *Cytogenet. Cell Genet.* **45**: 148–155.

Douglass EC, Rowe ST, Valentine M, Parham DM, Berkow R, Bowman WP, Maurer HM. (1991) Variant translocations of chromosome 13 in alveolar rhabdomyosarcoma. *Genes Chrom. Cancer* **3**: 480–482.

Douglass EC, Shapiro DN, Valentine M, Rowe ST, Carroll AJ, Raney RB, Ragab AH, Abella SM, Parham DM. (1993) Alveolar rhabdomyosarcoma with the t(2;13): cytogenetic findings and clinicopathologic correlations. *Med. Pediatr. Oncol.* **21**: 83–87.

Engel R, Ritterbach J, Schwabe D, Lampert F. (1988) Chromosome translocation (2;13)(q37;q14) in a disseminated alveolar rhabdomyosarcoma. *Eur. J. Pediatr.* **148**: 69–71.

Epstein DJ, Vekemans M, Gros P. (1991) Splotch (SP2H), a mutation affecting development of the mouse neural tube shows a deletion within the paired homeodomain of Pax-3. *Cell* **67**: 767–774.

Epstein DJ, Vogan KJ, Trasler DG, Gros P. (1993) A mutation within intron 3 of the Pax-3 gene produces aberrantly spliced mRNA transcripts in the splotch (Sp) mouse mutant. *Proc. Natl Acad. Sci. USA* **90**: 532–536.

Farrer LA, Grundfast KM, Amos J, Arnos KS, Asher JH Jr, Beighton P, Diehl SR, Fex J, Foy C, Friedman TB, *et al.* (1992) Waardenburg syndrome (WS) type I is caused by defects at multiple loci, one of which is near ALPP on chromosome 2: first report of the WS consortium. *Am. J. Hum. Genet.* **50:** 902–913.

Galili N, Davis RJ, Fredericks WJ, Mukhopadhyay S, Rauscher FJ, Emanuel BS, Rovera G, Barr FG. (1993) Fusion of a fork head domain gene to PAX3 in the solid tumour alveolar rhabdomyosarcoma. *Nature Genet.* **5:** 230–235.

Goulding MD, Chalepakis G, Deutsch U, Erselius J, Gruss P. (1991) Pax-3, a novel murine DNA binding protein expressed during early neurogenesis. *EMBO J.* **10:** 1135–1147.

Grossniklaus U, Pearson RK, Gehring WJ. (1992) The Drosophila sloppy paired locus encodes two proteins involved in segmentation that show homology to mammalian transcription factors. *Genes. Devel.* **6:** 1030–1051.

Gruss P, Walther C. (1992) Pax in development. *Cell* **69:** 719–722.

Hacker U, Grossniklaus U, Gehring WJ, Jackle H. (1992) Developmentally regulated Drosophila gene family encoding the fork head domain. *Proc. Natl Acad. Sci. USA* **89:** 8754–8758.

Hayashi Y, Sugimoto T, Horii Y, Hosoi H, Inazawa J, Kemshead JT, Inaba T, Hanada R, Yamamoto K, Gown AM, *et al.* (1990) Characterization of an embryonal rhabdomyosarcoma cell line showing amplification and over-expression of the N-myc oncogene. *Int. J. Cancer* **45:** 705–711.

Henry I, Bonaiti Pellie C, Chehensse V, Beldjord C, Schwartz C, Utermann G, Junien C. (1991) Uniparental parental disomy in a genetic cancer predisposing syndrome. *Nature* **351:** 665–667.

Henry I, van Heyningen V, Puech A, Scrable H, Augereau P, Boehm T, Rabbitts T, Mannens M, Rochefort H, Jones C, *et al.* (1993) Reassessment of breakpoints in chromosome 11p15. *Cytogenet. Cell Genet.* **62:** 52–53.

Horn RC, Enterline HT. (1958) Rhabdomyosarcoma: a clinicopathological study identification and classification. *Cancer* **11:** 181–199.

Hosoi H, Sugimoto T, Hayashi Y, Inaba T, Horii Y, Morioka H, Fushiki S, Hamazaki M, Sawada T. (1992) Differential expression of myogenic regulatory genes, MyoD1 and myogenin, in human rhabdomyosarcoma sublines. *Int. J. Cancer* **50:** 977–983.

Hoth CF, Milunsky A, Lipsky N, Sheffer R, Clarren SK, Baldwin CT. (1993) Mutation in the paired domain of the human PAX3 gene causes Klein-Waardenburg Syndrome (WS-III) as well as Waardenburg Syndrome Type I (WS-I) *Am. J. Hum. Genet.* **52:** 455–462.

Hromas R, Moore J, Johnston T, Socha C, Klemsz M. (1993) Drosophila forkhead homologues are expressed in a lineage-restricted manner in human hematopoietic cells. *Blood* **81:** 2854–2859.

Jackson RJ. (1993) Cytoplasmic regulation of mRNA function: the importance of the 3′ untranslated region. *Cell* **74:** 9–14.

Jostes B, Walther C, Gruss P. (1990) The murine paired box gene, Pax7, is expressed specifically during the development of the nervous and muscular system. *Mech. Devel.* **33:** 27–38.

Kaestner KH, Lee KH, Schlöndoff J, Hiemisch H, Monaghan AP, Schütz G. (1993) Six members of the mouse forkhead gene family are developmentally regulated. *Proc. Natl Acad. Sci. USA* **90:** 7628–7631.

Kastury K, Li J, Druck T, Su H, Vogt PK, Croce CM, Huebner K. (1994) The human homologue of the retroviral oncogene qin maps to chromosome 14q13. *Proc. Natl Acad. Sci. USA* **91**: 3616–3618.

Koi M, Johnson LA, Kalikin LM, Little PF, Nakamura Y, Feinberg AP. (1993) Tumor cell growth arrest caused by subchromosomal transferable DNA fragments from chromosome 11. *Science* **260**: 361–364.

Koufos A, Hansen MF, Copeland NG, Jenkins NA, Lampkin BC, Cavanee WK. (1985) Loss of heterozygosity in three embryonal tumors suggests a common pathogenetic mechanism. *Nature* **316**: 330–334.

Kubo K, Naoe T, Utsumi KR, Ishiguro Y, Ueda K, Shiku H, Yamada K. (1991) Cytogenetic and cellular characteristics of a human embryonal rhabdomyosarcoma cell line, RMS–YM. *Br. J. Cancer* **63**: 879–884.

Lai E, Prezioso VR, Tao W, Chen WS, Darnell JE. (1991) Hepatocyte nuclear factor 3α belongs to a gene family in mammals that is homologous to the Drosophila homeotic gene fork head. *Genes Devel.* **5**: 416–427.

Li J, Vogt PK. (1993) The retroviral oncogene qin belongs to the transcription factor family that includes the homeotic gene fork head. *Proc. Natl Acad. Sci. USA* **90**: 4490–4494.

Li L, Olson EN. (1992) Regulation of muscle cell growth and differentiation by the MyoD family of helix-loop-helix proteins. *Adv. Cancer Res.* **58**: 95–119.

Lizard–Nacol S, Mugneret F, Volk C, Turc–Carel C, Favrot M, Philip T. (1987) Translocation (2;13)(q37;q14) in alveolar rhabdomyosarcoma: a new case [letter]. *Cancer Genet. Cytogenet.* **25**: 373–374.

Loh WE Jr, Scrable HJ, Livanos E, Arboleda MJ, Cavanee WK, Oshimura M, Weissman BE. (1992) Human chromosome 11 contains two different growth suppressor genes for embryonal rhabdomyosarcoma. *Proc. Natl Acad. Sci. USA* **89**: 1755–1759.

Magnani I, Faustinella F, Nanni P, Nicoletti G, Larizza L. (1991); Karyotypic characterization of a new human embryonal rhabdomyosarcoma cell line. *Cancer Genet. Cytogenet.* **54**: 83–89.

Mathieu MC, Niggli F, Vielh P, Oberlin O, Stevens M, Boccon–Gibod L, Flamant F. (1994) Prognostic value of flow cytometric DNA ploidy in childhood rhabdomyosarcomas enrolled in SIOP–MMT 89 study. *Med. Pediatr. Oncol.* **23**: 223.

Maulbecker CC, Gruss P. (1993) The oncogenic potential of Pax genes. *EMBO J.* **12**: 2361–2367.

Miller LM, Gallegos ME, Morisseau BA, Kim SK. (1993) lin-31, a *Caenorhabditis elegans* HNF-3/fork head transcription factor homolog, specifies three alternative cell fates in vulval development. *Genes Devel.* **7**: 933–947.

Mitchell CD, Ventris JA, Warr TJ, Cowell JK. (1991) Molecular definition in a somatic cell hybrid of a specific 2:13 translocation breakpoint in childhood rhabdomyosarcoma. *Oncogene* **6**: 89–92.

Mitchell PJ, Tjian R. (1989) Transcriptional regulation in mammalian cells by sequence-specific DNA binding proteins. *Science* **245**: 371–378.

Molenaar WM, Dam–Meiring A, Kamps WA, Cornelisse CJ. (1988) DNA-aneuploidy in rhabdomyosarcomas as compared with other sarcomas of childhood and adolescence. *Hum. Pathol.* **19**: 573–579.

Mulligan LM, Matlashewski GJ, Scrable HJ, Cavanee WK. (1990) Mechanisms of p53 loss in human sarcomas. *Proc. Natl Acad. Sci. USA* **87**: 5863–5867.

Murphy DB, Wiese S, Burfeind P, Schmundt D, Mattei MG, Schultz–Schaeffer W, Thies U. (1994) Human brain factor 1, a new member of the fork head gene family. *Genomics* **21**: 551–557.

Newsham I, Claussen U, Ludecke HJ, Mason M, Senger G, Horsthemke B, Cavenee W. (1991) Microdissection of chromosome band 11p15.5: characterization of probes mapping distal to the HBBC locus. *Genes Chrom. Cancer* **3**: 108–116.

Niggli FK, Powell JE, Parkes SE, Ward K, Raafat F, Mann JR, Stevens MC. (1994) DNA ploidy and proliferative activity (S-phase) in childhood soft-tissue sarcomas: their value as prognostic indicators. *Br. J. Cancer* **69**: 1106–1110.

Nojima T, Abe S, Yamaguchi H, Matsuno T, Inoue K. (1990) A case of alveolar rhabdomyosarcoma with a chromosomal translocation, t(2;13)(q37;q14). *Virchows Arch. A. Pathol. Anat. Histopathol.* **417**: 357–359.

Noll M. (1993) Evolution and role of Pax genes. *Curr. Biol.* **3**: 595–605.

Olegard C, Mandahl N, Heim S, Willen H, Leifsson B, Mitelman F. (1992) Embryonal rhabdomyosarcoma with 100 chromosomes but no structural aberrations. *Cancer Genet. Cytogenet.* **60**: 198–201.

Olson EN. (1990) MyoD family: a paradigm for development? *Genes. Devel.* **4**: 1454–1461.

Olson EN, Klein WH. (1994) bHLH factors in muscle development: dead lines and commitments, what to leave in and what to leave out. *Genes Devel.* **8**: 1–8.

Pani L, Overdier DG, Porcella A, Qian X, Lai E, Costa RH. (1992). Hepatocyte nuclear factor 3 beta contains two transcriptional activation domains, one of which is novel and conserved with the Drosophila fork head protein. *Mol. Cell Biol.* **12**: 3723–3732.

Pappo AS, Crist WM, Kuttesch J, Rowe S, Ashmun RA, Maurer HM, Newton WA, Asmar L, Luo X, Shapiro DN. (1993) Tumor-cell DNA content predicts outcome in children and adolescents with clinical group III embryonal rhabdomyosarcoma. *J. Clin. Oncol.* **11**: 1901–1905.

Parry P, Wei Y, Evans G. (1994) Cloning and characterization of the t(X;11) breakpoint from a leukemic cell line identify a new member of the forkhead gene family. *Genes Chrom. Cancer* **11**: 79–84.

Pierrou S, Hellqvist M, Samuelsson L, Enerback S, Carlsson P. (1994) Cloning and characterization of seven human forkhead proteins: binding site specificity and DNA bending. *EMBO J.* **13**: 5002–5012.

Potluri VR, Gilbert F. (1985) A cytogenetic study of embryonal rhabdomyosarcoma. *Cancer Genet. Cytogenet.* **14**: 169–173.

Rainier S, Johnson LA, Dobry CJ, Ping AJ, Grundy PE, Feinberg AP. (1993) Relaxation of imprinted genes in human cancer. *Nature* **362**: 747–749.

Raney BR, Hays DM, Tefft M, Triche TJ. (1989) Rhabdomyosarcoma and the undifferentiated sarcomas. In: *Principles and Practice of Pediatric Oncology* (eds PA Pizzo, DG Poplack). J.B. Lippincott, Philadelphia, pp. 635–658.

Rowe D, Gerrard M, Gibbons B, Malpas JS. (1987) Two further cases of t(2;13) in alveolar rhabdomyosarcoma indicating a review of the published chromosome breakpoints. *Br. J. Cancer* **56**: 379–380.

Sakai T, Toguchida J, Ohtani N, Yandell DW, Rapaport JM, Dryja TP. (1991) Allele-specific hypermethylation of the retinoblastoma tumor-suppressor gene. *Am. J. Hum. Genet.* **48**: 880–888.

Schurr E. (1990) Mapping of Col3a1 and Col6a3 to proximal murine chromosome 1 identifies conserved image of structural protein genes between murine chromosome 1 and human chromosome 2q. *Genomics* **8**: 477–486.

Scrable H, Cavenee W, Ghavimi F, Lovell M, Morgan K, Sapienza C. (1989a) A model for embryonal rhabdomyosarcoma tumorigenesis that involves genome imprinting. *Proc. Natl Acad. Sci. USA* **86**: 7480–7484.

Scrable H, Witte D, Shimada H, Seemayer T, Sheng WW, Soukup S, Koufos A, Houghton P, Lampkin B, Cavenee W. (1989b) Molecular differential pathology of rhabdomyosarcoma. *Genes. Chrom. Cancer* **1**: 23–35.

Scrable HJ, Witte DP, Lampkin BC, Cavenee WK. (1987) Chromosomal localization of the human rhabdomyosarcoma locus by mitotic recombination mapping. *Nature* **329**: 645–647.

Scrable HJ, Johnson DK, Rinchik EM, Cavenee WK. (1990) Rhabdomyosarcoma-associated locus and MYOD1 are syntenic but separate loci on the short arm of human chromosome 11. *Proc. Natl Acad. Sci. USA* **87**: 2182–2186.

Seidal T, Mark J, Hagmar B, Angervall L. (1982) Alveolar rhabdomyosarcoma: a cytogenetic and correlated cytological and histological study. *Acta. Pathol. Microbiol. Immunol. Scand.* **90**: 345–354.

Shapiro DN, Parham DM, Douglass MD, Ashmun R, Webber BL, Newton WA, Hancock ML, Maurer HM, Look AT. (1991) Relationship of tumor cell ploidy to histologic subtype and treatment outcome in children and adolescents with unresectable rhabdomyosarcoma. *J. Clin. Oncol.* **9**: 159–166.

Shapiro DN, Valentine MB, Sublett JE, Sinclair AE, Tereba AM, Scheffer H, Buys CH, Look AT. (1992) Chromosomal sublocalization of the 2;13 translocation breakpoint in alveolar rhabdomyosarcoma. *Genes. Chrom. Cancer* **4**: 241–249.

Shapiro DN, Sublett JE, Li B, Downing JR, Naeve CW. (1993) Fusion of PAX3 to a member of the forkhead family of transcription factors in human alveolar rhabdomyosarcoma. *Cancer Res.* **53**: 5108–5112.

Shapiro DN, Jones BG, Shapiro LH, Dias P, Houghton PJ. (1994) Antisense-mediated reduction in insulin-like growth factor-I receptor expression suppresses the malignant phenotype of a human alveolar rhabdomyosarcoma. *J. Clin. Invest.* **94**: 1235–1242.

Sorensen PH, Lessnick SL, Lopez-Terrada D, Liu XF, Triche TJ, Denny CT. (1994) A second Ewing's sarcoma translocation, t(21;22), fuses the EWS gene to another ETS-family transcription factor, ERG. *Nature Genet.* **6**: 146–151.

Strachan T, Read AP. (1994) PAX genes. *Curr. Opin. Genet. Devel.* **4**: 427–438.

Stratton MR, Fisher C, Gusterson BA, Cooper CS. (1989) Detection of point mutations in N-ras and K-ras genes of human embryonal rhabdomyosarcomas using oligonucleotide probes and the polymerase chain reaction. *Cancer Res.* **49**: 6324–6327.

Tao W, Lai E. (1992) Telencephalon-restricted expression of BF-1, a new member of the HNF – 3/fork head gene family, in the developing rat brain. *Neuron* **8**: 957–966.

Tapscott SJ, Weintraub H. (1991) MyoD and the regulation of myogenesis by helix-loop-helix proteins. *J. Clin. Invest.* **87**: 1133–1138.

Tapscott SJ, Thayer MJ, Weintraub H. (1993) Deficiency in rhabdomyosarcomas of a factor required for MyoD activity and myogenesis. *Science* **259**: 1450–1453.

Tassabehji M, Read AP, Newton VE, Harris R, Balling R, Gruss P, Strachan T. (1992) Waardenburg's syndrome patients have mutations in the human homologue of the Pax-3 paired box gene [see Comments]. *Nature* **355**: 635–636.

Tonin PN, Scrable H, Shimada H. (1991) Muscle-specific gene expression in rhabdomyosarcoma, and stages of human fetal skeletal muscle development. *Cancer Res.* 51: 5100–5106.

Treisman J, Gonczy P, Vashishtha M, Harris E, Desplan C. (1989) A single amino acid can determine the DNA binding specificity of homeodomain proteins. *Cell* 59: 553–562.

Tremblay P, Gruss P. (1994) Pax: genes for mice and men. *Pharmacol. Ther.* 61: 205–226.

Trent J, Casper J, Meltzer P, Thompson F, Fogh J. (1985) Nonrandom chromosome alterations in rhabdomyosarcoma. *Cancer Genet. Cytogenet.* 16: 189–197.

Turc-Carel C, Lizard-Nacol S, Justrabo E, Favrot M, Philip T, Tabone E. (1986) Consistent chromosomal translocation in alveolar rhabdomyosarcoma. *Cancer Genet. Cytogenet.* 19: 361–362.

Tycko B. (1994) Genomic imprinting: mechanisms and role in human pathology. *Am. J. Pathol.* 144: 431–443.

Valentine M, Douglass EC, Look AT. (1989) Closely linked loci on the long arm of chromosome 13 flank a specific 2;13 translocation breakpoint in childhood rhabdomyosarcoma. *Cytogenet. Cell Genet.* 52: 128–132.

Voullaire LE, Petrovic V, Sheffield LJ, Campbell P. (1991) Two forms of ring 13 in a child with rhabdomyosarcoma. *Am. J. Med. Genet.* 39: 285–287.

Walther C, Guenet J, Simon D, Deutsch U, Jostes B, Goulding MD, Plachov D, Balling R, Gruss P. (1991) Pax: a murine multigene family of paired box-containing genes. *Genomics* 11: 424–434.

Weigel D, Jackle H. (1990) The fork head domain: a novel DNA binding motif of eukaryotic transcription factors? [letter]. *Cell* 63: 455–456.

Weigel D, Jurgens G, Kuttner F, Seifert E, Jackle H. (1989) The homeotic gene fork head encodes a nuclear protein and is expressed in the terminal regions of the Drosophila embryo. *Cell* 57: 645–658.

Weigel D, Seifert E, Reuter D, Jackle H. (1990) Regulatory elements controlling expression of the Drosophila homeotic gene fork head. *EMBO J.* 9: 1199–1207.

Weintraub H, Davis R, Tapscott S, Thayer M, Krause M, Benezra R, Blackwell TK, Turner D, Rupp R, Hollenberg S, et al. (1991) The myoD gene family: nodal point during specification of the muscle cell lineage. *Science* 251: 761–766.

Weissman BE, Saxon PJ, Pasquale SR, Jones GR, Geiser AG, Stanbridge EJ. (1987) Introduction of a normal human chromosome 11 into a Wilms' tumor cell line controls its tumorigenic expression. *Science* 236: 175–180.

Whang–Peng J, Knutsen T, Theil K, Horowitz ME, Triche T. (1992) Cytogenetic studies in subgroups of rhabdomyosarcoma. *Genes Chrom. Cancer* 5: 299–310.

Wijnaendts LC, van der Linden JC, van Diest P, van Unnik AJ, Delemarre JF, Voute PA, Meijer CJ. (1993) Prognostic importance of DNA flow cytometric variables in rhabdomyosarcomas. *J. Clin. Pathol.* 46: 948–952.

Zucman J, Melot T, Desmaze C, Ghysdael J, Plougastel B, Peter M, Zucker JM, Triche TJ, Sheer D, Turc–Carel C. et al. (1993) Combinatorial generation of variable fusion proteins in the Ewing family of tumours. *EMBO J.* 12: 4481–4487.

222

Knock-out mice for the *RET* oncogene

Vassilis Pachnis, Pascale Durbec, Lena Larsson-Blomberg,
Baljinder Mankoo, Anita Schuchardt, Vivette D'Agati and
Frank Costantini

10.1 Introduction

One of the underlying themes from the preceding chapters is that it is the homozygous inactivation of critical genes which results in tumour initiation. The function of these genes can be diverse, from DNA binding proteins to DNA repair genes. However, in many cases their primary role is to control the normal differentiation of either a single tissue or a very limited set of tissues. Detailed analysis of inactivating mutations has told us a great deal about the functional domains of these proteins but, ultimately, in order to study their role in normal development and tumorigenesis, model animal systems must be created which can be easily manipulated. The most successful approach has been to use transgenic mouse technology either to express constitutively the gene of interest (transgenic mice) or, perhaps more relevant to the recessive oncogenes, to produce mice with homozygous inactivation (null) of the particular gene, which are referred to as 'knock-out' mice. Simplistically, if the gene in question is responsible for tumorigenesis, then mice lacking this gene should develop the specific tumour associated with it. Although *p53* knock-outs appear to develop the expected range of tumours, this is the exception rather than the rule. However, since these *p53* knock-outs seem to develop normally it appears that this gene is not essential for embryonic development. For the majority of the recessive oncogenes, however, homozygous knock-outs die during development demonstrating the critical role of these genes in embryogenesis and thereby precluding an analysis of their role in tumorigenesis. Whether these animal models are going to be useful in the analysis of tumour development, however, is not clear since mice carrying heterozygous mutations in the *RB1* and *WT1* genes, for example, do not develop tumours in the retina or kidneys as in the human situation. It is not clear either why we should expect that they would since mice do not develop retinoblastoma or Wilms' tumour sporadically.

Although knock-out mice have not fully lived up to the expectations of the oncologists, because the genes in question are often critical for normal development, they have provided an exciting model for developmental biologists. Thus, knock-outs for *WT1* demonstrate the critical role of this gene in normal development of not only the kidneys and gonads but also the mesothelium, disruption of which appears to be the cause of embryonic death in the null mice. The *RB1* knock-outs do not develop retinoblastoma but, rather, tumours of the pituitary which is not part of the spectrum of tumours seen in patients carrying heterozygous constitutional *RB1* mutations. In the developing embryo, the null *RB1* mice show defects in the production of mature erythrocytes in the liver and extensive cell death in the central nervous system.

An intriguing observation concerning oncogenes is that the dominantly acting mutant form of the *RET* oncogene is also responsible for the development of tumours in patients with multiple endocrine neoplasia type 2A (MEN2A). The generation of knock-out mice for this gene has also led to some unexpected observations in normal development of tissues not related to the MEN2 phenotype. In this chapter, the role of *RET* in normal mouse development is discussed and illustrates how models designed to dissect the development of the malignant phenotype can unexpectedly provide models for other human genetic disorders.

10.2 Receptor tyrosine kinase

Receptor tyrosine kinases (RTKs) constitute a superfamily of cell surface molecules that serve as receptors for a variety of hormones, growth factors and neurotrophic factors (Yarden and Ullrich, 1988; Schlessinger and Ullrich, 1992). All RTKs have a similar topology: they possess an extracellular ligand binding domain, a trans-membrane segment and a cytoplasmic segment containing the catalytic tyrosine kinase domain (Yarden and Ullrich, 1988). Binding of the ligand leads to dimerization of the receptor and activation of the kinase domain, manifested as autophosphorylation of the receptor and phosphorylation of specific substrates which mediate the intracellular signalling (Ullrich and Schlessinger, 1990). Until recently, the vast majority of studies concerned the role of RTKs in cellular growth and proliferation. However, over the last few years, it has been increasingly realized that RTKs, in addition to their roles as mediators of cell growth and proliferation, are essential for diverse developmental processes in a wide range of organisms, ranging from flies to mammals and humans (Pawson and Bernstein, 1990). The best evidence that RTKs are key regulators of embryogenesis is derived from genetic studies in both invertebrates and vertebrates (Hafen and Basler, 1990; Pawson and Bernstein, 1990). In *Drosophila melanogaster*, several well-characterized developmental mutations have been shown to reside in genes encoding RTKs. One of the best studied examples is the *sevenless* locus, which encodes an RTK that interacts with a cell surface molecule encoded at the *boss* locus. Mutations at either of these loci prevent the differentiation of R7 photoreceptors and lead to abnormal patterning of the *D. melanogaster* eye. In mice, mutations in the *Dominant White Spotting* (*W*) and *Steel* (*Sl*) loci, which encode the kit RTK and its ligand, respectively, cause similar defects in the haematopoietic, germ cell and melanocytic lineages (Green, 1989). Finally, RTKs of the trk subfamily have been shown to function as the high affinity receptors for the nerve growth factor (NGF)-related neurotrophins (Kaplan *et al.*, 1991a, b; Klein *et al.*, 1991a, b; Lamballe *et al.*, 1991; Soppet *et al.*, 1991). Several studies, including immunoablation experiments, *in vitro* culture of neuronal cells and their precursors, as well as genetic ablation of members of the *trk* and the neurotrophin gene family, have established that these signalling molecules are involved in critical developmental processes related to the proliferation, differentiation and survival of a wide range of lineages of the peripheral nervous system (PNS) (Martinez *et al.*, 1985; Levi-Montalcini, 1987).

The *RET* proto-oncogene encodes a member of the RTK superfamily (Takahashi *et al.*, 1988; Iwamoto *et al.*, 1993; Pachnis *et al.*, 1993), whose extracellular (and putative ligand binding) domain displays no homology with other RTKs, and whose cognate ligand remains to be identified. *RET* was originally cloned because of the transforming ability of rearranged forms of the gene (*RET* oncogenes), in which the kinase domain is constitutively activated as a result of loss of the normal extracellular domain (Takahashi *et al.*, 1985; Takahashi and Cooper, 1987). *In vivo*, somatic rearrangements generating *RET* oncogenes have been observed in human thyroid papillary carcinomas (Santoro *et al.*, 1990, 1992), and similarly rearranged forms of the *RET* gene have been associated with various tumours when expressed ectopically in transgenic mice (Iwamoto *et al.*, 1990, 1991a, b). Recently, several laboratories have identified germline mutations at the *RET* locus in patients with multiple endocrine neoplasia type 2A (MEN2A) (Donis-Keller *et al.*, 1993; Mulligan *et al.*, 1993) and type 2B (MEN2B) (Carlson *et al.*, 1994; Hofstra *et al.*, 1994). These are dominantly inherited cancer syndromes characterized by overgrowth or tumours of derivatives of the neural crest, such as medullary thyroid carcinoma, phaeochromocytomas and ganglioneuromas of the intestinal tract. Although the role of *RET* mutations in the generation of the MEN2 syndromes has been firmly established, the mechanism of action of the mutated locus is still a matter of debate. For example, it is not known whether the phenotype of the MEN2A and MEN2B syndromes results (in the cell types affected by these mutations) from constitutive activation of the RET receptor (rendering it ligand independent), or from inactivation of its kinase catalytic domain. Finally, the identification of a variety of mutations in the coding region of the *RET* locus in patients with Hirschsprung's disease (HSCR, congenital megacolon) (Edery *et al.*, 1994; Romeo *et al.*, 1994) strongly suggests that *RET* plays a critical role in the development of the enteric nervous system in mammals.

10.3 Expression of the *RET* proto-oncogene during mouse embryogenesis

A common theme underlying the phenotype of the diseases caused by mutations in the *RET* proto-oncogene is that they are derived from a common origin, the neural crest. Analysis of the distribution of *RET* mRNA in developing mammalian and avian embryos has established that the gene is expressed in subsets of neural crest cells and their derivatives, as well as in various subsets of cells of the developing central nervous system (Pachnis *et al.*, 1993). *RET* transcripts were first detected in specific groups of cells in the 8.5–9.0-day-old mouse embryo. In the head, strong signal was localized to the cohesive group of neural crest cells emigrating from rhombomere 4 (r4) of the hindbrain. Other groups of neural crest cells emigrating from the neuroepithelium anterior or posterior to r4 were negative for *RET* transcripts.

Two components of the hindbrain neural crest have been described: the early migrating cells, which populate the ventrally located branchial arches and adopt a mesenchymal fate, and the late emerging cells, which remain closer to the neural tube and adopt a neural fate (Nichols, 1981; Nichols, 1986). The detection of *RET* mRNA in both the dorsal and ventral neural crest of r4 suggests that, during this stage of embryogenesis, both the neurogenic and the mesenchymal components of the r4 crest express *RET* mRNA at comparable levels. Interestingly, during the embryonic period under consideration, the brain neuroepithelium, including r4, was devoid of hybridization signal.

In addition to the migrating neural crest cells, *RET* is expressed in the 8.5–9.0-day-old embryo in a well-defined domain of cranial surface ectoderm and pharyngeal endoderm associated with the posterior branchial arches. The *RET*-positive domain of foregut endoderm is thought to contribute to the formation of the parathyroid glands. The *RET*-positive surface ectoderm includes the epibranchial placodes, which give rise to the sensory neurons of the inferior ganglia of the IXth and Xth cranial nerve ganglion complexes (Altman and Bayer, 1982).

In the heads of 9.5-day-old embryos, and consistent with the restricted expression of *RET* in r4 neural crest of earlier stage embryos, signal was localized in the condensing facioacoustic ganglion complex (VII–VIIIth) but not in the anlage of other cranial ganglia present anteriorly (Vth) or posteriorly (IXth, Xth). Despite the expression of *RET* in the anlage of the facioacoustic ganglion, no signal was detected in the second branchial arch which is populated to a large extent by the mesenchymal component of r4 crest, indicating that, by the time the r4 crest has completed its migration into the second branchial arch, the expression of *RET* is down-regulated. In 10.5-day-old embryos, in addition to the facioacoustic ganglion, the inferior ganglia of the IXth and Xth cranial nerve ganglion complexes are positive for *RET* transcripts, while the trigeminal ganglion (V) is still negative. Finally, in 13.5–14.5-day-old embryos, all cranial ganglia, including the trigeminal ganglion and the superior ganglia of the IXth and Xth cranial nerve ganglion complexes, are positive for *RET* mRNA. Overall, two distinct phases of *RET* expression can be discerned: an early, segment-specific phase (8.5–9.5-day-old embryo), during which *RET* mRNA is restricted to neural crest cells derived from r4 and the anlage of the facioacoustic ganglion, and a later phase (10.5–14.5-days) during which signal appears gradually in all cranial ganglia irrespective of the origin of the contributing neural crest cells along the anteroposterior axis of the brain neuroepithelium. The positionally restricted localization of *RET* mRNA in the cranial neural crest and the surface ectoderm and pharyngeal endoderm of the branchial arches suggests that expression of *RET* is related to the intrinsic mechanisms of segmentation of the vertebrate head. It has been suggested that the patterning information for the mesenchymal and neuronal components of each branchial arch is contained within the neural crest that populates that arch, and that this information has been imprinted on the neural crest by its origin along the anteroposterior axis of the brain neuroepithelium (Noden, 1983, 1988). It has also been postulated that the combinatorial expression of members of the *Hox* gene clusters provides at least part of the patterning information for the specification of individual rhombomeres of the hindbrain, and the neural crest cells derived from them (Hunt *et al.*, 1991). How might such a 'Hox code' in the cranial neural crest translate into different neuronal and mesenchymal structures in the head of the adult animal? The differential activation of downstream target genes could provide the means of translating such positional information into diverse developmental programmes. The restricted expression of the *RET* gene in the r4 neural crest of the 8.5–9.5-day-old embryo provides additional evidence that the various groups of cranial neural crest cells are molecularly distinct. Moreover, it raises the possibility that the RET receptor is part of the molecular cascade that leads to the establishment of the unique properties of the r4 crest derivatives.

Although the documented expression of *RET* in cell lines of neural crest origin is consistent with the neural crest origin of medullary thyroid carcinoma (C-cells of the thyroid) and phaeochromocytoma (sympathoadrenal progenitor), the occurrence of parathyroid hyperplasia in patients with MEN2A syndrome has been puzzling, as parathyroid cells are derived embryologically from the endoderm of the fourth pharyngeal pouch (Balinsky, 1970). This led to the suggestion that the tumours of the MEN2A syndrome may result from defects in closely linked genes that independently control neural crest or endocrine tissue development (Lairmore *et al.*, 1991). The pattern of expression of the *RET* proto-oncogene further supports the suggestion that the entire MEN2A syndrome is due to mutations of a single gene, by providing evidence that all lineages affected in this syndrome are likely to express *RET*. The precursors of the C-cells, originating from the posterior hindbrain, are transiently localized, along with the precursors to the enteric nervous system (ENS), in the mesenchyme of the posterior branchial arches (Le Douarin, 1982), an area where *RET* is strongly expressed. This, along with the high levels of *RET* expression in C-cell-derived medullary thyroid carcinomas (Santoro *et al.*, 1990) and preliminary experiments by myself and co-workers which show low levels of expression in the C-cells of adult mouse thyroid (Durbec *et al.*, unpublished data), suggests that the C-cell lineage expresses *RET* mRNA from the early embryonic stages. Despite lack of expression in the adrenal medulla, it is likely that the chromaffin cell precursors are also expressing *RET* mRNA, since they originate from the RET-positive cells of the sympathetic ganglia condensations (Pankratz, 1931). Finally, the parathyroid cell precursors are derived from the *RET*-positive endoderm of the posterior branchial arches. Overall, the common expression of *RET* reveals a developmental link between the cell types affected in the MEN2A syndrome, and provides further support for the hypothesis that germline mutations in *RET* are responsible for all the manifestations of the MEN2A syndrome.

10.3.1 RET *expression in the developing nervous system*

RET mRNA was also present in the migrating neural crest cells of the trunk, albeit at lower levels than in the r4 neural crest. In later stage embryos, i.e. embryonic day 11.5–13.5, intense punctate signal was observed in the dorsal root ganglia (DRG), indicating that expression of *RET* in the DRG is restricted to subsets of neurons and is absent from the glial cells of the ganglion. The apparent localization of *RET* mRNA in subsets of neurons of the sensory ganglia (cranial nerve ganglia and DRG) is consistent with previous studies indicating that distinct gene expression programmes exist in the various functional classes of the sensory neurons. Using monoclonal antibodies, Dodd and co-workers were able to identify subsets of primary sensory neurons in rat DRG expressing unique carbohydrate differentiation antigens (Dodd *et al.*, 1984; Dodd and Jessell, 1985). More recently, *in situ* hybridization analysis of the pattern of expression of members of the trk RTK subfamily has revealed that, similar to the RET receptor, individual members of the trk subfamily are expressed in subclasses of DRG cells (Carroll *et al.*, 1992). Furthermore, immune deprivation of NGF and genetic analysis of the function of various neurotrophins and their receptors using knock-out mice have indicated that each neurotrophin receptor signalling system is primarily responsible for the differentiation and survival of a specific subset of sensory neurons (Carroll *et al.*, 1992; Ruit *et al.*, 1992). Expression studies suggest that, as is the case for the trk subfamily of RTKs, the RET receptor plays a role in the differentiation and survival of a specific class of primary sensory neurons. The identity of this class and the particular sensory modality it conveys are currently under investigation.

High levels of *RET* mRNA are also expressed in the developing ENS. The ENS is composed of the myenteric and the submucosal ganglion complexes of the gut (Furness and Costa, 1987). The majority of the cells in the ganglia of the ENS are derived from the vagal crest originating from the post-otic hindbrain (corresponding to somites 1–7 of the avian embryo) (Yntema and Hammond,

1954; Le Douarin and Teillet, 1973). The presumptive enteric ganglioblasts of the vagal crest migrate first ventrally through the posterior branchial arches into the foregut mesenchyme, and then rostrocaudally inside the gut wall mesenchyme, where they eventually coalesce and form the enteric ganglia along the entire length of the gut (Tucker *et al.*, 1986; Baetge and Gershon, 1989; Kapur *et al.*, 1992; Gershon *et al.*, 1993). *RET*-positive cells have been identified in the migratory pathway of the presumptive enteric neuroblasts of the vagal crest, and in the myenteric ganglia of the gut (*Figure 10.1*). In the 9.0–9.5-day-old mouse embryo, strong expression was detected in mesenchymal cells of the third and fourth branchial arches, between the *RET*-positive pharyngeal endoderm and surface ectoderm. Twelve hours later, a stream of *RET*-positive cells could be seen emerging from the posterior branchial arch mesenchyme migrating towards the foregut. At subsequent stages (10.5–11.5 days) *RET*-positive cells were found at increasingly more caudal levels of the embryonic gut. Finally, in the 13.5–14.5-

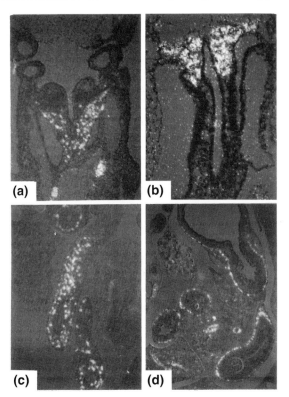

Figure 10.1. Expression of *RET* in the vagal neural crest and the derivative myenteric plexus of the embryonic gut. In the embryonic day (E)-9.5 mouse embryo, *RET*-positive neural crest cells have migrated into the foregut mesenchyme (a), while the midgut and hindgut are devoid of *RET*-expressing cells (b). In the E11.5 embryo, *RET*-positive cells have populated most of the gut (c), while in the E14.5 embryo, they have formed the typical ring-like structures of the myenteric plexus (d).

day-old embryo, in which the rostrocaudal migration of the ENS precursors has been completed, strong *RET* expression was detected in the myenteric plexus along the entire axis of the gut.

RET mRNA is also expressed in the autonomic ganglia of the PNS. At embryonic day 9.5, *RET* transcripts were detected in groups of cells by the dorsal aorta, where derivatives of the trunk neural crest coalesce to form the sympathetic ganglia anlage. Expression is maintained at later stages, e.g. 14.5 days, when sympathetic gangliogenesis has been essentially completed. Although, at this stage, no expression of *RET* was observed in the endocrine derivatives of the sympathoadrenal lineage, i.e. the adrenal chromaffin cells, we expect that *RET* mRNA is expressed in the sympathoadrenal precursors of these cells. Phenotypic analysis of the sympathoadrenal progenitors and the progenitors of the ENS in rodents has established that these two lineages share a number of independent molecular markers (Baetge and Gershon, 1989; Baetge *et al.*, 1990; Johnson *et al.*, 1990; Carnahan *et al.*, 1991; Lo *et al.*, 1991). This led to the suggestion that sympathoadrenal and enteric lineages are derived from progenitors that are either identical or closely related (Carnahan *et al.*, 1991), and that the particular phenotypes eventually acquired by these progenitors are determined by the specific microenvironment at their final destination. The concomitant expression of *RET* in the anlage of the sympathetic ganglia and the precursors of the ENS in the mouse embryo further supports their close relationship. Moreover, it raises the possibility that *RET* plays a role in the proliferation, migration, differentiation or survival of the cells of these lineages. The expression of *RET* mRNA in the migratory ENS progenitors is particularly interesting, since, despite the established importance of signals encountered along the migratory pathway of the vagal crest and in the microenvironment of the embryonic bowel for the differentiation of the enteric neurons (Gershon *et al.*, 1993), very little is known about the molecular nature of these signals. The expression studies presented here, in combination with genetic experiments involving RET-deficient mice (see below) clearly indicate that *RET* plays a critical role in the development of the ENS in mammals.

10.3.2 RET *expression in nephrogenesis*

In addition to its postulated role in the developing nervous system, the expression pattern of the *RET* gene suggests that the RET receptor could function during the development of the excretory system. *RET* mRNA was first detected at day 8.5 in the nephrotome, a region of lateral mesoderm from which the embryonic (pronephric and mesonephric) and permanent (metanephric) kidneys are formed and continued to be detected in the nephric (Wolffian) ducts during the development of the pronephric and mesonephric kidneys (day 9.5–10.5). The expression pattern of *RET* at this stage suggests that the *RET* receptor might play a role in the formation of the pronephros and mesonephros.

The development of the mammalian metanephric kidney has been extensively studied as a model system for the role of inductive tissue interactions during organogenesis (Saxen, 1987). Formation of the kidney is initiated on day 11 of mouse embryogenesis, when the ureteric bud emerges near the caudal end of the nephric duct and grows dorsally, eventually contacting the metanephric blastema, a specific condensate of mesenchymal cells. At this point, the ureteric bud begins to grow and branch repeatedly, dependent upon specific induction by the metanephric mesenchymal cells (Grobstein, 1953, 1955; Erickson, 1968), eventually giving rise to the entire renal collecting system (calyces, papillae and collecting ducts). At the same time, the tips of the branching ureteric bud induce the surrounding mesenchymal cells to condense into epithelial renal vesicles, which eventually differentiate to form the various segments of the nephron, including the proximal and distal tubules and glomeruli (Grobstein, 1955, 1956; Saxen, 1987). At a stage when the ureteric bud has first branched within the metanephric mesenchyme, we observed that *RET* was strongly expressed in the ureteric bud epithelium but was not expressed at detectable levels in the surrounding undifferentiated mesenchyme (*Figure 10.2*). At later stages of renal organogenesis, growth and branching of the ureteric derivatives, as well as the induction of new nephrons, is limited to a narrow 'nephrogenic' zone at the perimeter of the kidney (Saxen, 1987). At these stages (day 13.5–17.5), *RET* mRNA was observed only in the tips of the growing collecting ducts in the outer nephrogenic zone, and not in the previously formed and more centrally located collecting ducts (*Figure 10.2*). Like the undifferentiated metanephric mesenchyme, both the newly condensed renal vesicles and the more mature mesenchymal derivatives continued to be negative for *RET* expression.

Based on this restricted pattern of expression, it appears likely that the RET receptor could play a role in the growth and development of the ureteric bud and its derivative structures in the kidney. More specifically, RET might serve as the

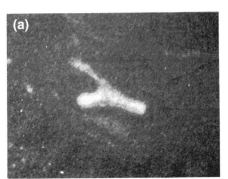 

Figure 10.2. *RET* expression in the excretory system of the mouse embryo. In E11.5 mouse embryos, high levels of *RET* mRNA are detected in the ureteric bud and its branches (a). In later stage embryos (E16.5), signal is restricted to the outer zone of the kidney, where induction of new nephrons is still taking place (b).

receptor for an inductive factor, produced by the metanephric mesenchymal cells, which stimulates the growth and/or branching of the ureteric bud epithelium. Support for the hypothesis that the *RET* gene serves an important function in renal organogenesis has recently been obtained by the production of mice carrying a targeted, loss-of-function mutation of the *RET* gene (see below).

10.4 *RET* knock-out mice

The role of the RET receptor in mammalian development was tested directly by generating mice homozygous for a targeted mutation in the *RET* locus. The mouse *RET* proto-oncogene was mutated by inserting a *neo*^R gene in the sequences encoding the tyrosine kinase domain of RET and simultaneously deleting 0.8 kb of genomic sequences (*Figure 10.3*). Since these sequences include the part of the receptor encoding the ATP binding site of the kinase domain, we anticipated that any potential protein product of the mutant locus would have no tyrosine kinase activity. Therefore, we named the mutation *ret-k⁻*. At birth, all of the pups generated from heterozygous intercrosses appeared normal in behaviour and external appearance. By 12 h after birth, however, one quarter of the pups started to become cyanotic and less active. All of these animals died approximately 16–24 hours after birth and, when genotyped, were found to be homozygous for the *ret-k⁻* allele. Of the mice that survived to 5 weeks, all were wild-types or *ret-k⁻* heterozygotes. Therefore, the perinatal lethal phenotype of the *ret-k⁻* mice appeared to be due to a single recessive mutation with complete penetrance.

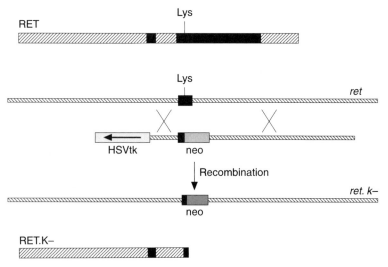

Figure 10.3. Targeting strategy of the murine *RET* locus. The mutant locus (*ret-k⁻*) is expected to produce a protein that is capable of being anchored on to the cell membrane, but is inactive due to lack of the entire tyrosine kinase domain.

10.4.1 Defects in urogenital development

The first abnormality that became obvious upon dissection of the homozygous *ret-k⁻* pups was the lack of normal kidneys (*Figure 10.4*). When kidneys were present in the mutant animals, they were rudimentary organs that lacked the characteristic reniform shape and were approximately the size of the adrenal glands. Despite their greatly reduced size, these kidney rudiments were connected to the bladder by ureters, and appeared to be partially functional in some cases, based on the presence of urine in the bladder. Other mutant mice displayed one or two blind-ending ureters, which emerged from the bladder at the

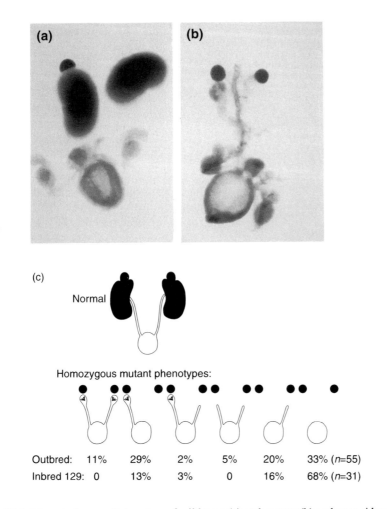

Figure 10.4. Dissected urogenital system of wild-type (a) and mutant (b) embryos. Also shown in (c) is the frequency of kidney rudiments and blind-ending ureters among homozygous *ret-k⁻* newborn mice.

appropriate position, but terminated half-way between the adrenal and bladder and had no apparent renal tissue at the tip. The most severe defect observed, and the most frequent, was the complete absence of ureter and kidney (renal agenesis), which occurred either unilaterally or bilaterally. The extreme variability in ureteric/renal development may be partially due to genetic background effects, since the mice examined were of mixed genetic background. However, since the degree of development often differed between the left and right side of an individual mouse, there appears to be a stochastic component to the developmental events underlying these phenotypes.

Histological analysis of several mutant kidney rudiments revealed severe dysplasia. The normal newborn mouse kidney contains three distinct zones, the medulla, the cortex and the nephrogenic zone where new nephrons continue to be formed (Saxen, 1987). The mutant kidneys contained reduced numbers of recognizable nephric elements, including proximal and distal tubules, glomerular tufts and vessels, which were randomly distributed with no recognizable medulla, cortex or nephrogenic zone. In addition, Bowman's capsules and tubules were focally dilated, and the mutant kidneys contained large regions of undifferentiated mesenchyme not seen in normal kidneys at this stage. Sections taken at the level of entry of the ureter showed reduced branchings of the ureter, numbering four or less, without formation of mature collecting ducts or intervening vasa recta.

Based on these histological findings, it was suggested that the RET protein transduces a mesenchyme-derived signal, which stimulates the formation as well as the subsequent growth and branching of the ureteric bud. This hypothesis is consistent with the expression pattern of the normal *RET* gene and with the *ret-k⁻* mutant phenotype. *RET* mRNA is expressed in the ureteric bud and its derivatives (specifically in the distal tips of the growing collecting ducts within the nephrogenic zone, where branching and elongation is taking place), but not in the mesenchymal compartment (Pachnis *et al.*, 1993). Furthermore, the histology of the rudimentary kidneys that develop in some *ret-k⁻* mice appeared consistent with a failure of the ureteric bud to develop appropriately. Branching of the ureter was greatly reduced, and the renal histoarchitecture, which is normally dictated by the complex branching pattern of the ureteric epithelium, was highly disorganized. In contrast, the fact that the mutant kidney rudiments contain some differentiated mesenchymal derivatives indicates that the mutant ureteric epithelium is capable of inducing tubulogenesis, and that the mutant metanephric mesenchyme is competent to respond to this stimulus.

Even more common than the formation of a kidney rudiment was the unilateral or bilateral absence of ureter and kidney in the newborn animal. This phenotype appears to result from failure of the ureteric bud to form, based on preliminary analyses of homozygous *ret-k⁻* embryos (unpublished data). Recent observations of mice with a mutation in the Wilms' tumour gene (*WT1*) implied that a signal from the metanephric mesenchyme is required to induce the initial formation of the ureteric bud (Kreidberg *et al.*, 1993). Our results, combined with the expression of *RET* in the posterior Wolffian duct before the outgrowth of the ureteric bud, support this model and suggest that *RET* encodes the receptor for this signal.

As is the case in human bilateral renal agenesis, some newborn mutant mice displayed underdeveloped and/or collapsed lungs (data not shown), which is likely to be a secondary effect of oligohydramnios, a deficiency in amniotic fluid production by the kidneys.

Overall, these findings suggest that mutations in the human *RET* gene may be responsible for hereditary defects of the excretory system. Renal agenesis or dysgenesis occurs at a frequency of approximately 1/3000 to 1/6000 births, including both sporadic and familial cases. While no genetic loci associated with these defects have yet been identified, the critical role of the *RET* proto-oncogene in mouse kidney development suggest that mutations or deletions at this locus could be responsible for a fraction of these cases.

10.4.2 Defects in the PNS

The second major abnormality observed in all the mutant homozygous pups was a failure of milk to progress from the stomach into the small intestine, indicating a defect in gastrointestinal peristalsis. As the normal *RET* gene is highly expressed in the developing enteric ganglia (Pachnis *et al.*, 1993), which control gut motility, this phenotype suggested that the mutation may have affected the development of the ENS. We therefore examined histological sections from various levels of the gastrointestinal tract, for the presence of these ganglia. This analysis revealed that the neurons of the myenteric plexus, normally located between the longitudinal and circular muscle layers and clearly visible in normal samples, were in each case absent from the entire bowel. To confirm these light microscopic observations, immunohistochemical studies were performed using antibodies against neuron-specific enolase (NSE) and peripherin, two proteins expressed in the neurons of enteric ganglia (Bishop *et al.*, 1985; Troy *et al.*, 1990). While many NSE- and peripherin-positive cells were observed in the myenteric plexus of the small and large intestine from heterozygous (*ret-k$^{-/+}$*) animals, no such positive cells were observed in the corresponding region of the intestine of homozygous mutant animals (*Figure 10.5*). Peripherin-positive neurons of the submucosal plexus could also be visualized in the large intestine of heterozygotes, while none were observed in the homozygous mutant large intestine. These experiments further indicate that all ganglia of the ENS are absent from the mutant bowel.

The majority of the neurons and glia of the ENS are derived from the vagal neural crest, which migrates from the post-otic hindbrain along predetermined pathways to populate the entire length of the bowel (Yntema and Hammond, 1954; Le Douarin and Teillet, 1973). Despite the unique properties of the ENS, neural crest cells that do not normally generate enteric neurons or glia have the

+ / + + / −   − / −

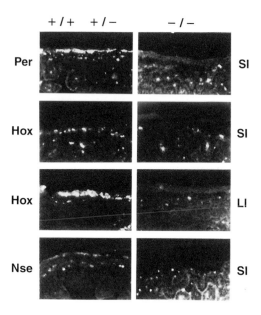

Figure 10.5. Absence of neurons from the *ret-k⁻* mutant gut wall. Sections of small intestine (SI) and large intestine (LI) of wild-type or heterozygous (+/+ or +/−) and homozygous mutant (−/−) newborn mice stained with antibodies against peripherin (Per), Hoxb-4 (Hox) and neuron-specific enolase (Nse).

potential to form a normal and functional ENS, containing bowel-appropriate neuronal and glial phenotypes, when grafted to ectopic sites in chimeric avian embryos (Le Douarin *et al.*, 1975; Fontaine-Perus *et al.*, 1982). These results demonstrate that the microenvironment of the embryonic bowel is capable of directing uncommitted neural crest precursors towards ENS-specific phenotypes (Gershon *et al.*, 1993). None of the molecules mediating such effects have been identified so far. We demonstrate here that *RET* plays a critical role in the establishment of the enteric neuronal lineage. Although the mechanism of action of *RET* is currently unknown, the nature of its product suggests that it encodes the receptor for a signalling molecule required for the migration or differentiation of the vagal neural crest and its derivatives.

The most common congenital defect (1/5000 births) affecting the development of the ENS in humans is Hirschsprung's disease (HSCR, aganglionic mega-colon). HSCR is characterized by the absence of enteric ganglia from the hindgut and, although the histopathological abnormalities are usually restricted to the sigmoid colon, some cases also involve the proximal colon and even the distal small intestine (Passarge, 1973; Meier-Ruge, 1974). Recent genetic linkage and physical mapping studies, as well as mutational analysis, have suggested that a certain percentage of cases of HSCR are due to mutations of the *RET* locus

(Edery *et al.*, 1994; Romeo *et al.*, 1994). These genetic studies are consistent with our analysis of the effect of the *ret-k⁻* mutation on the ENS, and indicate that the *ret-k⁻* mutant mice represent a useful animal model of the developmental defects underlying HSCR.

How does the apparently recessive *ret-k⁻* mutation relate to the dominantly inherited germline mutations that lead to the MEN2 syndromes? All MEN2 mutations described so far result in non-conservative substitutions in the ligand binding domain (MEN2A) or kinase domain (MEN2B), which presumably result in the constitutive activation of the tyrosine kinase. Since all cell types affected in MEN2A and MEN2B express *RET* mRNA not only during embryogenesis but into early postnatal life as well (Pachnis *et al.*, 1993; and unpublished observations), it is likely that the various tumours associated with the MEN2 syndromes result from the constitutive, i.e. ligand-independent, activation of the mutant RET receptor beyond the embryonic period during which it performs a developmental function. In contrast to the tumorigenic dominant mutations, the inactivating *ret-k⁻* mutation leads to major cellular deficiencies in the affected organs. The opposite effects of these mutations are clearly manifested in the ENS, where MEN2B results in hyperplasia of the enteric ganglia while the *ret-k⁻* mutation in mice results in elimination of all enteric neurons. This model is, however, complicated by the description of families in which a single MEN2A mutation co-segregates with a tumour phenotype (e.g. medullary thyroid carcinoma) and a cell loss phenotype (congenital megacolon due to absence of enteric neurons). In this respect, it is interesting that several oncogenes have been described which are capable of inducing cellular transformation or apoptotic cell death depending on the genotype of the cell or the conditions of the extracellular environment (Tanaka *et al.*, 1994). It is therefore possible that activating mutations of the *RET* proto-oncogene can lead, in different cell lineages, to either tumour formation or cell death. The *RET* locus therefore provides a unique opportunity to study the effects of gain-of-function and loss-of-function alleles of a *RTK* gene in tumour formation and normal mammalian development.

References

Altman J, Bayer S. (1982) Development of the cranial nerve ganglia and related nuclei in the rat. *Adv. Anat. Embryol. Cell Biol.* **74**.

Baetge G, Gershon MD. (1989) Transient catecholaminergic (TC) cells in the vagus nerves and bowel of fetal mice: relationship to the development of enteric neurons. *Devel. Biol.* **132**: 189–211.

Baetge G, Pintar JE, Gershon MD. (1990) Transiently catecholaminergic (TC) cells in the bowel of the fetal rat: precursors of noncatecholaminergic enteric neurons. *Devel. Biol.* **141**: 353–380.

Balinsky BI. (1970) *An Introduction to Embryology.* W.B. Saunders Co., Philadelphia.

Bishop AE, Carlei F, Lee V, Trojanowski JQ, Marangos PJ, Dahl D, Polak JM. (1985) Combined immunostaining of neurofilaments, neuron specific enolase, GFAP, and S-100: a possible means for assessing the morphological and functional status of the enteric nervous system. *Histochemistry* **82**: 93–97.

Carlson KM, Dou S, Chi D, Scavarda N, Toshima K, Jackson CE, Wells SAJ, Goodfellow PJ, Donis-Keller H. (1994) Single missense mutation in the tyrosine kinase catalytic domain of the RET protooncogene is associated with multiple endocrine neoplasia type 2B. *Proc. Natl Acad. Sci. USA* **91**: 1579–1583.

Carnahan JF, Anderson DJ, Patterson PH. (1991) Evidence that enteric neurons may derive from the sympathoadrenal lineage. *Devel. Biol.* **148**: 552–561.

Carroll SL, Silos-Santiago I, Frese SE, Ruit KG, Milbrandt J, Snider WD. (1992) Dorsal root ganglion neurons expressing trk are selectively sensitive to NGF deprivation in utero. *Neuron* **9**: 779–788.

Dodd J, Solter D, Jessell TM. (1984) Monoclonal antibodies against carbohydrate differentiation antigens identify subsets of primary sensory neurons. *Nature* **311**: 469–472.

Dodd J, Jessell TM. (1985) Lactoseries carbohydrates specify subsets of dorsal root ganglion neurons projecting to the superficial dorsal horn of rat spinal cord. *J. Neurosci.* **5**: 3278–3294.

Donis-Keller H, Dou S, Chi D, Carlson KM, Toshima K, Lairmore TC, Howe JR, Moley JF, Goodfellow P, Wells SAJ. (1993) Mutations in the RET proto-oncogene are associated with MEN 2A and FMTC. *Hum. Mol. Genet.* **2**: 851–856.

Edery P, Lyonnet S, Mulligan LM, Pelet A, Dow E, Abel L, Holder S, Nihoul Fekete C, Ponder BA, Munnich A. (1994) Mutations of the RET proto-oncogene in Hirschsprung's disease. *Nature* **367**: 378–380.

Erickson RA. (1968) Inductive interactions in the development of the mouse metanephros. *J. Exp. Zool.* **169**: 33–42.

Fontaine-Perus J, Chanconie M, Le Douarin N. (1982) Differentiation of peptidergic neurons in quail-chick chimaeric embryos. *Cell Differ.* **11**: 183–193.

Furness J, Costa M. (1987) *The Enteric Nervous System.* Churchill Livingstone, New York.

Gershon MD, Chalazonitis A, Rothman TP. (1993) From neural crest to bowel: development of the enteric nervous system. *J. Neurobiol.* **24**: 199–214.

Green MC. (1989) Catalog of mutant genes and polymorphic loci. In *Genetic Variants and Strains of the Laboratory Mouse* (eds MF Lyon, AG Searle). Oxford University Press, Oxford, pp. 12–403.

Grobstein C. (1953) Inductive epithelio-mesenchymal interaction in cultured organ rudiments of the mouse. *Science* **118**: 52–55.

Grobstein C. (1955) Inductive interaction in the development of the mouse metanephros. *J. Exp. Zool.* **130**: 319–340.

Grobstein C. (1956) Trans-filter induction of tubules in mouse metanephrogenic mesenchyme. *Exp. Cell Res.* **10**: 424–440.

Hafen E, Basler K. (1990) Role of receptor tyrosine kinases during Drosophila development. *Ciba Found. Symp.* **150**: 191–204.

Hofstra RM, Landsvater RM, Ceccherini I, et al. (1994) A mutation in the RET proto-oncogene associated with multiple endocrine neoplasia type 2B and sporadic medullary thyroid carcinoma. *Nature* **367**: 375–376.

Hunt P, Wilkinson D, Krumlauf R. (1991) Patterning the vertebrate head: murine Hox 2 genes mark distinct subpopulations of premigratory and migrating cranial neural crest. *Development* **112**: 43–50.

Iwamoto T, Takahashi M, Ito M, Hamaguchi M, Isobe K, Misawa N, Asai J, Yoshida T, Nakashima I. (1990) Oncogenicity of the ret transforming gene in MMTV/ret transgenic mice. *Oncogene* 5: 535–542.

Iwamoto T, Pu M, Ito M, Takahashi M, Isobe K, Nagase F, Kawashima K, Ichihara M, Nakashima I. (1991a) Preferential development of pre-B lymphomas with drastically down-regulated N-myc in the E mu-ret transgenic mice. *Eur J. Immunol.* 21: 1809–1814.

Iwamoto T, Takahashi M, Ito M, Hamatani K, Ohbayashi M, Wajjwalku W, Isobe K, Nakashima I. (1991b) Aberrant melanogenesis and melanocytic tumour development in transgenic mice that carry a metallothionein/ret fusion gene. *EMBO J.* 10: 3167–3175.

Iwamoto T, Taniguchi M, Asai N, Ohkusu K, Nakashima I, Takahashi M. (1993) cDNA cloning of mouse ret proto-oncogene and its sequence similarity to the cadherin superfamily. *Oncogene* 8: 1087–1091.

Johnson JE, Birren SJ, Anderson DJ. (1990) Two rat homologues of Drosophilia achaete-scute specifically expressed in neuronal precursors. *Nature* 346: 858–861.

Kaplan DR, Hempstead BL, Martin-Zanca D, Chao MV, Parada LF. (1991a) The trk proto-ocogene product: a signal transducing receptor for nerve growth factor. *Science* 252: 554–558.

Kaplan DR, Martin-Zanca D, Parada LF. (1991b) Tyrosine phosphorylation and tyrosine kinase activity of the trk proto-oncogene product induced by NGF. *Nature* 350: 158–160.

Kapur RP, Yost C, Palmiter RD. (1992) A transgenic model for studying development of the enteric nervous system in normal and aganglionic mice. *Development* 116: 167–175.

Klein R, Jing SQ, Nanduri V, O'Rourke E, Barbacid M. (1991a) The trk proto-oncogene encodes a receptor for nerve growth factor. *Cell* 65: 189–197.

Klein R, Nanduri V, Jing SA, Lamballe F, Tapley P, Bryant S, Cordon-Cardo C, Jones KR, Reichardt LF, Barbacid M. (1991b) The trkB tyrosine protein kinase is a receptor for brain-derived neurotrophic factor and neurotrophin-3. *Cell* 66: 395–403.

Kreidberg JA, Sariola H, Loring JM, Maeda M, Pelletier J, Housman D, Jaenisch R (1993) WT-1 is required for early kidney development. *Cell* 74: 679–691.

Lairmore TC, Howe JR, Korte JA, Dilley WG, Aine L, Aine E, Wells SAJ, Donis-Keller H. (1991) Familial medullary thyroid carcinoma and multiple endocrine neoplasia type 2B map to the same region of chromosome 10 as multiple endocrine neoplasia type 2A. *Genomics* 9: 181–192.

Lamballe F, Klein R, Barbacid M. (1991) trkC, a new member of the trk family of tyrosine protein kinases, is a receptor for neurotrophin-3. *Cell* 66: 967–979.

Le Douarin N, Renaud D, Teillet MA, Le Douarin GH. (1975) Cholinergic differentiation of presumptive adrenergic neuroblasts in interspecific chimaeras after heterotopic transplantations. *Proc. Natl Acad. Sci. USA* 72: 728–732.

Le Douarin N, Teillet MA. (1973) The migration of neural crest cells to the wall of the digestive tract in avian embryo. *J. Embryol. Exp. Morphol.* 30: 31–48.

Levi-Montalcini R. (1987) The nerve growth factor 35 years later. *Science* 237: 1154–1162.

Lo LC, Johnson JE, Wuenschell CW, Saito T, Anderson DJ. (1991). Mammalian achaete-scute homolog 1 is transiently expressed by spatially restricted subsets of early neuroepithelial and neural crest cells. *Genes Devel.* 5: 1524–1537.

Martinez HJ, Dreyfus CF, Jonakait GM, Black IB. (1985) Nerve growth factor promotes cholinergic development in brain atrial cultures. *Proc. Natl Acad. Sci. USA* 82: 7777–7781.

Meier-Ruge W. (1974) Hirschsprung's disease: its aetiology, pathogenesis and differential diagnosis. *Curr. Top. Pathol.* 59: 131–179.

Mulligan LM, Kwok JB, Healey CS, *et al.* (1993) Germ-line mutations of the RET proto-oncogene in multiple endocrine neoplasia type 2A. *Nature* 363: 458–460.

Noden DM. (1983) The role of the neural crest in patterning of avian cranial skeletal, connective, and muscle tissue. *Devel. Biol.* 96: 144–165.

Noden DM. (1988) Interactions and fates of avian craniofacial mesenchyme. *Development* 103 (Suppl.): 121–140.

Pachnis V, Mankoo B, Costantini F. (1993) Expression of the c-ret proto-oncogene during mouse embryogenesis. *Development* 119: 1005–1017.

Pankratz DS. (1931) The development of the suprarenal gland in the albino rat. *Anat. Rec.* 49: 31–39.

Passarge E. (1973) Genetics of Hirschsprung's disease. *Clin. Gastroenterol.* 2: 507–513.

Pawson T, Bernstein A. (1990) Receptor tyrosine kinases: genetic evidence for their role in Drosophila and mouse development. *Trends Genet.* 6: 350–356.

Robertson E. (1987) *Teratocarcinomas and Embryonic Stem Cells: a Practical Approach.* IRL Press, Oxford.

Romeo G, Ronchetto P, Luo Y, *et al.* (1994) Point mutations affecting the tyrosine kinase domain of the RET proto-oncogene in Hirschsprung's disease. *Nature* 367: 377–378.

Ruit KG, Elliott JL, Osborne PA, Yan Q, Snider WD. (1992) Selective dependence of mammalian dorsal root ganglion neurons on nerve growth factor during embryonic development. *Neuron* 8: 573–587.

Santoro M, Rosati R, Grieco M, Berlingieri MT, D'Amato GL, de-Franciscis V, Fusco A. (1990) The ret proto-oncogene is consistently expressed in human pheochromocytomas and thyroid medullary carcinomas. *Oncogene* 5: 1595–1598.

Santoro M, Carlomagno F, Hay ID, *et al.* (1992) Ret oncogene activation in human thyroid neoplasms is restricted to the papillary cancer subtype. *J. Clin. Invest.* 89: 1517–1522.

Saxen L. (1970) Failure to demonstrate tubule formation in a heterologous mesenchyme. *Devel. Biol.* 23: 511–523.

Saxen L. (1987) *Organogenesis of the Kidney.* Cambridge University Press, Cambridge.

Schlessinger J, Ullrich A. (1992) Growth factor signaling by receptor tyrosine kinases. *Neuron* 9: 383–391.

Schuchardt A, D'Agati V, Larsson-Blomberg L, Costantini F, Pachnis V. (1994) Defects in the kidney and enteric nervous system of mice lacking the tyrosine kinase receptor Ret. *Nature* 367: 380–383.

Soppet D, Escandon E, Maragos J, *et al.* (1991) The neurotrophic factors brain-derived neurotrophic factor and neurotrophin-3 are ligands for the trkB tyrosine kinase receptor. *Cell* 65: 895–903.

Takahashi M, Ritz J, Cooper GM. (1985) Activation of a novel human transforming gene, ret, by DNA rearrangement. *Cell* 42: 581–588.

Takahashi M, Buma Y, Iwamoto T, Inaguma Y, Ikeda H, Hiai H. (1988) Cloning and expression of the ret proto-oncogene encoding a tyrosine kinase with two potential transmembrane domains. *Oncogene* 3: 571–578.

Takahashi M, Cooper GM. (1987) ret transforming gene encodes a fusion protein homologous to tyrosine kinases. *Mol. Cell Biol.* 7: 1378–1385.

Tanaka N, Ishihara M, Kitagawa M, Harada H, Kimura T, Matsuyama T, Lamphier MS, Aizawa S, Mak TW, Taniguchi, T. (1994) Cellular commitment to oncogene-induced transformation or apoptosis is dependent on the transcription factor IRF-1. *Cell* 77: 829–839.

Troy CM, Brown K, Greene LA, Shelanski ML. (1990) Ontogeny of the neuronal intermediate filament protein, peripherin, in the mouse embryo. *Neuroscience* **36**: 217–237.

Tucker G, Ciment G, Thiery JP. (1986) Pathways of avian neural crest cell migration in the developing gut. *Devel. Biol.* **116**: 439–450.

Yarden Y, Ullrich A. (1988) Growth factor receptor tyrosine kinases. *Annu. Rev. Biochem.* **57**: 443–478.

Yntema CL, Hammond WS. (1954) The origin of intrinsic ganglia of trunk viscera from vagal neural crest in the chick embryo. *J. Comp. Neurol.* **101**: 515–542.

Index